T0269020

Rotating Flow

Rotating Flow

Peter R.N. Childs
FIMechE, FRSA, FHEA, mem.ASME, MIED,
BSc(hons), DPhil, CEng
Imperial College London

AMSTERDAM • BOSTON • HEIDELBERG • LONDON •NEW YORK • OXFORD
PARIS • SAN DIEGO •SAN FRANCISCO • SINGAPORE • SYDNEY • TOKYO
Butterworth-Heinemann is an imprint of Elsevier

ELSEVIER

Butterworth-Heinemann is an imprint of Elsevier
30 Corporate Drive, Suite 400, Burlington, MA 01803, USA
Linacre House, Jordan Hill, Oxford, OX2 8DP, UK

Library of Congress Cataloging-in-Publication Data
Childs, Peter R. N.
Rotating flow / Peter R. N. Childs.
    p.   cm.
Includes bibliographical references and index.
ISBN 978-0-12-810212-1
1. Vortex-motion.   2. Rotational motion.   I. Title.
QC159.C47   2011
532'.0595–dc22

                                                            2010028787

British Library Cataloguing-in-Publication Data
A catalogue record for this book is available from the British Library.

ISBN: 978-0-12-810212-1

For information on all Butterworth–Heinemann publications
visit our Website at www.books.elsevier.com

Printed and bound in China
11 12 13   10 9 8 7 6 5 4 3 2 1

The subject of rotating flows is important for many engineers, mathematicians and physicists. The earth's weather system is controlled by the combined effects of solar radiation and rotation. The jet stream and ocean circulations occur as a direct result of the earth's rotation; hurricanes and tornadoes are extreme phenomena that owe their existence to rotation-induced swirl.

There are many examples of flow near rotating machines, the most important of which is the gas turbine. Rotating-disc systems are used to model—theoretically, experimentally and computationally—the flow and heat transfer associated with the internal-air systems of gas turbines, where discs rotate close to a rotating or a stationary surface. The engine designer uses compressed air to cool the turbine discs: too little air could result in catastrophic failure; too much is wasteful and increases the fuel consumption and CO_2 production of the engine. Small improvements to the cooling system can result in significant savings, but the optimum design requires an understanding of the principles of rotating flows and the development and solution of the appropriate equations.

In this book, Professor Childs draws on his extensive experience to cover the basic theory of rotating flows. He shows how the Navier-Stokes equations can be used to derive the boundary-layer and Ekman-layer equations for rotating surfaces, and how—with appropriate assumptions—these equations can be applied to a disc rotating in a stationary fluid (the classic free disc) and to discs rotating close to stationary surfaces (rotor-stator systems) or close to another rotating disc (rotating cavities). The solutions of the equations are compared with experimental measurements, and the numerical examples should be helpful to students and practising engineers alike.

The student familiar with conventional non-rotating flows has, like Alice, to go through the looking glass into a strange world that is counter-intuitive to the uninitiated. This book provides a guide to this important world, and the reader who is prepared to put in the effort and follow the sign-posted paths will emerge better informed and a lot wiser. It's a book that I would recommend to postgraduate students, practising engineers and scientists who wish to know more about this fascinating subject.

J Michael Owen

Professor Emeritus
University of Bath, UK

Contents

Swirling, whirling, and rotating flow has proved fascinating and challenging throughout the ages. From sink vortices to the swirling motion seen in a cornfield as the wind blows across it, the subject provides a talking point and a level of complexity that often defies simple explanation. Atmospheric and oceanic flow is also significantly affected by rotation, including cyclonic and anticyclonic flow circulations, intense atmospheric vortices, and ocean circulations. The observed flow phenomena often do not match our expectations and intuition, and it is through the insight revealed by detailed observation and modeling that significant satisfaction in the study of rotating flow can result.

Modeling of rotating flow is critically important across a wide range of scientific, engineering, and product design applications, providing design capability for products such as jet engines, pumps, and vacuum cleaners and modeling capability for geophysical flows. Even for applications where rotation is not initially evident, the subject is often fundamental to understanding and modeling the details of the flow physics. Examples include the vorticity produced in flow along a channel, the secondary flow produced for flow around a bend, and the wing-tip vortices produced downstream of a wing.

There are fundamental differences between rotating and linear flows, and this text explores these differences, their physical manifestations, and modeling. The text includes both anecdotal explanations and in-depth development of modeling and analytical methods. This book introduces examples of rotating flow in technology, science, and nature. The fundamental methods of modeling such flows are explored in conjunction with simplified techniques that enable approximate solutions to practical applications. Simple explanations of complex flows can be very useful in developing understanding and flow models, even though they sometimes neglect important details. The in-depth proofs and advanced models provide an indication of the validity of approach and additional analytical capability.

Specific subjects covered include:

- an introduction to rotating flow
- fundamental equations
- vorticity and vortices
- rotating disc flow
- flow around rotating cylinders and in a rotating annulus
- flow in rotating cavities
- an introduction to atmospheric and oceanic circulations

This text has been written principally to assist engineers, technologists, physicists, and meteorologists in developing their understanding and modeling of the fluid mechanics associated with rotating flow. The text has been developed to be of value to practitioners, researchers, and students. It has been assumed that the reader will have already developed some skills in the field of fluid mechanics, typically compatible with first- and second-year undergraduate level. The text has been used for both short and extended courses in rotating flow and for augmenting undergraduate, graduate, and continuing professional development courses in fluid mechanics.

The book has principally arisen from over 20 years of research into rotating fluids and associated heat transfer at the Thermo-Fluid Mechanics Research Centre, University of Sussex, the author's former affiliation. Much of this work has been focused on the internal air system for gas turbine engines where rotating discs, cavities, and channels are common features, as well as flow visualization experiments. The research has been sponsored by a wide range of industrial companies and funding bodies, including Rolls-Royce plc, Alstom, Siemens, Ruston Gas Turbines, Turbomeca, Motor-Turbieren Union, Volvo, Fiat, Snecma, Industrial Turbinas Propulsores, ABB, Turbocam, the European Union, and the Engineering and Physical Sciences Research Council. Their support and partnership in the research programs is gratefully acknowledged. I have also been fortunate to have worked in the same laboratory as Professor Fred Bayley, Professor Mike Owen, Professor Alan Turner, and Dr. Christopher Long, all of whom have been influential in developing the science of rotating flow.

The book is organized into eight principal chapters. Chapter 1 provides an overview of rotating flow phenomena, setting the context for subsequent development of the subject. Chapter 2 provides an overview and detailed introduction to the fundamental mathematical means of modeling fluid flow, with a specific emphasis on the conservation equations of continuity, momentum, and energy. These enable detailed modeling of fluid flow and are presented in forms directly applicable to rotating flow. The subjects of vorticity and vortices are introduced and developed in Chapter 3.

Chapters 4, 5, 6, and 7 concentrate on rotating flow in specific geometric configurations. Chapters 4 and 5 focus on the flow in rotating disc applications important to a range of engineering applications, such as gas turbine engines and pumps, and provide a relatively simple geometric configuration for the modeling of, in some cases, complex three-dimensional and time-dependent flow. The subject of a plain rotating disc is considered in detail in Chapter 4, serving to provide an in-depth development of understanding of the flow physics and modeling approach for both laminar and turbulent flow. Chapter 5 builds on this treatment, extending consideration to the cavity formed between a stationary and a rotating disc. Chapter 6 is concerned with flows associated with rotating cylinders and spheres. This chapter also addresses the flow in a fluid-filled annulus where the inner or outer cylinder rotates and there may also be an axial flow through the annulus. The flow in a rotating annulus can, under certain conditions, result in the production of toroidal vortices called Taylor vortices, which arise from instabilities in the flow field. The flow in a cavity formed between two co-rotating discs is considered in Chapter 7. This subject is particularly challenging as the flow physics depends on the thermal and geometric boundary conditions and tends to be three-dimensional and time dependent and sometimes intractable even to the most sophisticated of current modeling techniques. The subjects of vortex flow and of rotating disc and rotating cavity applications have particular elements of commonality with geophysical flows.

Flow in the Earth's atmosphere and its oceans is significantly influenced by the Earth's rotation. Examples include the major circulations in the atmosphere and oceans, cyclones and anticyclones, and large-scale vortices. Chapter 8 provides a brief description of the atmosphere and oceans and the principal flow structures, with a specific focus on the associated flow circulations and the influence of the Earth's rotation.

About the Author

Peter Childs is the Professorial Lead of Engineering Design in the Faculty of Engineering at Imperial College, having taken up this chair in 2008. He was formerly the director of the Rolls-Royce-supported University Technology Centre for Aero-Thermal Systems, director of InQbate, the HEFCE funded Centre of Excellence in Teaching and Learning in Creativity, and a professor at the University of Sussex where he worked for 21 years. His general interests include creativity; the application of creative methods; sustainable energy component, concept, and system design; fluid flow in rotating applications; and heat transfer. He has undertaken extensive research activities principally focused on the internal air system for gas turbine engines. He has written several books on mechanical design, fluid flow, and temperature measurement and is a former winner of the American Society of Mechanical Engineers—International Gas Turbine Institute John P. Davis award for exceptional contribution to the literature of gas turbine technology.

Peter Childs

Email p.childs@imperial.ac.uk

Notation and Units

An attempt has been made to use a nomenclature system that is consistent throughout the text. This is challenging in a subject dealing with a variety of applications and specific geometries. The letter u has been selected for velocity and components identified by subscripts. This has the advantage that the reader can readily identify the direction concerned, but it is more complex and prone to error than, say, a separate letter being used for each. The selection of u for velocity allows the choice for volume to be V. A distinction has been made where possible between velocity components in a stationary frame of reference denoted, in the case of cylindrical coordinates, by u_r, u_ϕ, and u_z, and velocity components in a rotating frame of reference denoted by u, v, and w.

Where possible emboldened letters have been used for vectors.

Many rotating flow applications involve consideration of inner and outer radii. It is common practice in disc geometries to use the letter a to denote an inner radius and the letter b to denote the outer radius. Similarly, for an annulus the letter a can be used for the radius of the inner cylindrical surface and the letter b for the radius of the outer cylindrical surface. The letter r can then be used for the local radius.

The relationship between fluid flow and heat transfer has required consideration of notation relevant to both fields. The letter t has been selected for time and the capital T for temperature.

There are many dimensionless groups that are useful in fluid mechanics. It is possible for ambiguities to arise between, for example, the Rossby number, Ro, and the multiplication between two variables R and o. It is hoped that the context will resolve such an ambiguity if it arises.

In accordance with the SI convention, a space has been left between numerals and units and between units.

a	radius (m)
A	area (m^2), coefficient (dimensionless)
A_{gap}	cross-sectional area of annular gap (m^2)
a_z	acceleration in z direction (m/s^2)
b	outer radius (m)

b_o	characteristic length (m)
B	coefficient (dimensionless)
c	clearance (m), swirl ratio (dimensionless)
c_{eff}	effective swirl ratio (dimensionless)
c_p	specific heat capacity at constant pressure (J/kg K)
c_{p*}	dimensionless specific heat capacity at constant pressure (dimensionless)
c_v	specific heat capacity at constant volume (J/kg K)
c_{wave}	wave speed (m/s)
C	vortex strength, coefficient
C_1	constant (1/s), constant (dimensionless), first radiation constant (W·m^2)
C_2	constant (m^2/s), constant (dimensionless), second radiation constant (m K)
C_d	discharge coefficient (dimensionless)
C_D	drag coefficient (dimensionless)
C_f	skin friction coefficient (dimensionless)
C_L	lift coefficient (dimensionless)
C_m	moment coefficient (dimensionless)
C_{mc}	moment coefficient for a rotating cylinder (dimensionless)
C_p	pressure coefficient (dimensionless)
C_{rs}	empirical constant
Cw	nondimensional flow rate (dimensionless)
$Cw_{free\ disc}$	nondimensional flow rate entrained by a free disc (dimensionless)
Cw_{local}	local nondimensional flow rate (dimensionless)
Cw_{min}	minimum nondimensional flow rate to prevent ingress (dimensionless)
d	diameter (m)
d_{bolt}	exposed bolt head diameter (m)
d_h	hydraulic diameter (m)
D	journal diameter (m)
e	roughness grain size (m), eccentricity (m)
E	energy (J)
$E_{\lambda,b}$	spectral emissive power for a blackbody (W/m^3)
Ek	Ekman number (dimensionless)
Eu	Euler number (dimensionless)
f	friction factor (dimensionless), coefficient of friction (dimensionless), Coriolis parameter (1/s)
F	force magnitude (N)
$\mathbf{F}$	body force (N/m^3)
$F_{centrifugal}$	centrifugal force (N)
$F_{Coriolis}$	Coriolis force (N)
F_g	geometrical factor (dimensionless)
F_L	lift force magnitude (N)
F_r	body force in the r direction (N/m^3)
Fr	Froude number (dimensionless)
F_r'	body force in the r direction (N/m^3)
F_x	body force component in the x direction (N/m^3)
F_y	body force component in the y direction (N/m^3)
F_z	body force component in the z direction (N/m^3)
F_z'	body force component in the z direction (N/m^3)

F_ϕ	body force in the azimuthal direction (N/m^3)
F_ϕ'	body force in the azimuthal direction (N/m^3)
g	acceleration due to gravity (m/s^2)
G	gap ratio (dimensionless)
G_1	coefficient
G_2	coefficient
G_3	coefficient
G_4	coefficient
G_5	coefficient
G_c	shroud clearance ratio (dimensionless)
Gr	Grashof number (dimensionless)
Gr_b	Grashof number (dimensionless)
G_λ	spectral radiation
h	film thickness (m)
h_o	minimum film thickness (m)
H	total head (m), height of bolt head (m)
I	moment of inertia (kg m^2)
k	thermal conductivity (W/m K)
k_*	dimensionless thermal conductivity (dimensionless)
k_1	constant (dimensionless)
k_2	constant (dimensionless)
k_3	constant
k_4	constant
k_5	constant
k_o	characteristic thermal conductivity (W/m K)
k_s	sand-grain roughness (m)
k_s'	sand-grain size (m)
K	coefficient (dimensionless)
K_1	velocity shape factor (dimensionless)
K_2	velocity shape factor (dimensionless)
K_{bl}	boundary layer constant (dimensionless)
K_f	form drag correction factor (dimensionless)
K_L	loss coefficient (dimensionless)
K_u	velocity shape factor (dimensionless)
K_v	velocity shape factor (dimensionless)
K_w	stress flow rate parameter (dimensionless)
l_o	characteristic length (m)
l_{pitch}	pitch (m)
L	characteristic length, length (m)
M	Mach number (dimensionless), molecular mass (kg/mol)
m	mass (kg)
$m_{element}$	mass of an arbitrary fluid element (kg)
$\dot{m}$	mass flow rate (kg/s)
$\dot{m}_d$	mass flow rate in the boundary layer (kg/s)
$\dot{m}_{in}$	inflow mass flow rate (kg/s)
$\dot{m}_o$	mass flow rate in the rotor boundary layer (kg/s)
$\dot{m}_{out}$	outflow mass flow rate (kg/s)
$\dot{m}_{ref}$	reference mass flow rate (kg/s)

$\dot{m}_s$	mass flow rate in the stator boundary layer (kg/s)
$\dot{m}_{superposed}$	superposed mass flow rate (kg/s)
M_z	axial Mach number (dimensionless)
n	exponent (dimensionless)
$\boldsymbol{n}$	outward normal unit vector (m)
N	number of blades (integer), speed (rpm)
N_s	rotational speed (revolutions per second)
P	load capacity (N/m^2)
p	static pressure (Pa)
p'	fluctuating component of static pressure (Pa)
p_∞	static pressure in free stream (Pa)
p_*	dimensionless pressure (dimensionless)
p_a	external static pressure (Pa)
p_{cav}	rotor-stator cavity static pressure (N/m^2)
p_{max}	maximum annulus static pressure (N/m^2)
p_{min}	minimum annulus static pressure (N/m^2)
p_o	characteristic pressure (Pa)
$p_{reduced}$	reduced pressure (Pa)
p_{sat}	saturation vapor pressure (Pa)
Pr	Prandtl number (dimensionless)
q	heat flux (W/m^2)
q_x	volumetric flow rate per unit width in the x direction (m^2/s)
q_y	volumetric flow rate per unit width in the y direction (m^2/s)
$\dot{q}$	flow rate (m^3/s)
Q	flow rate (m^3/s), heat flux (W)
Q_s	side flow rate (m^3/s)
R	common radius (m), radius (m), specific gas constant (J/kg K)
r	distance along the r-axis (m)
$\boldsymbol{r}$	position vector (m)
r_e	radial extent of source region (m)
r_f	final radius (m)
r_i	inner radius (m)
r_m	mean radius (m)
r_o	outer radius (m)
r_s	shaft radius (m)
R_a	centre line average roughness (m)
$\mathfrak{R}$	molar gas constant (J/mol K)
Ra	Rayleigh number (dimensionless)
Re	Reynolds number (dimensionless)
$Re_{critical}$	critical Reynolds number (dimensionless)
Re_d	Reynolds number based on pipe diameter (dimensionless)
Re_x	Reynolds number based on x component of velocity (dimensionless)
Re_y	Reynolds number based on y component of velocity (dimensionless)
Re_z	external flow Reynolds number (dimensionless), Reynolds number based on z component of velocity (dimensionless)
Re_ϕ	rotational Reynolds number (dimensionless)
Re_ϕ'	modified rotational Reynolds number (dimensionless)
$Re_{\phi,local}$	local rotational Reynolds number (dimensionless)

$Re_{\phi m}$	rotational Reynolds number based on annulus gap (dimensionless)
Ro	Rossby number (dimensionless)
s	gap (m)
s_c	clearance between shroud and rotor (m)
S	Taylor vortex critical speed factor (dimensionless), Sommerfield number (dimensionless)
S_o	solar constant (W/m^2)
St	Stanton number (dimensionless)
t	time (s)
t_o	characteristic timescale (s)
t_*	dimensionless time ratio (dimensionless)
T	temperature (K)
T_{av}	average temperature (K)
T_e	emission temperature (K)
T_o	characteristic temperature (K)
T_q	moment, torque (N m)
T_w	wall temperature (K)
T_*	dimensionless temperature (dimensionless)
Ta	Taylor number (dimensionless)
Ta_m	Taylor number based on mean radius (dimensionless)
$Ta_{m,cr}$	critical Taylor number based on mean annulus radius (dimensionless)
u	relative radial velocity (m/s), relative velocity in the x direction (m/s)
$\boldsymbol{u}$	velocity vector in a stationary frame of reference (m/s)
$\boldsymbol{u}'$	fluctuating component of the velocity vector in a stationary frame of reference (m/s)
u_∞	free-stream velocity component (m/s)
u_{dm}	mixed out radial velocity (m/s)
u_{max}	maximum velocity (m/s)
u_r	velocity component in the r direction (m/s)
u_r'	instantaneous radial velocity component (m/s)
u_{r*}	dimensionless radial velocity component (dimensionless)
$u_{resultant}$	resultant velocity magnitude (m/s)
u_x	velocity component in the x direction (m/s)
u_{x*}	dimensionless velocity component in the x direction (dimensionless)
$u_{x,o}$	characteristic velocity component in the x direction (m/s)
u_y	velocity component in the y direction (m/s)
u_{y*}	dimensionless velocity component in the y direction (dimensionless)
$u_{y,o}$	characteristic velocity component in the y direction (m/s)
u_z	velocity component in the z direction (m/s)
u_z'	instantaneous axial velocity component (m/s)
u_{z*}	dimensionless axial velocity component (dimensionless)
$u_{z,\infty}$	free-stream velocity component (m/s)
$u_{z,o}$	characteristic velocity component in the z direction (m/s)
$\bar{u}_z$	average axial component of velocity for the external flow (m/s)
u_τ	friction velocity (m/s)
u_ϕ	velocity component in the tangential direction (m/s)
u_ϕ'	instantaneous tangential velocity component (m/s)
$u_{\phi*}$	dimensionless tangential velocity component (dimensionless)

$u_{\phi,\infty}$	velocity component in the circumferential direction in the free stream (m/s)
$u_{\phi,1}$	velocity component in the tangential direction at the inner radius (m/s)
$u_{\phi,2}$	velocity component in the tangential direction at the outer radius (m/s)
$u_{\phi,c}$	velocity component in the circumferential direction in the inviscid core (m/s)
$u_{\phi,o}$	velocity component in the circumferential direction at the rotor (m/s)
$\bar{u}$	average velocity (m/s)
$\bar{\boldsymbol{u}}$	average velocity vector (m/s)
U	amplitude of velocity (m/s)
U_o	characteristic velocity (m/s), reference velocity (m/s)
v	relative tangential velocity (m/s)
v_c	inviscid rotating core relative tangential velocity (m/s)
v_o	relative rotor tangential velocity (m/s)
v_∞	relative tangential velocity in the free-stream (m/s)
V	volume (m^3)
$V_{paraboloid}$	volume of the paraboloid (m^3)
w	relative axial velocity (m/s)
$\boldsymbol{w}$	relative velocity vector (m/s)
w_*	dimensionless relative velocity vector (dimensionless)
W	load (N)
x	distance along the x-axis (m), ratio of local to disc outer radius (dimensionless)
x_*	dimensionless location (dimensionless)
x_a	ratio of inner to outer radius (dimensionless)
x_e	ratio of local to disc outer radius defining extent of source region (dimensionless)
$x_{e,downstream}$	ratio of local to disc outer radius defining extent of source region (dimensionless)
X	arbitrary vector (m)
y	distance along the y-axis (m), wall distance (m)
y_*	dimensionless location (dimensionless)
y_+	universal coordinate (dimensionless)
$y_{sublayer}$	sublayer thickness (m)
z	distance along the z-axis (m)
z_*	dimensionless axial location (dimensionless)
z_p	upward unit vector in pressure coordinates
α	ratio of radial and tangential components of stress (dimensionless), coefficient (dimensionless)
α_o	coefficient for the rotor (dimensionless)
α_r	absorptivity (dimensionless)
α_s	coefficient for the stator (dimensionless)
β	swirl fraction or velocity ratio (dimensionless), coefficient (dimensionless)
β_v	volumetric expansion coefficient (m^3/m^{3}K)
β_ω	relative vorticity (1/s)
β^*	velocity ratio when the superposed flow rate is zero (dimensionless)
δ	boundary layer thickness (m)
ε	eccentricity ratio (dimensionless), emissivity
ε_m	coefficient (dimensionless)

ε_m'	coefficient (dimensionless)
ε_M	coefficient (dimensionless)
η	mass flow ratio (dimensionless)
ζ	dimensionless distance (dimensionless)
ϕ	azimuth angle (rad)
ϕ_{h_o}	angular position of minimum film thickness ($^\circ$)
$\phi_{p_{max}}$	angular position of maximum pressure ($^\circ$)
ϕ_{p_o}	film termination angle ($^\circ$)
ϕ_*	nondimensional circumferential location (dimensionless)
ϕ'	relative azimuth angle (rad)
γ	coefficient (dimensionless), isentropic index (dimensionless)
γ_o	coefficient for the rotor (dimensionless)
γ_s	coefficient for the stator (dimensionless)
κ	thermal diffusivity (m^2/K)
λ	wavelength (m), longitude ($^\circ$)
λ_t	turbulent flow parameter (dimensionless)
λ_m	rotation parameter (dimensionless)
μ	viscosity (Pa s)
μ_e	eddy viscosity (Pa s)
μ_o	reference viscosity (Pa s)
μ_*	dimensionless viscosity (dimensionless)
ν	kinematic viscosity (m^2/s)
ρ	density (kg/m^3)
ρ_d	partial density of dry air (kg/m^3)
ρ_o	reference density (kg/m^3)
ρ_r	reflectivity (dimensionless)
ρ_v	partial density of water vapour (kg/m^3)
$\tilde{\rho}$	density (kg/m^3)
ρ_*	dimensionless density (dimensionless)
σ	normal stress (N/m^2), Stefan-Boltzmann constant ($Wm^{-2}K^{-4}$)
σ_{im}	ratio of mean axial and tangential components of flow (dimensionless)
σ_x	normal stress in the x direction (N/m^2)
σ_r	normal stress in the r direction (N/m^2)
σ_s	squeeze number (dimensionless)
σ_y	normal stress in the y direction (N/m^2)
σ_z	normal stress in the z direction (N/m^2)
σ_φ	normal stress in the φ direction (N/m^2)
τ	shear stress (N/m^2)
τ_e	wind shear stress in the direction of flow (N/m^2)
τ_o	shear stress at the rotor (N/m^2)
τ_r	radial component of shear stress (N/m^2)
$\tau_{r,o}$	radial shear stress at the rotor (N/m^2)
τ_{rz}	viscous shear stress acting in the z direction, on a plane normal to the r direction (N/m^2)
$\tau_{r\phi}$	viscous shear stress acting in the φ direction, on a plane normal to the r direction (N/m^2)
τ_s	surface shear stress (N/m^2)

τ_{xy}	viscous shear stress acting in the y direction, on a plane normal to the x direction (N/m^2)
τ_{xz}	viscous shear stress acting in the z direction, on a plane normal to the x direction (N/m^2)
τ_{yx}	viscous shear stress acting in the x direction, on a plane normal to the y direction (N/m^2)
τ_{yz}	viscous shear stress acting in the z direction, on a plane normal to the y direction (N/m^2)
τ_{zr}	viscous shear stress acting in the r direction, on a plane normal to the z direction (N/m^2)
τ_{zx}	viscous shear stress acting in the x direction, on a plane normal to the z direction (N/m^2)
τ_{zy}	viscous shear stress acting in the y direction, on a plane normal to the z direction (N/m^2)
$\tau_{z\phi}$	viscous shear stress acting in the φ direction, on a plane normal to the z direction (N/m^2)
τ_{ϕ}	tangential component of shear stress (N/m^2)
$\tau_{\phi,o}$	tangential shear stress at the rotor (N/m^2)
$\tau_{\phi r}$	viscous shear stress acting in the r direction, on a plane normal to the ϕ direction (N/m^2)
$\tau_{\phi z}$	viscous shear stress acting in the z direction, on a plane normal to the ϕ direction (N/m^2)
ω	vorticity (1/s), frequency (Hz), angular velocity (rad/s)
$\boldsymbol{\omega}$	vorticity vector (1/s)
ω_r	radial vorticity component (1/s)
ω_x	vorticity component about the x-axis (1/s)
ω_y	vorticity component about the y-axis (1/s)
ω_z	vorticity component about the z-axis (1/s)
ω_{φ}	azimuthal vorticity component (1/s)
ξ	nondimensional decay rate (dimensionless)
ψ	stream function (m^2/s)
$\Delta C_{m,form}$	form drag moment coefficient (dimensionless)
ΔC_p	nondimensional static pressure difference (Pa)
$\Delta H_{vaporisation}$	enthalpy of vaporization (J/mol)
Δp	static pressure difference (Pa)
Δp_s	circumferentially averaged pressure drop across the shroud (Pa)
ΔT	temperature difference (K)
Δz	depression of surface (m)
Φ	velocity potential function (m^2/s)
Γ	circulation (m^2/s)
Ω	angular velocity magnitude (rad/s)
$\boldsymbol{\Omega}$	angular velocity vector (rad/s)
Ω'	angular velocity of rotating frame of reference (rad/s)
Ω_{cr}	critical angular velocity magnitude (rad/s)
Ω_x	angular velocity component about the x-axis (rad/s)
Ω_y	angular velocity component about the y-axis (rad/s)
Ω_z	angular velocity component about the z-axis (rad/s)

Chapter 1

Introduction to Rotating Flow

1.1. INTRODUCTION

Swirling, whirling, rotating flow has proved fascinating and challenging through-out the ages. Whether it be the vortex formed as water exits the bathtub or the eddying motion seen in a cornfield as wind blows across it, the subject provides a talking point and a level of complexity that often defy simple explanation. Rotating flow is critically important across a wide range of scientific, engineering, and product design applications. The subject provides a means of modeling, and as a result, design capability for products such as jet engines, pumps, and vacuum cleaners.

Even for applications where rotation is not initially evident, the subject is often fundamental to understanding and modeling the details of the flow physics. An example is the very strong wing tip vortices shed from the wings of an aircraft as it flies, as illustrated in Figures 1.1 and 1.2.

In Figure 1.1 the massive vortices shed by the wings of a Boeing 727 are made visible by smoke emitted by smoke generators attached to the wing tips. This smoke becomes entrained into the vortices as the plane moves. These vortices can take a few minutes to dissipate, and the flow disturbance caused in the wake of a plane is therefore one of the factors that needs to be considered in order to ensure adequate intervals between different aircraft occupying the same airspace. The vortex that is shed from one of the wings of an agricultural spray plane is illustrated in Figure 1.2. Here the flow is made visible by using colored smoke rising from the ground. This kind of vortex flow is similar to the whirling motion seen in the vicinity of an oar for a boat as it is rowed along. For both the wing tip and paddle vortices examples, where the action disturbing the flow is a linear motion, the result is a flow with significant rotation.

It is the much weaker forces that hold the particles together in a fluid, in comparison to those in a solid, that give rise to their more complex behavior. A fluid does not offer lasting resistance to displacement of one layer of particles over another. If a fluid experiences a shear force, the fluid particles will move in response with a permanent change in their relative position, even when the force is removed. By comparison, a solid—provided the shear force does not exceed elastic limits—will adopt its previous shape

Rotating Flow, DOI: 10.1016/B978-0-12-382098-3.00001-9

FIGURE 1.1 Typical wing tip vortices for a passenger aircraft. Photograph courtesy of NASA.

FIGURE 1.2 A photograph of the vortex that is shed from one of the wings of an agricultural spraying plane. The wing tip vortex has been made visible by means of coloured smoke rising from the ground. Photograph courtesy of NASA.

when the force is removed. In a fluid, shear forces are possible only when relative movement between layers is taking place, that is, the fluid is flowing. The purpose of this text is to introduce the subject of flow in applications where the flow rotates or swirls and the associated physical phenomena and principles.

1.2. GEOMETRIC CONFIGURATIONS

There are many subtle interactions between fluids and associated structures and boundaries that produce vorticity and secondary flows. Vorticity involves the rotation of a flow element, representing a chunk of fluid particles, as it moves through a flow field and can be visualized by a physical or virtual cork in a flow. If the cork rotates, then the flow has vorticity, whereas if the cork just translates through the flow field, then the flow does not have vorticity. A secondary flow is a flow pattern superimposed on the primary flow path. For example, when a viscous fluid moves through a bend in a pipe or channel, the differences in pressure and velocity that arise between the central core and the boundary layer flow near the surfaces set in motion a complex secondary flow pattern of spiralling vortices (Figure 1.3). These secondary flows are responsible for additional losses in pressure within a pipeline and increased erosion and scouring of surfaces.

An interest in the fluid flow associated with machinery may stem from the need to know the losses associated with windage in, say, an electric motor where the armature is spinning in air (Figure 1.4) or from use of fluid as a means of transferring energy as in, for example, a centrifugal compressor (Figure 1.5), or a rotary pump.

The jet engine has become the principal prime mover for aviation transport, and the industrial gas turbine engine is responsible for the production of a significant proportion of base load and standby power. It is also the prime

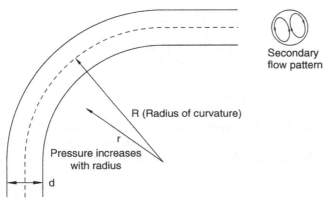

FIGURE 1.3 The secondary flow associated with flow around a pipe bend, where r is a local radius, R is the radius of curvature, d is the internal diameter of the pipe.

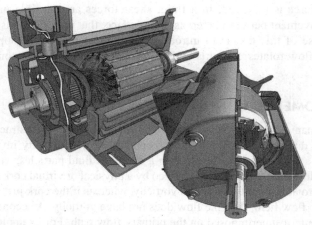

FIGURE 1.4　Electric motor armature. Figure courtesy of Loctite.

FIGURE 1.5　Centrifugal compressor from a turboprop engine.

mover of choice for certain types of ship and pumping applications. Gas turbine engines typically comprise a compressor, combustor, and turbine. Flow is compressed by diffusing the flow through alternating rows of stationary and rotating blades in a compressor; energy is added to the flow in the combustor; and the power to drive the compressor is produced by expanding the flow through the alternating rows of stationary and rotating turbine blades. The rotating blades are typically supported on discs. Rather than use completely

FIGURE 1.6 Schematic cutaway of the Rolls-Royce Trent 1000 engine for the Boeing 787 Dreamliner, illustrating the bladed discs of the compressor and turbine. Drawing courtesy of Rolls-Royce plc.

solid rotors, compressor and turbine discs tend to have profiled cross sections that are designed to dissipate the local high stresses and minimize the mass of material used. As illustrated in Figure 1.6, a typical gas turbine engine cross section will reveal that cylindrical cavities are formed between adjacent discs. These cavities will typically be filled with air. The cavity formed may involve a rotating disc adjacent to a stationary disc, which is known as a wheelspace or rotor-stator disc cavity. Alternatively, if both coaxial discs are rotating, this is referred to as a rotating cavity.

In both aircraft and electric power generation gas turbine engines, it is common practice for the cavity formed between a rotating turbine disc and a stationary disc to be purged with coolant air. This air reduces the thermal load on the disc and prevents the ingress of hot mainstream gas from the blade path into the cavity between the rotating disc and the adjacent stationary casing. However, use of this air is detrimental to engine cyclic performance, and turbine efficiency can be adversely affected by the seal air efflux into the main annulus. The lifetime of the rotor, and the cyclic and component performance of the engine are therefore dependent on the efficiency with which the cavity is purged of hot gases. The combined requirements to be able to determine the power required to overcome frictional drag, local flow characteristics, and associated heat transfer have provided a sustained impetus to investigate a number of rotor-stator disc configurations. The range of practical applications for disc flow extends to computer memory disc drives (as illustrated in Figure 1.7), centrifuges, cutting discs and saws, gears, and brakes.

Rotating closed cylindrical and annular cavities are a common feature in compressor and turbine rotor assemblies. The flow associated with a rotating

FIGURE 1.7 Photograph of a computer memory disc drive.

cavity depends on the radial and axial temperature distribution and the presence of geometrical features such as shrouds and the presence of any superposed flow. The flow can be highly complex with buoyancy-driven time-dependent flow features that provide a challenge to the most advanced analytical and numerical models.

Further applications in which rotating flow is important are in cyclone and U-shaped separators, as illustrated in Figures 1.8 and 1.9. In cyclone separators, a raw mixture of substances at high velocity is introduced into a stationary mechanical structure that deflects the mixture into a circular path. In the conical cyclone separator, the flow's angular velocity increases as the mixture moves

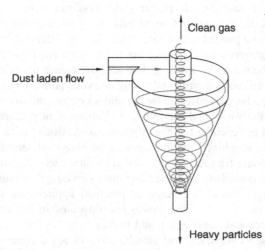

FIGURE 1.8 A conical cyclone separator.

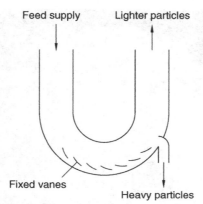

FIGURE 1.9   A U shaped separator.

radially inward due to the conservation of angular momentum. Because of inertia, the heavier components in the flow resist being deflected more than lighter particles, and the heavier particles therefore pass preferentially toward the outside of the flow. Secondary flows in the conical section assist in transporting heavier particles toward the bottom outlet, and a second outtake can be used for lighter particles.

In the simple U-shaped geometry of Figure 1.9, the fluid is caused to change direction by 180°, and an outtake located partway along the tube is used to separate the flows. This type of design is reasonably effective with fluid and solid mixtures, where the solids are much more dense than the fluid and have a low enough drag so that the solid particles can move radially through the fluid during the short residence time in which the mixture is in the U shaped section of the separator. U-shaped separators are generally located vertically, so that gravity can be used to aid the collection process of the denser particles.

1.3. GEOPHYSICAL FLOW

Theories for modeling rotating flow originally developed from attempts to understand fluid flow phenomena on the Earth's surface and its oceans. The fluid motions in the Earth's atmosphere are driven by a combination of thermal convection, Coriolis forces, and vorticity produced in regions of wind shear. The energy source for this motion is thermal radiation from the Sun, which is absorbed by the Earth and emitted to space. The massive amounts of energy involved, combined with the energy differentials between the poles and the equator, the influence of the Earth's rotation, and flow instability, result in major atmospheric circulations and under certain conditions intense atmospheric vortices. The views of the Earth from space illustrated in Figures 1.10 and 1.11 provide an indication of the variety of cloud formations in our atmosphere.

FIGURE 1.10   Composite image of the Earth from space. Courtesy of NASA.

FIGURE 1.11   Composite image of the Earth from space. Courtesy of NASA.

Over the oceans, shear stresses caused by winds blowing across an ocean surface can cause currents of water up to 100 m in depth or more to flow. These currents can be thousands of kilometers wide and move at speeds of several kilometers per day. Rotational patterns known as gyres are caused by the combination of wind-induced shear stresses and Coriolis forces due to the Earth's rotation in each of the five major oceans. The net effect is that in the northern hemisphere the general direction of rotation of the gyres is clockwise, and in the southern hemisphere it is anticlockwise.

Examples of intense atmospheric vortices include tropical cyclones and tornadoes. Tropical cyclones are called hurricanes over the Atlantic Ocean and typhoons over the Pacific Ocean. They can be several hundred kilometers in diameter. Tropical cyclones are caused by large-scale convection over a region of warm water. The convection is augmented by the release of latent heat during the condensation of rain. The cyclone's angular momentum is derived from the Earth's rotation during convergence of air toward the cyclone's low-pressure center. Because the angular momentum is derived from the Earth's rotation, the direction of rotation for a tropical cyclone is the same as the Earth's, which is anticlockwise when viewed from an observer above the northern hemisphere and clockwise in the southern hemisphere.

Figure 1.12 shows a photograph of Hurricane Gordon taken at 18:15:36 Greenwich Mean Time on September 17, 2006 by one of the crew members aboard the Space Shuttle *Atlantis*. The center of the storm was located near a latitude of 34° degrees north and a longitude of 53° west, while moving north-northeast. At the time the photo was taken, the sustained winds were 36 m/s

FIGURE 1.12   Hurricane Gordon taken at 18:15:36 Greenwich Mean Time on the 17th of September, 2006 by one of the crew members aboard the Space Shuttle Atlantis. Photograph courtesy of NASA.

with gusts to 44 m/s (almost 100 mph). Figure 1.13 shows a satellite image of Hurricane Rita approaching the Florida coastline, and Figure 1.14 shows an image of three typhoons over the western Pacific Ocean taken on August 7, 2006. In all three figures the spiral nature of the hurricane or typhoon is apparent with a series of rainbands that surround the low-pressure "eye" at the center.

In comparison to cyclones, tornadoes are much smaller in scale but can involve more intense and destructive vorticity. Tornadoes have captured the interest of observers for centuries because of their destructive nature on a localized scale, flow intricacies, and the difficulty in predicting their occurrence and path. A tornado is a columnar vortex formed in the atmosphere with a height of about 3 km and is typically 10 to 100 m in diameter. A tornado involves concentration of angular momentum and kinetic energy into a low-altitude cloud sometimes called a rotor cloud. Rotating columns of air called funnels extend down from the cloud toward the ground, and if they have enough energy and momentum, they reach the Earth's surface. If the funnel reaches the Earth's surface, the central pressure will be low enough for the radial pressure gradient across the sheath of air forming the funnel to balance the centrifugal force in the rotating sheath of air. A secondary flow is super-posed on the rotation of the funnel, and the flow pattern associated with a tornado is in the form of a spiral, with a radial inward flow toward the eye of the tornado near the ground. The local winds resulting from a tornado can reach

FIGURE 1.13   Hurricane Rita, 21st September 2005. GOES Project, NASA-GSFC. Image courtesy of NASA.

FIGURE 1.14   Three typhoons over the western Pacific Ocean on the 7th August 2006. Image courtesy of NASA.

speeds of up to 200 or even 300 m/s. A tornado column will become visible if water vapor condenses within it or if its motion results in it picking up dust and debris. The presence of a low-pressure region at the center of a tornado accounts for some of the destructive capability. If the eye of the tornado where the pressure is lowest passes over a hollow building, the pressure difference between the air inside the building and the low pressure at the center of the tornado can result in very large forces on the walls and roof and can cause the building to literally explode. (A "small" tornado off the author's hometown in the United Kingdom is illustrated in Figure 1.15.) Tornadoes derive their energy and vorticity from other phenomena such as hurricanes, squall lines, thunderstorms, volcanoes, and firestorms. If a cyclone provides the vorticity required for the formation of a tornado, as a result, the tornado will tend to rotate counterclockwise, although anticyclonic tornadoes with clockwise rotation do occur. When a tornado forms over water, it is referred to as a water spout.

Atmospheric circulations can occur on any planet with a fluid atmosphere. For example, Jupiter's atmosphere is well known for its Great Red Spot, illustrated in Figure 1.16. This is a swirling mass of gas resembling a hurricane or massive anticyclone. The immense size of this nearly two-dimensional vortex, its persistence, and its very high turbulence have provided a source of fascination to observers and scientists for centuries. The current length of the elongated spot in the east-west direction is circa 20,000 km and is about three

FIGURE 1.15   Tornado off the coast of Brighton, England, 2nd October, 2006. Photograph by Jean Hopkins, Brighton.

FIGURE 1.16   Composite image of Jupiter taken by the narrow angle camera onboard NASA's Cassini spacecraft on December 29, 2000, during its closest approach to the giant planet at a distance of approximately 10 million kilometers (6.2 million miles). Image courtesy of NASA.

times the diameter of Earth. The width of the spot is about 12,000 km in the north-south direction, and its depth is believed to be somewhere between 20 km (Marcus 1993) and 200 km (Irwin 2003). The length of the spot varies, and 100 years ago estimates indicate that it had a length of 46,000 km (Irwin 2003). The color of the spot usually varies from brick-red to slightly brown, and the spot has been known to fade entirely. Its color may be due to small amounts of sulfur and phosphorus in the ammonia crystals. The edge of the Great Red Spot circulates at a speed of about 100 m/s (225 miles per hour). The spot remains at the same distance from the equator but drifts slowly east and west. The zones, belts, and the Great Red Spot are much more stable than similar circulation systems on Earth, but the mechanisms responsible for its formation (Marcus 1993) have similarities to those of the free vortex introduced in Chapter 3 and large-scale atmospheric vortices on Earth as considered in Chapter 8.

Figure 1.16 shows a variety of cloud features, as well as the Great Red Spot, including parallel dark and white bands, multilobed chaotic regions, white ovals, and many small vortices. Many clouds appear in streaks and waves due to continual stretching and folding by Jupiter's winds and turbulence. The gray features along the north edge of the central bright band are equatorial "hotspots." Small bright spots within the orange band north of the equator are lightning-bearing thunderstorms. The polar regions shown here are less visible because the Cassini spacecraft viewed them at an angle and through thicker atmospheric haze.

The parallel dark and white bands, the white ovals, and the Great Red Spot persist over many years or centuries, despite intense atmospheric turbulence. The most energetic features are the small, bright clouds to the left of the Great Red Spot and in similar locations in the northern half of the planet. These clouds grow and disappear over a few days and generate lightning. Streaks form as clouds are sheared apart by Jupiter's intense jet streams that run parallel to the colored bands. The prominent dark band in the northern half of the planet is the location of Jupiter's fastest jet stream, with eastward winds of about 130 m/s (300 miles per hour). Jupiter's diameter is 11 times that of Earth (142,984 km). Thus the smallest storms in Figure 1.16 are comparable in size to the largest hurricanes on Earth.

Unlike Earth, where only water condenses to form clouds, Jupiter's clouds are made of ammonia, hydrogen sulfide, water, and methane. The convective atmospheric updrafts and downdrafts bring different mixtures of these sub-stances up from lower elevations, leading to stratified clouds at different heights. It is the combination of shear, convection and capture, and the absorp-tion of smaller eddies that characterizes the atmospheric interactions on Jupiter. The brown and orange colors of its atmosphere may be due to trace chemicals dredged up from deeper levels of the atmosphere, or they may be by-products of chemical reactions driven by ultraviolet light from the Sun. Bluish areas, such as the small features just north and south of the equator, are areas of reduced cloud cover, where one can see deeper into the atmosphere (NASA 2003).

1.4. CONCLUSIONS

This chapter has introduced a series of industrial applications as well as examples of flows in the atmosphere and oceans in order to illustrate some of the principal phenomena and complexity associated with rotating flow. This book aims to describe the subject of rotating flow and associated phenomena and provide the reader with insights into modeling both general and specific rotating flow applications. A particular emphasis has been placed on those applications that involve disc and cylindrical geometries. In addition, a chapter is devoted to the modeling of the geophysical flows associated with the Earth's atmosphere and its oceans.

Chapters 2 and 3 provide an overview and detailed introduction to the fundamental mathematical means of modeling fluid flow, with a specific emphasis on the conservation equations of continuity, momentum, and energy. These enable detailed modeling of fluid flow and are presented in forms that are directly applicable to rotating flow. Several relatively simple applications of these equations are included in Chapter 3 illustrating the modeling of vortex flow. Chapters 4 and 5 focus on the flow in rotating disc applications important to a range of engineering applications such as gas turbine engines and pumps, and in some cases they provide a relatively simple geometric configuration for the modeling of complex three-dimensional and time-dependent flow. Chapter 6 is concerned with flows associated with rotating cylinders and spheres. This chapter also addresses the flow in a fluid-filled annulus where the inner or outer cylinder rotates, and there may also be an axial flow through the annulus. The flow in a cavity formed between two co-rotating discs is considered in Chapter 7. This subject is particularly challenging, as the flow physics depends on the thermal and geometric boundary conditions and tends to be three-dimensional and time-dependent and sometimes intractable even to the most sophisticated of current modeling techniques. The subjects of vortex flow and rotating disc and rotating cavity applications have particular elements of commonality with geophysical flows. These subjects are introduced in Chapter 8, with a particular focus on Earth's weather systems and, to a lesser extent, on oceanic circulations.

The subject of rotating flow has been addressed in a number of monographs and textbooks. The text by Dorfman (1963) has served as a standard reference for engineering researchers interested in rotating flow. This work has more recently been complemented by the two monographs by Owen and Rogers (1989, 1995), the textbooks by Vanyo (1993) and a monograph on rotating disc applications by Shevchuk (2009). A mathematical treatment of rotating flows is given by Greenspan (1968). In addition, valuable sections on rotating flow are developed in the books by Batchelor (1967) and Tritton (1988). Finally, a brief summary of the modeling methods for rotating discs and cylinders is provided by ESDU (2007).

REFERENCES

Batchelor, G.K. *An introduction to fluid dynamics*. Cambridge University Press, 1967.

Dorfman, L.A. *Hydrodynamic resistance and the heat loss of rotating solids*. Oliver & Boyd, 1963.

ESDU 07004. *Flow in Rotating Components*. Engineering Sciences Data Unit, 2007.

Greenspan, H.P. *The theory of rotating fluids*. Cambridge University Press, 1968.

Irwin, P.G.J. *Giant planets of our solar system*. Springer-Praxis, 2003.

Marcus, P.S. Jupiter's great red spot and other vortices. *Annual Review of Astronomy and Astrophysics*, 31, pp. 523–573, 1993.

NASA, 2003. http://photojournal.jpl.nasa.gov/catalog/PIA04866 accessed 010207.

Owen, J.M., & Rogers, R.H. *Flow and heat transfer in rotating-disc systems*. Volume 1—*Rotor-stator systems*. Research Studies Press, 1989.

Owen, J.M., & Rogers, R.H. *Flow and heat transfer in rotating-disc systems*. Volume 2—*Rotating cavities*. Research Studies Press, 1995.

Shevchuk, I.V. *Convective heat and mass transfer in rotating disk systems*. Springer, 2009.

Tritton, D.J. *Physical fluid dynamics*, 2nd ed. Oxford Science Publications, 1988.

Vanyo, J.P. *Rotating fluids in engineering and science*. Butterworth Heinemann, 1993.

Chapter 2

Laws of Motion

The laws of motion that apply to solids are valid for all matter including liquids and gases. When considering flow involving circular or swirling motion the laws of motion serve as a useful starting point.

2.1. INTRODUCTION

A principal difference between fluids and solids is that fluids distort without limit. For example, if a shear stress is applied to a fluid, as illustrated in Figure 2.1, then layers of fluid particles will move relative to each other and the particles will not return to their original position if application of the shear force is stopped. Analysis of a fluid needs to take account of such distortions.

One of the pre-eminent historical methods for analysing motion that has been applied involves identifying specific particles or objects and following their path. This method follows Newton's approach for particle motion and projectiles and is traditionally known as the Lagrangian formulation of particle mechanics. In this approach the position of a particle is the principal variable and particle velocity, and acceleration can be obtained by the first and second derivatives of the position vector. For example if x, y, z define the coordinates of the vector of a particle P, then the component velocities are given by $u_x = dx/dt$, $u_y = dy/dt$ and $u_z = dz/dt$. The Lagrangian formulation is illustrated for the case of a particle P in Figure 2.2.

An alternative approach is to identify fixed locations in space and analyse the motion of whatever fluid is in the fixed location at a particular time. This approach was developed by Euler (1755, 1759). The fluid domain can be divided up in to a grid of coordinates as illustrated in Figure 2.3. If x, y, z are the coordinates of a fixed location forming an Eulerian grid then $dx/dt = 0$, $dy/dt = 0$ and $dz/dt = 0$. In an Eulerian formulation, the principal variables are the velocity or momentum of the fluid and the pressure and density at all points of space occupied by the fluid for all instants of time.

Both Lagrangian and Eulerian approaches can be used to analyse the behaviour of a fluid. In general, however, the Eulerian approach has proved much simpler because of the large number of particles involved in a typical

Rotating Flow, DOI: 10.1016/B978-0-12-382098-3.00002-0
17

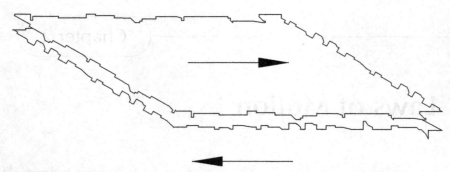

FIGURE 2.1   Relative motion of one layer of fluid over another. If the force causing the motion is removed then the fluid particles will not return to their original positions.

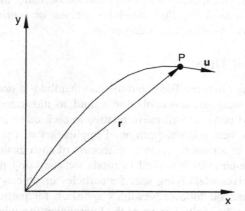

FIGURE 2.2   The Lagrangian formulation of particle mechanics for a particle P.

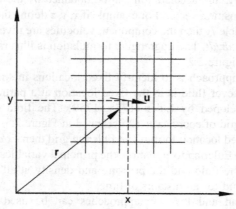

FIGURE 2.3   The Eulerian formulation of particle mechanics illustrating the division of a flow domain using a grid.

fluid flow and as a result is much more common. To give an idea of the magnitude of the problem that would be involved in tracking every molecule, in each cubic meter of air at room temperature and pressure there are approximately 2.549×10^{25} molecules. This represents a formidable challenge even to the largest current computational facilities.

A fluid particle will respond to a force in the same way that a solid particle will. If a force is applied to a particle, an acceleration will result, as governed by Newton's second law of motion, which states that the rate of change of momentum of a body is proportional to the unbalanced force acting on it and takes place in the direction of the force. It is useful to consider the forces that a fluid particle can experience. These include:

- body forces such as gravity and electromagnetism
- forces due to pressure
- forces due to viscous action
- forces due to rotation

The usual laws of mechanics, including conservation of mass and Newton's laws of motion, can be applied to a fluid. The resulting formulations provide the mathematical basis for modeling fluid flow and developing understanding and insight into fluid flow phenomena. The equations necessary for modelling a given flow are called the continuity and the Navier-Stokes equations, or the laws of motion. In the case of a flow involving a change in the total energy of a flow, the energy equation must also be taken into account. Although familiar from courses in fluid mechanics and standard text books such as Douglas et al. (2005) and White (2003), the Navier-Stokes equations are developed in Section 2.2 and the continuity equation in Section 2.3 for completeness and to allow the reader to develop an understanding of the context of the terms concerned. Methods for using these equations are discussed in Section 2.4. For many rotating flow applications, it is useful to model the flow using a rotating frame of reference. This subject and the associated equations of motion are introduced in Section 2.5. A facet of fluid mechanics is that by non-dimensionalizing the governing equations, the solution to a flow in a specified geometry will depend only on the values of the relevant non-dimensionalized groups that appear in the transformed equations and the boundary conditions. A solution to the governing equations will apply to any geometrically similar flow that has the same values for the dimensionless groups; and this subject is introduced in Section 2.6.

2.2. NAVIER-STOKES EQUATIONS

Assuming that the shear rate in a fluid is linearly related to shear stress and that the fluid flow is laminar, Cloaude-Louis Navier (1823) derived the equations of motion for a viscous fluid from molecular considerations. George Stokes (1845) also derived the equations of motion for a viscous fluid in a slightly different form and the basic equations that govern fluid flow are now generally known as

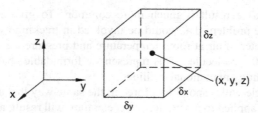

FIGURE 2.4 Fluid element.

the Navier-Stokes equations of motion. The Navier-Stokes equations can also be used for turbulent flow, with appropriate modifications.

Quantities of interest in engineering applications can typically be expressed in terms of integrals. Examples include volumetric flow rate, which can be determined by the integral of the velocity over an area; force, which is the integral of a stress over an area; and heat transfer, which is the integral of heat flux over an area. In order to determine an integral quantity, the integrand must be known, or sufficient information available so that a realistic estimate for it can be made. Integration can then be performed to give the integral quantity. If the integrand is not known and cannot be estimated, then the appropriate differential equations can be solved to give the required integrands. The equations of fluid mechanics that relate the integral quantities are known as the integral equations or control-volume formulation of fluid mechanics. Here, however, it is assumed that the integrands are not known and the relevant differential equations are developed.

The Navier-Stokes equations can be derived by considering the dynamic equilibrium of a fluid element. A fluid element is an arbitrary volume of fluid. Here a rectangular prism of fluid will be considered as illustrated in Figure 2.4, although any elemental volume of fluid could be used, albeit with different levels of algebraic complexity depending on the shape selected. The dimensions and location of this element of fluid is arbitrary and could be anywhere in the continuum of the fluid, in order to ensure that the results are generally applicable. The forces acting on the fluid element that need to be considered are surface forces, resulting from shear and normal stresses, body forces, such as those due to gravity or electromagnetism, and inertia forces. Body and surface forces arise as a result of the underlying structure of matter and the four fundamental forces of the Universe of which we are aware, namely gravitational forces, electromagnetic forces, and strong and weak intermolecular forces.

2.2.1 Surface Forces

Surface forces are short-range forces that act on a fluid element through physical contact between the fluid element and its surroundings. Examples of surface forces are those resulting from pressure or shear stress acting on an area. As the magnitude of a surface force is proportional to the contact area between the

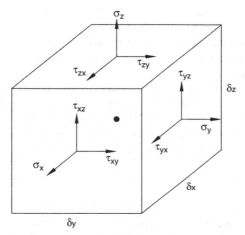

FIGURE 2.5 Stresses on three surfaces of a fluid element.

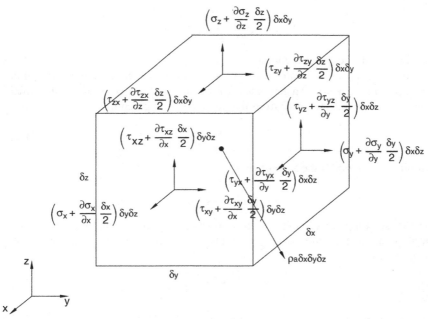

FIGURE 2.6 Forces acting on an infinitesimal fluid element.

fluid and its surroundings, surface forces tend to be expressed in terms of the force per unit area with units of N/m².

The stresses on the surface of a fluid element consist of normal and shear stresses. These are illustrated in Figure 2.5, and the corresponding forces are shown in Figure 2.6. It is common practice to identify the shear stress components by using two subscripts. The first denotes the direction of the outward normal, the second the direction of the stress component. For example, the component τ_{yx}, is the viscous shear stress acting in the x direction, on a plane normal to the y direction. The normal stress components require only one subscript to be unambiguous.

It is frequently useful to be able to estimate the conditions of a fluid at a small distance away from a point where the conditions are actually known. A Taylor series expansion can be used to do this as it allows estimation of a function in terms of a known value and differentials of the conditions at that value. The Taylor series expansion in a general form (Stroud, 2001) is given by

$$f(x+h) = f(x) + h.f'(x) + \frac{h^2}{2!}f''(x) + \frac{h^3}{3!}f'''(x) + \dots \qquad (2.1)$$

where $f'(x)=d(f(x))/dx$, $f''(x)=d^2(f(x))/dx^2$, $f'''(x)=d^3(f(x))/dx^3$, and so on.

If, for example, the force at the point (x, y, z) is known, an estimate of the force at a location $\delta x/2$, represented by h in Equation 2.1, away can be made, in the case of the x component using

$$f\left(x+\frac{\delta x}{2}\right) = f(x) + \frac{\delta x}{2}f'(x) \qquad (2.2)$$

if higher order terms are ignored.

In the analysis presented here, it is assumed that conditions at the point concerned are known. An estimate of, for example, the x component of normal stress, σ_x, at a distance $\delta x/2$ from the point x, y, z is given by

$$\sigma_x + \frac{\delta x}{2}\frac{\partial \sigma_x}{\partial x} \qquad (2.3)$$

Similarly at $x-(\delta x/2)$ the x component of stress is given by

$$\sigma_x - \frac{\delta x}{2}\frac{\partial \sigma_x}{\partial x} \qquad (2.4)$$

For these components of stress, the corresponding forces can be evaluated by multiplying by the appropriate area, in this case $\delta y\delta z$. Hence the force at $x+(\delta x/2)$ is given by

$$\left(\sigma_x + \frac{\delta x}{2}\frac{\partial \sigma_x}{\partial x}\right)\delta y\delta z \qquad (2.5)$$

The corresponding force at $x-(\delta x/2)$ is given by

$$\left(\sigma_x - \frac{\delta x}{2}\frac{\partial \sigma_x}{\partial x}\right)\delta y\delta z \qquad (2.6)$$

The component forces are illustrated in Figure 2.6 for three of the faces. The corresponding forces on the opposite faces have been omitted to prevent further crowding of the figure.

For equilibrium of the moments on the fluid element, the stresses must be symmetric. The shear stresses on adjacent planes must therefore be equal as follows

$$\tau_{xy} = \tau_{yx} \tag{2.7}$$

$$\tau_{xz} = \tau_{zx} \tag{2.8}$$

$$\tau_{yz} = \tau_{zy} \tag{2.9}$$

The hydrostatic pressure in the fluid element, p, is taken as the average of the three normal stresses.

$$p = -\frac{1}{3}(\sigma_x + \sigma_y + \sigma_z) \tag{2.10}$$

A negative sign is used here because hydrostatic forces are compressive, whereas as positive stresses are tensile.

In order to provide additional information necessary to match the number of unknowns and equations, a relationship between stress and rate of strain for the particular fluid concerned is required. This relationship forms part of the so-called constitutive model for a fluid, which provides the necessary relationships between the components of the stress tensor and the fluid velocity field to give a set of governing equations that are well posed and solvable. The constitutive model uses parameters that reflect the physical and molecular structure of the fluid and its thermodynamic state.

The magnitude of the shear stresses depends on the rate at which the fluid is being distorted and taken, in Cartesian coordinates, to be

$$\tau_{xy} = \tau_{yx} = \mu\left(\frac{\partial u_y}{\partial x} + \frac{\partial u_x}{\partial y}\right) \tag{2.11}$$

$$\tau_{yz} = \tau_{zy} = \mu\left(\frac{\partial u_z}{\partial y} + \frac{\partial u_y}{\partial z}\right) \tag{2.12}$$

$$\tau_{zx} = \tau_{xz} = \mu\left(\frac{\partial u_x}{\partial z} + \frac{\partial u_z}{\partial x}\right) \tag{2.13}$$

The magnitude of the normal stresses are taken as (Schlichting 1979)

$$\sigma_x = -p - \frac{2}{3}\mu\left(\frac{\partial u_x}{\partial x} + \frac{\partial u_y}{\partial y} + \frac{\partial u_z}{\partial z}\right) + 2\mu\frac{\partial u_x}{\partial x} \tag{2.14}$$

$$\sigma_y = -p - \frac{2}{3}\mu\left(\frac{\partial u_x}{\partial x} + \frac{\partial u_y}{\partial y} + \frac{\partial u_z}{\partial z}\right) + 2\mu\frac{\partial u_y}{\partial y} \tag{2.15}$$

$$\sigma_z = -p - \frac{2}{3}\mu\left(\frac{\partial u_x}{\partial x} + \frac{\partial u_y}{\partial y} + \frac{\partial u_z}{\partial z}\right) + 2\mu\frac{\partial u_z}{\partial z} \tag{2.16}$$

The term

$$\frac{\partial u_x}{\partial x} + \frac{\partial u_y}{\partial y} + \frac{\partial u_z}{\partial z} \qquad (2.17)$$

represents the divergence of the velocity vector, or dilation, and provides an indication of the rate at which fluid is flowing out from each point, that is, the expansion of the fluid.

Normal and shear stresses tend to move the fluid element in the x, y and z directions. The surface forces resulting from these stresses can be expressed in the form

$$\frac{\partial \sigma_x}{\partial x} \delta x \delta y \delta z \qquad (2.18)$$

$$\frac{\partial \sigma_y}{\partial y} \delta x \delta y \delta z \qquad (2.19)$$

$$\frac{\partial \sigma_z}{\partial z} \delta x \delta y \delta z \qquad (2.20)$$

and

$$\frac{\partial \tau_{xy}}{\partial y} \delta x \delta y \delta z \qquad (2.21)$$

$$\frac{\partial \tau_{zx}}{\partial x} \delta x \delta y \delta z \qquad (2.22)$$

$$\frac{\partial \tau_{yz}}{\partial z} \delta x \delta y \delta z \qquad (2.23)$$

$$\frac{\partial \tau_{yx}}{\partial x} \delta x \delta y \delta z \qquad (2.24)$$

$$\frac{\partial \tau_{xz}}{\partial z} \delta x \delta y \delta z \qquad (2.25)$$

$$\frac{\partial \tau_{zy}}{\partial y} \delta x \delta y \delta z \qquad (2.26)$$

2.2.2 Body Forces

Body forces are long-range forces that act on an elemental body in such a way that the magnitude of the body force is proportional to the mass of the elemental body. As the mass of the elemental body is equal to the product of its volume and density, the magnitude of the body force is proportional to the volume of the elemental body. Body forces therefore tend to be expressed in terms of the force per unit volume with units of N/m^3. Using the definitions given here, the body force is given by F, or in component form in Cartesian coordinates by F_x, F_y and F_z. The component forces acting on an element are given by

$$F_x \delta x \delta y \delta z \tag{2.27}$$

$$F_y \delta x \delta y \delta z \tag{2.28}$$

$$F_z \delta x \delta y \delta z \tag{2.29}$$

2.2.3 Inertia Forces

Velocity is a function of space of time and be defined by the vector

$$\boldsymbol{u} = f(x, y, z, t) \tag{2.30}$$

with components of velocity in Cartesian coordinates defined by u_x, u_y and u_z.
From the 'rates of change' rule of partial differentiation,

$$\frac{du_x}{dt} = \frac{\partial u_x}{\partial t} + \frac{\partial u_x}{\partial x}\frac{dx}{dt} + \frac{\partial u_x}{\partial y}\frac{dy}{dt} + \frac{\partial u_x}{\partial z}\frac{dz}{dt} \tag{2.31}$$

In the limit as dt tends to zero, dx/dt tends to u_x, dy/dt tends to u_y and dz/dt tends to u_z. So Equation 2.31 can be written as

$$\frac{du_x}{dt} = \frac{\partial u_x}{\partial t} + u_x\frac{\partial u_x}{\partial x} + u_y\frac{\partial u_x}{\partial y} + u_z\frac{\partial u_x}{\partial z} \tag{2.32}$$

This form is commonly represented by the total derivative, D/Dt, also known as the substantive derivative or material derivative. Its name arises because the derivative involves following a particular fluid particle, material, or substance. It can be applied to other dependent variables as well as velocity. For example, DT/Dt would represent the change of temperature of a fluid particle as the particle flows along a path.
For the case of the x component of velocity,

$$\underbrace{\frac{Du_x}{Dt}}_{\text{Total derivative}} = \underbrace{\frac{\partial u_x}{\partial t}}_{\text{Local derivative}} + \underbrace{u_x\frac{\partial u_x}{\partial x} + u_y\frac{\partial u_x}{\partial y} + u_z\frac{\partial u_x}{\partial z}}_{\text{Convective differential terms}} \tag{2.33}$$

Similarly, for the y and z components of velocity,

$$\frac{Du_y}{Dt} = \frac{\partial u_y}{\partial t} + u_x\frac{\partial u_y}{\partial x} + u_y\frac{\partial u_y}{\partial y} + u_z\frac{\partial u_y}{\partial z} \tag{2.34}$$

$$\frac{Du_z}{Dt} = \frac{\partial u_z}{\partial t} + u_x\frac{\partial u_z}{\partial x} + u_y\frac{\partial u_z}{\partial y} + u_z\frac{\partial u_z}{\partial z} \tag{2.35}$$

The total derivative, D/Dt thus gives a measure of the change of velocity of one fluid element as it moves about in space. The term $\partial/\partial t$ gives the variation of velocity at a fixed point and is known as the local derivative. The remaining three terms on the right-hand side of Equations 2.33, 2.34 and 2.35 are grouped together and known as the convective terms or convective differential.

The forces required to accelerate a fluid element are given by

$$\rho \frac{Du_x}{Dt} \delta x \delta y \delta z \tag{2.36}$$

$$\rho \frac{Du_y}{Dt} \delta x \delta y \delta z \tag{2.37}$$

$$\rho \frac{Du_z}{Dt} \delta x \delta y \delta z \tag{2.38}$$

2.2.4 Equilibrium

The inertial forces acting on a fluid element are balanced by the surface and body forces. These have been defined for each component direction in Sections 2.2.1, 2.2.2, and 2.2.3. An equation can now be formulated equating these terms. For the x component,

$$\rho \frac{Du_x}{Dt} \delta x \delta y \delta z = \frac{\partial \sigma_x}{\partial x} \delta x \delta y \delta z + \frac{\partial \tau_{xy}}{\partial y} \delta x \delta y \delta z + \frac{\partial \tau_{xz}}{\partial z} \delta x \delta y \delta z + F_x \delta x \delta y \delta z \tag{2.39}$$

Dividing through by $\delta x \delta y \delta z$, gives

$$\rho \frac{Du_x}{Dt} = \frac{\partial \sigma_x}{\partial x} + \frac{\partial \tau_{xy}}{\partial y} + \frac{\partial \tau_{xz}}{\partial z} + F_x \tag{2.40}$$

Similarly, for the y and z components

$$\rho \frac{Du_y}{Dt} = \frac{\partial \sigma_y}{\partial y} + \frac{\partial \tau_{yx}}{\partial x} + \frac{\partial \tau_{yz}}{\partial z} + F_y \tag{2.41}$$

$$\rho \frac{Du_z}{Dt} = \frac{\partial \sigma_z}{\partial z} + \frac{\partial \tau_{zx}}{\partial x} + \frac{\partial \tau_{zy}}{\partial y} + F_z \tag{2.42}$$

Using the relationships between shear stress and strain developed in Equations 2.11–2.16, Equations 2.40–2.42 become

$$\rho \left(\frac{\partial u_x}{\partial t} + u_x \frac{\partial u_x}{\partial x} + u_y \frac{\partial u_x}{\partial y} + u_z \frac{\partial u_x}{\partial z} \right) = -\frac{\partial p}{\partial x} - \frac{2}{3} \frac{\partial}{\partial x} \left[\mu \left(\frac{\partial u_x}{\partial x} + \frac{\partial u_y}{\partial y} + \frac{\partial u_z}{\partial z} \right) \right]$$

$$+ 2 \frac{\partial}{\partial x} \left(\mu \frac{\partial u_x}{\partial x} \right) + \frac{\partial}{\partial y} \left[\mu \left(\frac{\partial u_x}{\partial y} + \frac{\partial u_y}{\partial x} \right) \right] + \frac{\partial}{\partial z} \left[\mu \left(\frac{\partial u_x}{\partial z} + \frac{\partial u_z}{\partial x} \right) \right] + F_x$$

$$\tag{2.43}$$

$$\rho \left(\frac{\partial u_y}{\partial t} + u_x \frac{\partial u_y}{\partial x} + u_y \frac{\partial u_y}{\partial y} + u_z \frac{\partial u_y}{\partial z} \right) = -\frac{\partial p}{\partial y} - \frac{2}{3}\frac{\partial}{\partial y}\left[\mu \left(\frac{\partial u_x}{\partial x} + \frac{\partial u_y}{\partial y} + \frac{\partial u_z}{\partial z} \right) \right]$$

$$+2\frac{\partial}{\partial y}\left(\mu \frac{\partial u_y}{\partial y} \right) + \frac{\partial}{\partial x}\left[\mu \left(\frac{\partial u_x}{\partial y} + \frac{\partial u_y}{\partial x} \right) \right] + \frac{\partial}{\partial z}\left[\mu \left(\frac{\partial u_y}{\partial z} + \frac{\partial u_z}{\partial y} \right) \right] + F_y$$

$$(2.44)$$

$$\rho \left(\frac{\partial u_z}{\partial t} + u_x \frac{\partial u_z}{\partial x} + u_y \frac{\partial u_z}{\partial y} + u_z \frac{\partial u_z}{\partial z} \right) = -\frac{\partial p}{\partial z} - \frac{2}{3}\frac{\partial}{\partial z}\left[\mu \left(\frac{\partial u_x}{\partial x} + \frac{\partial u_y}{\partial y} + \frac{\partial u_z}{\partial z} \right) \right]$$

$$+2\frac{\partial}{\partial z}\left(\mu \frac{\partial u_z}{\partial z} \right) + \frac{\partial}{\partial x}\left[\mu \left(\frac{\partial u_x}{\partial z} + \frac{\partial u_z}{\partial x} \right) \right] + \frac{\partial}{\partial y}\left[\mu \left(\frac{\partial u_y}{\partial z} + \frac{\partial u_z}{\partial y} \right) \right] + F_z$$

$$(2.45)$$

Equations 2.43–2.45 are a statement of the Navier-Stokes equations. The Navier-Stokes equations are an expression of Newton's second law of motion for a fluid of constant density. The terms in Equations 2.43 to 2.45 on the left hand side represent inertia effects and those on the right hand side, in order, pressure gradient, viscous terms, and body force.

Equations 2.43–2.45 are valid for viscous, compressible flow with varying viscosity. If the viscosity is assumed to be constant, the Navier-Stokes equations can be simplified to

$$\rho \left(\frac{\partial u_x}{\partial t} + u_x \frac{\partial u_x}{\partial x} + u_y \frac{\partial u_x}{\partial y} + u_z \frac{\partial u_x}{\partial z} \right) = -\frac{\partial p}{\partial x} + \mu \left(\frac{\partial^2 u_x}{\partial x^2} + \frac{\partial^2 u_x}{\partial y^2} + \frac{\partial^2 u_x}{\partial z^2} \right)$$

$$+\frac{\mu}{3}\frac{\partial}{\partial x}\left(\frac{\partial u_x}{\partial x} + \frac{\partial u_y}{\partial y} + \frac{\partial u_z}{\partial z} \right) + F_x \qquad (2.46)$$

$$\rho \left(\frac{\partial u_y}{\partial t} + u_x \frac{\partial u_y}{\partial x} + u_y \frac{\partial u_y}{\partial y} + u_z \frac{\partial u_y}{\partial z} \right) = -\frac{\partial p}{\partial y} + \mu \left(\frac{\partial^2 u_y}{\partial x^2} + \frac{\partial^2 u_y}{\partial y^2} + \frac{\partial^2 u_y}{\partial z^2} \right)$$

$$+\frac{\mu}{3}\frac{\partial}{\partial y}\left(\frac{\partial u_x}{\partial x} + \frac{\partial u_y}{\partial y} + \frac{\partial u_z}{\partial z} \right) + F_y \qquad (2.47)$$

$$\rho \left(\frac{\partial u_z}{\partial t} + u_x \frac{\partial u_z}{\partial x} + u_y \frac{\partial u_z}{\partial y} + u_z \frac{\partial u_z}{\partial z} \right) = -\frac{\partial p}{\partial z} + \mu \left(\frac{\partial^2 u_z}{\partial x^2} + \frac{\partial^2 u_z}{\partial y^2} + \frac{\partial^2 u_z}{\partial z^2} \right)$$

$$+\frac{\mu}{3}\frac{\partial}{\partial z}\left(\frac{\partial u_x}{\partial x} + \frac{\partial u_y}{\partial y} + \frac{\partial u_z}{\partial z} \right) + F_z \qquad (2.48)$$

If the density is also constant, then the dilation is zero and Equations 2.46–2.48 become

$$\rho\left(\frac{\partial u_x}{\partial t}+u_x\frac{\partial u_x}{\partial x}+u_y\frac{\partial u_x}{\partial y}+u_z\frac{\partial u_x}{\partial z}\right)=-\frac{\partial p}{\partial x}+\mu\left(\frac{\partial^2 u_x}{\partial x^2}+\frac{\partial^2 u_x}{\partial y^2}+\frac{\partial^2 u_x}{\partial z^2}\right)+F_x$$

$$(2.49)$$

$$\rho\left(\frac{\partial u_y}{\partial t}+u_x\frac{\partial u_y}{\partial x}+u_y\frac{\partial u_y}{\partial y}+u_z\frac{\partial u_y}{\partial z}\right)=-\frac{\partial p}{\partial y}+\mu\left(\frac{\partial^2 u_y}{\partial x^2}+\frac{\partial^2 u_y}{\partial y^2}+\frac{\partial^2 u_y}{\partial z^2}\right)+F_y$$

$$(2.50)$$

$$\rho\left(\frac{\partial u_z}{\partial t}+u_x\frac{\partial u_z}{\partial x}+u_y\frac{\partial u_z}{\partial y}+u_z\frac{\partial u_z}{\partial z}\right)=-\frac{\partial p}{\partial z}+\mu\left(\frac{\partial^2 u_z}{\partial x^2}+\frac{\partial^2 u_z}{\partial y^2}+\frac{\partial^2 u_z}{\partial z^2}\right)+F_z$$

$$(2.51)$$

The Navier-Stokes equation can equally well be expressed using vector notation and is given in the case of a constant density and constant viscosity fluid by

$$\rho\underbrace{\frac{D\mathbf{u}}{Dt}}_{\textit{inertial terms}}=-\underbrace{\nabla p}_{\substack{\textit{pressure}\\\textit{gradient}}}+\underbrace{\mu\nabla^2\mathbf{u}}_{\substack{\textit{viscous}\\\textit{terms}}}+\underbrace{\mathbf{F}}_{\substack{\textit{body}\\\textit{forces}}} \qquad (2.52)$$

In Equation 2.52 ∇ is the del operator (sometimes also referred to as the gradient, grad, or nabla, operator), and $\nabla^2=\nabla\cdot\nabla$, the Laplacian. In Cartesian coordinates, the del operator, ∇, and the Laplacian, ∇^2, are determined respectively, by

$$\nabla=\frac{\partial}{\partial x}\mathbf{i}+\frac{\partial}{\partial y}\mathbf{j}+\frac{\partial}{\partial z}\mathbf{k} \qquad (2.53)$$

$$\nabla^2=\nabla\cdot\nabla=\frac{\partial^2}{\partial x^2}+\frac{\partial^2}{\partial y^2}+\frac{\partial^2}{\partial z^2} \qquad (2.54)$$

It is often convenient to use cylindrical polar coordinates, r, ϕ, z, for rotating flows, particularly, for example, in fluid machinery applications where, say, a rotor is spinning in a cylindrical casing as is the case in turbochargers and gas turbine engines. The definition of cylindrical coordinates is illustrated in Figure 2.7, and the transformation between Cartesian coordinates is defined by

$$x=r\cos\phi \qquad (2.55)$$

$$y=r\sin\phi \qquad (2.56)$$

$$z=z \qquad (2.57)$$

The transformations given in Equations 2.55–2.57 can be used to give the Navier-Stokes equations for constant viscosity and density in cylindrical coordinates as given in Equations 2.58–2.60.

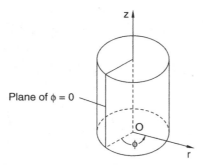

FIGURE 2.7   Cylindrical polar coordinates.

$$\rho\left(\frac{\partial u_r}{\partial t} + u_r\frac{\partial u_r}{\partial r} + \frac{u_\phi}{r}\frac{\partial u_r}{\partial \phi} + u_z\frac{\partial u_r}{\partial z} - \frac{u_\phi^2}{r}\right) = -\frac{\partial p}{\partial r}$$

$$+\mu\left(\frac{\partial^2 u_r}{\partial r^2} + \frac{1}{r}\frac{\partial u_r}{\partial r} - \frac{u_r}{r^2} + \frac{1}{r^2}\frac{\partial^2 u_r}{\partial \phi^2} + \frac{\partial^2 u_r}{\partial z^2} - \frac{2}{r^2}\frac{\partial u_\phi}{\partial \phi}\right) + F_r$$

(2.58)

$$\rho\left(\frac{\partial u_\phi}{\partial t} + u_r\frac{\partial u_\phi}{\partial r} + \frac{u_r u_\phi}{r} + \frac{u_\phi}{r}\frac{\partial u_\phi}{\partial \phi} + u_z\frac{\partial u_\phi}{\partial z}\right) = -\frac{1}{r}\frac{\partial p}{\partial \phi}$$

$$+\mu\left(\frac{\partial^2 u_\phi}{\partial r^2} + \frac{1}{r}\frac{\partial u_\phi}{\partial r} - \frac{u_\phi}{r^2} + \frac{1}{r^2}\frac{\partial^2 u_\phi}{\partial \phi^2} + \frac{\partial^2 u_\phi}{\partial z^2} + \frac{2}{r^2}\frac{\partial u_r}{\partial \phi}\right) + F_\phi$$

(2.59)

$$\rho\left(\frac{\partial u_z}{\partial t} + u_r\frac{\partial u_z}{\partial r} + \frac{u_\phi}{r}\frac{\partial u_z}{\partial \phi} + u_z\frac{\partial u_z}{\partial z}\right) = -\frac{\partial p}{\partial z}$$

$$+\mu\left(\frac{\partial^2 u_z}{\partial r^2} + \frac{1}{r}\frac{\partial u_z}{\partial r} + \frac{1}{r^2}\frac{\partial^2 u_z}{\partial \phi^2} + \frac{\partial^2 u_z}{\partial z^2}\right) + F_z$$

(2.60)

Equations 2.40–2.42, 2.43–2.45, 2.46–2.48, 2.49–2.51, 2.52 and 2.58–2.60 are valid for laminar flow. If the flow is turbulent, it is necessary to account for both the velocity and pressure by an average and fluctuating component. Replacing $\boldsymbol{u}$ by $\bar{\boldsymbol{u}} + \boldsymbol{u}'$ and p by $\bar{p} + p'$ where the primes represent the fluctuating components, and assuming the average values of the derivatives of u' and p' are zero, the steady state momentum equation, using vector notation becomes

$$(\bar{\boldsymbol{u}}.\nabla)\bar{\boldsymbol{u}} = -\frac{1}{\rho}\nabla\bar{p} + \nu\nabla^2\bar{\boldsymbol{u}} - \overline{(\boldsymbol{u}'.\nabla)\boldsymbol{u}'}$$

(2.61)

The term $\overline{(\boldsymbol{u}'.\nabla)\boldsymbol{u}'}$ is equivalent to an extra stress that affects the average motion in turbulent flow and is usually referred to as the Reynolds stress. In component form, the Navier-Stokes equations in cylindrical coordinates can be stated as

$$\rho\left(\frac{\partial u_r}{\partial t} + u_r\frac{\partial u_r}{\partial r} + \frac{u_\phi}{r}\frac{\partial u_r}{\partial \phi} + u_z\frac{\partial u_r}{\partial z} - \frac{u_\phi^2}{r}\right) = -\frac{\partial p}{\partial r} + \mu\left(\nabla^2 u_r - \frac{u_r}{r^2} - \frac{2}{r^2}\frac{\partial u_\phi}{\partial \phi}\right)$$

$$+ \frac{1}{r}\frac{\partial}{\partial r}(-\rho\overline{u_r'^2}) + \frac{1}{r}\frac{\partial}{\partial \phi}(-\rho\overline{u_r' u_\phi'}) + \frac{\partial}{\partial z}(-\rho\overline{u_r' u_z'}) - \frac{1}{r}(-\rho\overline{u_\phi'^2}) + F_r \quad (2.62)$$

$$\rho\left(\frac{\partial u_\phi}{\partial t} + u_r\frac{\partial u_\phi}{\partial r} + \frac{u_\phi}{r}\frac{\partial u_\phi}{\partial \phi} + u_z\frac{\partial u_\phi}{\partial z} + \frac{u_r u_\phi}{r}\right) = -\frac{\partial p}{\partial \phi} + \mu\left(\nabla^2 u_\phi + \frac{2}{r^2}\frac{\partial u_r}{\partial \phi} - \frac{u_\phi}{r}\right)$$

$$+ \frac{1}{r}\frac{\partial}{\partial \phi}(-\rho\overline{u_\phi'^2}) + \frac{\partial}{\partial r}(-\rho\overline{u_r' u_\phi'}) + \frac{\partial}{\partial z}(-\rho\overline{u_\phi' u_z'}) + \frac{2}{r}(-\rho\overline{u_r' u_\phi'}) + F_\phi \quad (2.63)$$

$$\rho\left(\frac{\partial u_z}{\partial t} + u_r\frac{\partial u_z}{\partial r} + \frac{u_\phi}{r}\frac{\partial u_z}{\partial \phi} + u_z\frac{\partial u_z}{\partial z}\right) = -\frac{\partial p}{\partial z} + \mu(\nabla^2 u_z)$$

$$+ \frac{1}{r}\frac{\partial}{\partial r}(-\rho\overline{u_r' u_z'}) + \frac{1}{r}\frac{\partial}{\partial \phi}(-\rho\overline{u_\phi' u_z'}) + \frac{\partial}{\partial z}(-\rho\overline{u_z'^2}) + F_z \quad (2.64)$$

where

$$\nabla^2 = \frac{\partial}{\partial r^2} + \frac{1}{r}\frac{\partial}{\partial r} + \frac{1}{r^2}\frac{\partial}{\partial \phi^2} + \frac{\partial}{\partial z^2} \quad (2.65)$$

In component form, the momentum equations for an axisymmetric flow can be stated as

$$\rho\left(u_r\frac{\partial u_r}{\partial r} + u_z\frac{\partial u_r}{\partial z} - \frac{u_\phi^2}{r}\right) = -\frac{\partial p}{\partial r} + \mu\left(\nabla^2 u_r - \frac{u_r}{r^2}\right) + \frac{1}{r}\frac{\partial}{\partial r}(-\rho\overline{u_r'^2})$$

$$+ \frac{\partial}{\partial z}(-\rho\overline{u_r' u_z'}) + \frac{\rho\overline{u_\phi'^2}}{r} + F_r \quad (2.66)$$

$$\rho\left(u_r\frac{\partial u_\phi}{\partial r} + u_z\frac{\partial u_\phi}{\partial z} + \frac{u_r u_\phi}{r}\right) = \mu\left(\nabla^2 u_\phi - \frac{u_\phi}{r^2}\right) + \frac{\partial}{\partial r}(-\rho\overline{u_r' u_\phi'})$$

$$+ \frac{\partial}{\partial z}(-\rho\overline{u_\phi' u_z'}) + \frac{2}{r}(-\rho\overline{u_r' u_\phi'}) + F_\phi \quad (2.67)$$

$$\rho\left(u_r\frac{\partial u_z}{\partial r} + u_z\frac{\partial u_z}{\partial z}\right) = -\frac{\partial p}{\partial z} + \mu\nabla^2 u_z + \frac{1}{r}\frac{\partial}{\partial r}(-\rho\overline{u_r' u_z'}) + \frac{\partial}{\partial z}(-\rho\overline{u_z'^2}) + F_z$$

$$(2.68)$$

2.3. CONTINUITY EQUATION

The continuity equation is an expression representing the idea that matter is conserved in a flow. Per unit volume, the sum of all masses flowing in and out per unit time must be equal to the change of mass due to change in density per unit time. This can be expressed in the form of an equation by

$$\dot{m}_{in} - \dot{m}_{out} = \frac{\partial}{\partial t} m_{element} \tag{2.69}$$

The mass flux through each face of an arbitrary fluid element is illustrated in Figure 2.8.

Summing the inflows and outflows and equating them to the change in mass gives

$$\left[\rho u_x - \frac{\delta x}{2}\frac{\partial(\rho u_x)}{\partial x}\right]\delta y \delta z - \left[\rho u_x + \frac{\delta x}{2}\frac{\partial(\rho u_x)}{\partial x}\right]\delta y \delta z + \left[\rho u_y - \frac{\delta y}{2}\frac{\partial(\rho u_y)}{\partial y}\right]\delta x \delta y$$

$$- \left[\rho u_y + \frac{\delta y}{2}\frac{\partial(\rho u_y)}{\partial y}\right]\delta x \delta y + \left[\rho u_z - \frac{\delta z}{2}\frac{\partial(\rho u_z)}{\partial z}\right]\delta x \delta y - \left[\rho u_z + \frac{\delta z}{2}\frac{\partial(\rho u_z)}{\partial z}\right]\delta x \delta y$$

$$= \frac{\partial}{\partial t}(\rho \delta x \delta y \delta z) \tag{2.70}$$

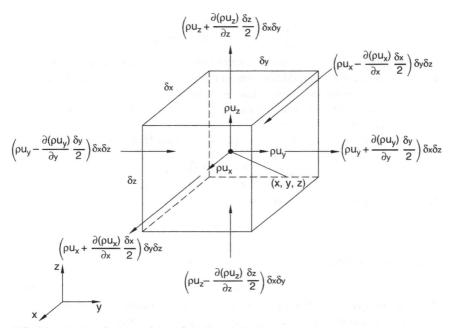

FIGURE 2.8 Mass flux through an infinitesimal fluid element control volume.

Collecting terms, dividing through by $\delta x \delta y \delta y$ and converting this to a partial differential equation gives

$$\frac{\partial(\rho u_x)}{\partial x} + \frac{\partial(\rho u_y)}{\partial y} + \frac{\partial(\rho u_z)}{\partial z} = -\frac{\partial \rho}{\partial t} \tag{2.71}$$

This can be restated in the form

$$\frac{\partial \rho}{\partial t} + u_x \frac{\partial \rho}{\partial x} + u_y \frac{\partial \rho}{\partial y} + u_z \frac{\partial \rho}{\partial z} + \rho\left(\frac{\partial u_x}{\partial x} + \frac{\partial u_y}{\partial y} + \frac{\partial u_z}{\partial z}\right) = 0 \tag{2.72}$$

Using the substantive derivative, Equation 2.72 is given by

$$\frac{D\rho}{Dt} + \rho\left(\frac{\partial u_x}{\partial x} + \frac{\partial u_y}{\partial y} + \frac{\partial u_z}{\partial z}\right) = 0 \tag{2.73}$$

or in vector form, using the del operator, the continuity equation for unsteady flows is given by

$$\frac{D\rho}{Dt} + \rho\nabla\cdot\mathbf{u} = 0 \tag{2.74}$$

The continuity equation applies to all fluids, compressible and incompressible flow, Newtonian and non-Newtonian fluids. It expresses the law of conservation of mass at each point in a fluid and must therefore be satisfied at every point in a flow field.

For incompressible flow the density of a fluid particle does not change as it travels, so

$$\frac{D\rho}{Dt} = \frac{\partial \rho}{\partial t} + u_x \frac{\partial \rho}{\partial x} + u_y \frac{\partial \rho}{\partial y} + u_z \frac{\partial \rho}{\partial z} = 0 \tag{2.75}$$

Note that the definition of incompressible flow is less restrictive than that of constant density. Constant density would require each term in Equation 2.75 to be zero. Incompressible flows that have density gradients are sometimes referred to as stratified flows, and examples of these are found in atmospheric and oceanic flows.

For an incompressible flow, with $D\rho/Dt = 0$, then from Equation 2.73

$$\rho\left(\frac{\partial u_x}{\partial x} + \frac{\partial u_y}{\partial y} + \frac{\partial u_z}{\partial z}\right) = 0 \tag{2.76}$$

Dividing through by the density gives

$$\frac{\partial u_x}{\partial x} + \frac{\partial u_y}{\partial y} + \frac{\partial u_z}{\partial z} = 0 \tag{2.77}$$

or in vector notation

$$\nabla\cdot\mathbf{u} = 0 \tag{2.78}$$

This equation is valid whether the velocity is time dependent or not.

In cylindrical coordinates, the continuity equation is given by

$$\frac{\partial \rho}{\partial t} + u_r \frac{\partial \rho}{\partial r} + \frac{u_\phi}{r}\frac{\partial \rho}{\partial \phi} + u_z \frac{\partial \rho}{\partial z} + \rho\left(\frac{1}{r}\frac{\partial(ru_r)}{\partial r} + \frac{1}{r}\frac{\partial u_\phi}{\partial \phi} + \frac{\partial u_z}{\partial z}\right) = 0 \qquad (2.79)$$

For incompressible flow, Equation 2.79 reduces to

$$\frac{1}{r}\frac{\partial}{\partial r}(ru_r) + \frac{1}{r}\frac{\partial u_\phi}{\partial \phi} + \frac{\partial u_z}{\partial z} = \frac{\partial u_r}{\partial r} + \frac{u_r}{r} + \frac{1}{r}\frac{\partial u_\phi}{\partial \phi} + \frac{\partial u_z}{\partial z} = 0 \qquad (2.80)$$

If the flow is turbulent, it is necessary to account for both the velocity and pressure by an average and fluctuating component. Replacing u by $\bar{u} + u'$ where the primes represent the fluctuating components, and assuming the average values of the derivatives of u' are zero, the continuity equation using vector notation becomes

$$\nabla\cdot\bar{u} = 0 \qquad (2.81)$$

In component form the continuity equation for an axisymmetric flow can be stated as

$$\frac{\partial \bar{u}_r}{\partial r} + \frac{\bar{u}_r}{r} + \frac{\partial \bar{u}_z}{\partial z} = 0 \qquad (2.82)$$

2.4. SOLUTION OF THE GOVERNING EQUATIONS OF FLUID MECHANICS

A solution of the governing equations of fluid mechanics provides a complete description of the flow at every point in the flow at each instant of time. So once available, a solution provides the engineer and scientist with data concerning the transport of mass, momentum and energy, the detailed flow structure, local velocities and fluid properties, the distributions of forces, and the interactions of the fluid with its surrounding environment. This level of data and knowledge is therefore very desirable in many instances. However, the governing equations of fluid mechanics are a complex set of nonlinear partial differential equations, and the number of exact analytical solutions available is limited. The available exact analytical solutions tend to involve simplifying assumptions to the flow structure and restrictive geometrical features. A number of these assumptions are included in this text, and an example is laminar flow contained in an annulus between two concentric cylinders with rotation of one, or both, of the cylinders; see Chapter 6, Section 6.3. In order to augment the limited solutions available, routine use is now made of computational methods to obtain approximate solutions to the relevant equations. This technique is known as computational fluid dynamics (CFD).

2.5. EQUATIONS OF MOTION IN A ROTATING COORDINATE SYSTEM

The mathematical modelling of rotating flow can be formulated, as seen by a stationary observer. All the boundary conditions governing the flow would, however, be specified in terms of the rotating frame, and it is therefore generally more convenient to modify the equations of motion so that they apply in a rotating frame of reference. This is particularly so in the case of many forms of rotating machinery, consisting say of a single rotor, rotating in a stationary casing, in which case it is advantageous to view the flow from a coordinate system fixed to the rotating parts. This provides flow motion that is predominantly steady relative to the rotating components, but the disadvantage is that the rotating system is not inertial and additional accelerations, including Coriolis and centrifugal accelerations, must be accounted for in order to apply Newton's second law to the rotating system.

The velocity vector, u, in a stationary frame of reference is related to the relative velocity vector w, in a rotating frame of reference by

$$u = w + (\Omega \times r) \tag{2.83}$$

where Ω is the angular velocity of the system and r is a position vector from the origin of rotation to the point of interest. The components of the relative velocity vector are u, v and w, which for the case of cylindrical coordinates are related to the stationary frame of reference component velocities by

$$u = u_r \tag{2.84}$$

$$v = u_\phi - \Omega r \tag{2.85}$$

$$w = u_z \tag{2.86}$$

For an arbitrary vector, X,

$$\left(\frac{dX}{dt}\right)_{stationary\ frame} = \left(\frac{dX}{dt}\right)_{rotating\ frame} + \Omega \times X \tag{2.87}$$

Similarly, for the substantive derivative

$$\left(\frac{DX}{Dt}\right)_{stationary\ frame} = \left(\frac{DX}{Dt}\right)_{rotating\ frame} + \Omega \times X \tag{2.88}$$

Differentiating Equation 2.83 using 2.88 gives

$$\left(\frac{Du}{Dt}\right)_{stationary\ frame} = \left(\frac{Du}{Dt}\right)_{rotating\ frame} + \Omega \times u \tag{2.89}$$

Eliminating u from the right-hand side of Equation 2.89 using Equation 2.83 gives

$$\left(\frac{Du}{Dt}\right)_{stationary\ frame} = \left(\frac{Dw}{Dt}\right)_{rotating\ frame} + \frac{D\Omega}{Dt} \times r + \Omega \times \frac{Dr}{Dt} + \Omega \times (w + \Omega \times r)$$

$$= \left(\frac{Dw}{Dt}\right)_{rotating\ frame} + \frac{D\Omega}{Dt} \times r + \Omega \times w + \Omega \times (w + \Omega \times r)$$

$$= \left(\frac{Dw}{Dt}\right)_{rotating\ frame} + \frac{D\Omega}{Dt} \times r + 2\Omega \times w + \Omega \times (\Omega \times r) \qquad (2.90)$$

If the equation is restricted to constant angular velocity, then, $D\Omega/Dt = 0$, giving

$$\left(\frac{Du}{Dt}\right)_{stationary\ frame} = \left(\frac{Dw}{Dt}\right)_{rotating\ frame} + 2\Omega \times w + \Omega \times (\Omega \times r) \qquad (2.91)$$

The momentum equation in a reference frame rotating at constant angular velocity is given by

$$\rho \left(\frac{Dw}{Dt}\right)_{rotating\ frame} = \rho \left[\frac{\partial w}{\partial t} + (w \cdot \nabla)w + 2\Omega \times w + \Omega \times (\Omega \times r)\right] \qquad (2.92)$$
$$= -\nabla p + \mu \nabla^2 w + F$$

In component form for a system of axes rotating at angular velocity Ω about the z axis, for cylindrical polar coordinates (r, ϕ', z) where $\phi' = \phi - \Omega t$, the momentum equation, 2.92, can be stated in component form as

$$\frac{\partial}{\partial t}(\rho u) + \frac{1}{r}\frac{\partial}{\partial r}(\rho r u^2) + \frac{1}{r}\frac{\partial}{\partial \phi'}(\rho u v) + \frac{\partial}{\partial z}(\rho u w) - \rho \left(\frac{v^2}{r} + 2\Omega v + \Omega^2 r\right) =$$
$$-\frac{\partial p}{\partial r} + \frac{1}{r}\frac{\partial}{\partial r}(r \sigma_r) - \frac{\sigma_\phi}{r} + \frac{1}{r}\frac{\partial \tau_{r\phi}}{\partial \phi'} + \frac{\partial \tau_{rz}}{\partial z} + F_r' \qquad (2.93)$$

$$\frac{\partial}{\partial t}(\rho v) + \frac{1}{r^2}\frac{\partial}{\partial r}(\rho r^2 u v) + \frac{1}{r}\frac{\partial}{\partial \phi'}(\rho v^2) + \frac{\partial}{\partial z}(\rho v w) + 2\rho \Omega u =$$
$$-\frac{1}{r}\frac{\partial p}{\partial \phi'} + \frac{1}{r^2}\frac{\partial}{\partial r}(r^2 \sigma_{\phi r}) + \frac{1}{r}\frac{\partial \sigma_\phi}{\partial \phi'} + \frac{\partial \tau_{\phi z}}{\partial z} + F_\phi' \qquad (2.94)$$

$$\frac{\partial}{\partial t}(\rho w) + \frac{1}{r}\frac{\partial}{\partial r}(\rho r u w) + \frac{1}{r}\frac{\partial}{\partial \phi'}(\rho v w) + \frac{\partial}{\partial z}(\rho w^2) =$$
$$-\frac{\partial p}{\partial z} + \frac{1}{r}\frac{\partial}{\partial r}(r \tau_{zr}) + \frac{1}{r}\frac{\partial \tau_{z\phi}}{\partial \phi'} + \frac{\partial \sigma_z}{\partial z} + F_z' \qquad (2.95)$$

where the viscous stress tensor is given by

$$\sigma_r = \mu\left[2\frac{\partial u}{\partial r} - \frac{2}{3}\left(\frac{\partial u}{\partial r} + \frac{u}{r} + \frac{1}{r}\frac{\partial v}{\partial \phi'} + \frac{\partial w}{\partial z}\right)\right] \tag{2.96}$$

$$\sigma_\phi = \mu\left[2\left(\frac{1}{r}\frac{\partial v}{\partial \phi'} + \frac{u}{r}\right) - \frac{2}{3}\left(\frac{\partial u}{\partial r} + \frac{u}{r} + \frac{1}{r}\frac{\partial v}{\partial \phi'} + \frac{\partial w}{\partial z}\right)\right] \tag{2.97}$$

$$\sigma_z = \mu\left[2\frac{\partial w}{\partial z} - \frac{2}{3}\left(\frac{\partial u}{\partial r} + \frac{u}{r} + \frac{1}{r}\frac{\partial v}{\partial \phi'} + \frac{\partial w}{\partial z}\right)\right] \tag{2.98}$$

$$\tau_{\phi z} = \tau_{z\phi} = \mu\left(\frac{\partial v}{\partial z} + \frac{1}{r}\frac{\partial w}{\partial \phi'}\right) \tag{2.99}$$

$$\tau_{zr} = \tau_{rz} = \mu\left(\frac{\partial w}{\partial r} + \frac{\partial u}{\partial z}\right) \tag{2.100}$$

$$\tau_{r\phi} = \tau_{\phi r} = \mu\left(\frac{1}{r}\frac{\partial u}{\partial \phi'} + \frac{\partial v}{\partial r} - \frac{v}{r}\right) \tag{2.101}$$

and the body force $\boldsymbol{F}$ is unchanged. $\boldsymbol{F} = (F_r', F_\phi', F_z')$. The body force components F_r', F_ϕ' are different from F_r and F_ϕ and $F_z' = F_z$.

The momentum equation in the form given in Equations 2.93–2.95 is valid for both laminar and turbulent flow when the stress tensor includes both viscous and Reynolds stresses.

For laminar flow, Equations 2.93–2.95 become

$$\frac{\partial}{\partial t}(\rho u) + \frac{1}{r}\frac{\partial}{\partial r}(\rho r u^2) + \frac{1}{r}\frac{\partial}{\partial \phi'}(\rho u v) + \frac{\partial}{\partial z}(\rho u w) - \rho\left(\frac{v^2}{r} + 2\Omega v + \Omega^2 r\right)$$

$$= -\frac{\partial p}{\partial r} + \mu\left[\nabla^2 u - \frac{u}{r^2} - \frac{2}{r^2}\frac{\partial v}{\partial \phi'} + \frac{1}{3}\frac{\partial}{\partial r}\left(\frac{\partial u}{\partial r} + \frac{u}{r} + \frac{1}{r}\frac{\partial v}{\partial \phi'} + \frac{\partial w}{\partial z}\right)\right]$$

$$+ \frac{\partial \mu}{\partial r}\left[2\frac{\partial u}{\partial r} - \frac{2}{3}\left(\frac{\partial u}{\partial r} + \frac{u}{r} + \frac{1}{r}\frac{\partial v}{\partial \phi'} + \frac{\partial w}{\partial z}\right)\right] + \frac{1}{r}\frac{\partial}{\partial \phi'}\left(\frac{1}{r}\frac{\partial u}{\partial \phi'} + \frac{\partial v}{\partial r} - \frac{v}{r}\right)$$

$$+ \frac{\partial \mu}{\partial z}\left(\frac{\partial u}{\partial z} + \frac{\partial w}{\partial r}\right) + F_r' \tag{2.102}$$

$$\frac{\partial}{\partial t}(\rho v) + \frac{1}{r^2}\frac{\partial}{\partial r}(\rho r^2 uv) + \frac{1}{r}\frac{\partial}{\partial \phi'}(\rho v^2) + \frac{\partial}{\partial z}(\rho vw) + 2\rho\Omega u =$$

$$-\frac{1}{r}\frac{\partial p}{\partial \phi'} + \mu\left[\nabla^2 v - \frac{v}{r^2} + \frac{2}{r^2}\frac{\partial u}{\partial \phi'} + \frac{1}{3}\frac{1}{r}\frac{\partial}{\partial \phi'}\left(\frac{\partial u}{\partial r} + \frac{u}{r} + \frac{1}{r}\frac{\partial v}{\partial \phi'} + \frac{\partial w}{\partial z}\right)\right]$$

$$+\frac{\partial \mu}{\partial r}\left(\frac{\partial v}{\partial r} - \frac{v}{r} + \frac{1}{r}\frac{\partial u}{\partial \phi'}\right) + \frac{1}{r}\frac{\partial \mu}{\partial \phi'}\left[\frac{2}{r}\frac{\partial v}{\partial \phi'} + \frac{2u}{r} - \frac{2}{3}\left(\frac{\partial u}{\partial r} + \frac{u}{r} + \frac{1}{r}\frac{\partial v}{\partial \phi'} + \frac{\partial w}{\partial z}\right)\right]$$

$$+\frac{\partial \mu}{\partial z}\left(\frac{1}{r}\frac{\partial w}{\partial \phi'} + \frac{\partial v}{\partial z}\right) + F_\phi' \tag{2.103}$$

$$\frac{\partial}{\partial t}(\rho w) + \frac{1}{r}\frac{\partial}{\partial r}(\rho ruw) + \frac{1}{r}\frac{\partial}{\partial \phi'}(\rho vw) + \frac{\partial}{\partial z}(\rho w^2) =$$

$$-\frac{\partial p}{\partial z} + \mu\left[\nabla^2 w + \frac{1}{3}\frac{\partial}{\partial z}\left(\frac{\partial u}{\partial r} + \frac{u}{r} + \frac{1}{r}\frac{\partial v}{\partial \phi'} + \frac{\partial w}{\partial z}\right)\right]$$

$$+\frac{\partial \mu}{\partial r}\left(\frac{\partial w}{\partial r} + \frac{\partial u}{\partial z}\right) + \frac{1}{r}\frac{\partial \mu}{\partial \phi'}\left(\frac{1}{r}\frac{\partial w}{\partial \phi'} + \frac{\partial v}{\partial z}\right)$$

$$+\frac{\partial \mu}{\partial z}\left[2\frac{\partial w}{\partial z} - \frac{2}{3}\left(\frac{\partial u}{\partial r} + \frac{u}{r} + \frac{1}{r}\frac{\partial v}{\partial \phi'} + \frac{\partial w}{\partial z}\right)\right] + F_z' \tag{2.104}$$

where

$$\nabla^2 = \frac{\partial^2}{\partial r^2} + \frac{1}{r}\frac{\partial}{\partial r} + \frac{1}{r^2}\frac{\partial^2}{\partial \phi'^2} + \frac{\partial^2}{\partial z^2} \tag{2.105}$$

If viscous forces are negligible then Equations 2.102–2.104 become

$$\frac{\partial}{\partial t}(\rho u) + \frac{1}{r}\frac{\partial}{\partial r}(\rho ru^2) + \frac{1}{r}\frac{\partial}{\partial \phi'}(\rho uv) + \frac{\partial}{\partial z}(\rho uw) - \rho\left(\frac{v^2}{r} + 2\Omega v + \Omega^2 r\right) = -\frac{\partial p}{\partial r} + F_r' \tag{2.106}$$

$$\frac{\partial}{\partial t}(\rho v) + \frac{1}{r^2}\frac{\partial}{\partial r}(\rho r^2 uv) + \frac{1}{r}\frac{\partial}{\partial \phi'}(\rho v^2) + \frac{\partial}{\partial z}(\rho vw) + 2\rho\Omega u = -\frac{1}{r}\frac{\partial p}{\partial \phi'} + F_\phi' \tag{2.107}$$

$$\frac{\partial}{\partial t}(\rho w) + \frac{1}{r}\frac{\partial}{\partial r}(\rho ruw) + \frac{1}{r}\frac{\partial}{\partial \phi'}(\rho vw) + \frac{\partial}{\partial z}(\rho w^2) = -\frac{\partial p}{\partial z} + F_z' \tag{2.108}$$

Equations 2.106–2.108 for an inviscid flow are also valid for turbulent flow as the Reynolds stresses are negligible.

In component form, Equation 2.92, gives rise to the terms $2\Omega u$ and $2\Omega v$. These are usually referred to as the Coriolis terms and are responsible in a fluid for the differences between the dynamics of non-rotating and rotating fluids.

The continuity equation in a rotating frame of reference is given in Equation 2.109, taking a similar form to that for a stationary frame of reference.

$$\frac{1}{\rho}\left(\frac{D\rho}{Dt}\right)_{rotating\ frame} + \nabla \cdot w = 0 \qquad (2.109)$$

If the flow is incompressible then $D\rho/Dt = 0$, and Equation 2.109 becomes $\rho \nabla \cdot w = 0$.

In component form, Equation 2.109 can be written as

$$\frac{\partial \rho}{\partial t} + \frac{1}{r}\frac{\partial}{\partial r}(\rho r u) + \frac{1}{r}\frac{\partial}{\partial \phi'}(\rho v) + \frac{\partial}{\partial z}(\rho w) = 0 \qquad (2.110)$$

2.6. DIMENSIONAL ANALYSIS AND SIMILARITY

The momentum and continuity equations, as presented in Section 2.2.4 and 2.3, are dimensional. These can be made dimensionless if we chose to redefine the dependent and independent variables by dividing them by constant reference properties appropriate to the flow.

It should be noted that dimensional analysis tends to benefit from previous experience and knowledge of the particular groupings of variables that are hoped for.

Appropriate reference constants is provided below.

- A reference velocity, U_o (e.g. the free-stream velocity) or $\mu_o/\rho_o L$ if there is no free stream (as in natural convection, for example).
- Free stream properties, $p_o, \mu_o, \rho_o, k_o, T_o$.
- A characteristic dimension. This is usually a characteristic length scale, L, derived from the geometry (e.g. body length or pipe diameter).
- A characteristic time scale, e.g. residence time L/U_o for steady flows.

The resulting dimensionless variables are commonly denoted using an asterisk. Recommended choices are follows.

$$x_* = \frac{x}{L} \qquad (2.111)$$

$$y_* = \frac{y}{L} \qquad (2.112)$$

$$z_* = \frac{z}{L} \qquad (2.113)$$

$$t_* = \frac{tU_o}{L} \qquad (2.114)$$

$$u_{x*} = \frac{u_x}{U_o} \tag{2.115}$$

$$u_{y*} = \frac{u_y}{U_o} \tag{2.116}$$

$$u_{z*} = \frac{u_z}{U_o} \tag{2.117}$$

$$p_* = \frac{p}{P_o} \tag{2.118}$$

where P_o is a characteristic pressure.

It is sometimes convenient to non-dimensionalise the pressure using the form given in Equation 2.119, where the pressure difference has been divided by twice the free-stream dynamic pressure. This idea is as a result of hindsight suggested by knowledge of Bernoulli's equation. Additional dimensionless variables for temperature and various other fluid properties are given in Equations 2.120 to 2.124.

$$p_* = \frac{p - p_o}{\rho U_o^2} \tag{2.119}$$

$$T_* = \frac{T - T_o}{T_w - T_o} \tag{2.120}$$

$$\mu_* = \frac{\mu}{\mu_o} \tag{2.121}$$

$$\rho_* = \frac{\rho}{\rho_o} \tag{2.122}$$

$$k_* = \frac{k}{k_o} \tag{2.123}$$

$$c_{p*} = \frac{c_p}{c_{p_o}} \tag{2.124}$$

The spatial derivatives are:

$$\frac{\partial()}{\partial x} = \frac{\partial()}{\partial x_*}\frac{\partial x_*}{\partial x} = \frac{1}{L}\frac{\partial()}{\partial x_*} \tag{2.125}$$

$$\frac{\partial()}{\partial y} = \frac{\partial()}{\partial y_*}\frac{\partial y_*}{\partial y} = \frac{1}{L}\frac{\partial()}{\partial y_*} \tag{2.126}$$

$$\frac{\partial()}{\partial z} = \frac{\partial()}{\partial z_*}\frac{\partial z_*}{\partial z} = \frac{1}{L}\frac{\partial()}{\partial z_*} \tag{2.127}$$

$$\frac{\partial^2()}{\partial x^2} = \frac{1}{L^2}\frac{\partial^2()}{\partial x_*} \tag{2.128}$$

$$\frac{\partial^2()}{\partial y^2} = \frac{1}{L^2}\frac{\partial^2()}{\partial y_*^2} \tag{2.129}$$

$$\frac{\partial^2()}{\partial z^2} = \frac{1}{L^2}\frac{\partial^2()}{\partial z_*^2} \tag{2.130}$$

The time derivative is

$$\frac{\partial()}{\partial t} = \frac{\partial()}{\partial t_*}\frac{\partial t_*}{\partial t} = \frac{1}{t_o}\frac{\partial()}{\partial t_*} \tag{2.131}$$

where $t_o = L/U_o$.

Substitution in the continuity equation gives

$$\frac{\partial u_x}{\partial x} + \frac{\partial u_y}{\partial y} + \frac{\partial u_z}{\partial z} = \frac{1}{L}\frac{\partial(u_{x*}U_o)}{\partial x_*} + \frac{1}{L}\frac{\partial(u_{y*}U_o)}{\partial y_*} + \frac{1}{L}\frac{\partial(u_{z*}U_o)}{\partial z_*} =$$
$$\frac{U_0}{L}\left(\frac{\partial u_{x*}}{\partial x_*} + \frac{\partial u_{y*}}{\partial y_*} + \frac{\partial u_{z*}}{\partial z_*}\right) = 0 \tag{2.132}$$

If this equation is now multiplied by L/U_o the non-dimensionalised form of the continuity equation becomes:

$$\left(\frac{\partial u_{x*}}{\partial x_*} + \frac{\partial u_{y*}}{\partial y_*} + \frac{\partial u_{z*}}{\partial z}\right) = 0 \tag{2.133}$$

A similar process can be applied to the Navier-Stokes equations, giving, in the case of a constant density, constant viscosity, Newtonian fluid where the only body force under consideration is gravity:

$$\left(\frac{L}{U_o t_o}\right)\frac{\partial u_{x*}}{\partial t_*} + u_{x*}\frac{\partial u_{x*}}{\partial x_*} + u_y^*\frac{\partial u_{x*}}{\partial y_*} + u_z^*\frac{\partial u_{x*}}{\partial z_*} =$$
$$-\left(\frac{p}{\rho U_o^2}\right)\frac{\partial p_*}{\partial x_*} + \left(\frac{\mu}{\rho U_o L}\right)\left(\frac{\partial^2 u_{x*}}{\partial x_*^2} + \frac{\partial^2 u_{x*}}{\partial y_*^2} + \frac{\partial^2 u_{x*}}{\partial z_*^2}\right) \tag{2.134}$$

$$\left(\frac{L}{U_o t_o}\right)\frac{\partial u_{y*}}{\partial t_*} + u_x^*\frac{\partial u_{y*}}{\partial x_*} + u_{y*}\frac{\partial u_{y*}}{\partial y_*} + u_z^*\frac{\partial u_{y*}}{\partial z_*} =$$
$$-\left(\frac{p}{\rho U_o^2}\right)\frac{\partial p_*}{\partial y_*} + \left(\frac{\mu}{\rho U_o L}\right)\left(\frac{\partial^2 u_{y*}}{\partial x_*^2} + \frac{\partial^2 u_{y*}}{\partial y_*^2} + \frac{\partial^2 u_{y*}}{\partial z_*^2}\right) \tag{2.135}$$

$$\left(\frac{L}{U_o t_o}\right)\frac{\partial u_{z^*}}{\partial t_*} + u_{x^*}\frac{\partial u_{z^*}}{\partial x_*} + u_{y^*}\frac{\partial u_{z^*}}{\partial y_*} + u_{z^*}\frac{\partial u_{z^*}}{\partial z_*} =$$

$$-\left(\frac{gL}{U_o^2}\right) - \left(\frac{p}{\rho U_o^2}\right)\frac{\partial p_*}{\partial z_*} + \left(\frac{\mu}{\rho U_o L}\right)\left(\frac{\partial^2 u_{z^*}}{\partial x_*^2} + \frac{\partial^2 u_{z^*}}{\partial y_*^2} + \frac{\partial^2 u_{z^*}}{\partial z_*^2}\right)$$

$$(2.136)$$

Four dimensionless groups evident in these equations:

$$\frac{L}{U_o t_o} \tag{2.137}$$

$$\frac{gL}{U_o^2} \tag{2.138}$$

$$\frac{p}{\rho U_o^2} \tag{2.139}$$

$$\frac{\mu}{\rho U_o L} \tag{2.140}$$

Some fluid flow phenomena involve a forced oscillation with a characteristic frequency, ω. The time scale in such a system is $1/\omega$, and the dimensionless group identified in (2.137) on substitution becomes $\omega L/U_o$. This group can be identified from previous experience of fluid flow as the Strouhal number, Equation 2.141. Forces, moments, friction and heat transfer in an oscillating flow are typically a function of the Strouhal number and also the Reynolds number. An example of an oscillating flow is the periodic shedding of vortices behind a blunt body immersed in a steady stream of velocity U_o. For the case of the alternating vortices shed from a circular cylinder immersed in steady cross flow, these are known as a Kármán vortex street after original investigations by von Kármán in 1912, see von Kármán (2004). Vortices are shed in the range $100 < Re < 10^7$ with a typical Strouhal number of about 0.21 allowing the shedding frequency to be readily calculated (see Roshko (1954) and Jones (1968) for detailed data). This can be important in, for example, ensuring that the shedding frequency is distinctly different to any structural vibration frequency in the design of components and structures.

$$St = \frac{\omega L}{U_o} \tag{2.141}$$

The dimensionless group in (2.138) can be identified as being related to the Froude number, Equation 2.142, and is the characterizing parameter in free surface flows but unimportant if there is no free surface.

$$Fr = \frac{U_o}{\sqrt{gL}} \tag{2.142}$$

The dimensionless group in (2.139) is related to a form of the Euler number, Equation 2.143. The Euler number is important in a liquid if the pressure drops low enough to cause the formation of vapour.

$$Eu = \frac{p - p_o}{0.5\rho U_o^2} \tag{2.143}$$

The last group (2.140) is recognizable as the inverse of the Reynolds number, Equation 2.144. The Reynolds number is the primary parameter characterizing the viscous behaviour of Newtonian fluids. Low Reynolds numbers for a flow indicate viscous creeping motion where inertia effects are negligible. Moderate Reynolds numbers are indicative of a smoothly varying laminar flow. High Reynolds numbers indicate turbulent flow where the flow is characterized by slowly varying mean velocities but with high frequency random oscillations of the local instantaneous velocity. Numerical values characterizing whether flow is laminar or turbulent depend on the particular geometrical configuration concerned.

$$\text{Re} = \frac{\rho U_o L}{\mu} \tag{2.144}$$

By introducing these dimensionless groups, the Navier-Stokes equations can be stated in non-dimensionalised form as

$$(St)\frac{\partial u_{x^*}}{\partial t_*} + u_{x^*}\frac{\partial u_{x^*}}{\partial x_*} + u_{y^*}\frac{\partial u_{x^*}}{\partial y_*} + u_{z^*}\frac{\partial u_{x^*}}{\partial z_*} =$$
$$- \left(\frac{Eu}{2}\right)\frac{\partial p_*}{\partial x_*} + \left(\frac{1}{Re}\right)\left(\frac{\partial^2 u_{x^*}}{\partial x_*^2} + \frac{\partial^2 u_{x^*}}{\partial y_*^2} + \frac{\partial^2 u_{x^*}}{\partial z_*^2}\right) \tag{2.145}$$

$$(St)\frac{\partial u_{y^*}}{\partial t_*} + u_{x^*}\frac{\partial u_{y^*}}{\partial x_*} + u_{y^*}\frac{\partial u_{y^*}}{\partial y_*} + u_{z^*}\frac{\partial u_{y^*}}{\partial z_*} =$$
$$- \left(\frac{Eu}{2}\right)\frac{\partial p_*}{\partial y_*} + \left(\frac{1}{Re}\right)\left(\frac{\partial^2 u_{y^*}}{\partial x_*^2} + \frac{\partial^2 u_{y^*}}{\partial y_*^2} + \frac{\partial^2 u_{y^*}}{\partial z_*^2}\right) \tag{2.146}$$

$$(St)\frac{\partial u_{z^*}}{\partial t_*} + u_x^*\frac{\partial u_{z^*}}{\partial x_*} + u_y^*\frac{\partial u_{z^*}}{\partial y_*} + u_{z^*}\frac{\partial u_{z^*}}{\partial z_*} =$$
$$- \left(\frac{1}{Fr^2}\right) - \left(\frac{Eu}{2}\right)\frac{\partial p_*}{\partial z_*} + \left(\frac{1}{Re}\right)\left(\frac{\partial^2 u_{z^*}}{\partial x_*^2} + \frac{\partial^2 u_{z^*}}{\partial y_*^2} + \frac{\partial^2 u_{z^*}}{\partial z_*^2}\right) \tag{2.147}$$

By non-dimensionalising the governing equations, the solution to a flow in a specified geometry will depend only on the values of the relevant non-dimensionalised groups that appear in the transformed equations and the boundary conditions. A solution to the governing equations will apply to any geometrically similar flow that has the same values for the dimensionless groups. For example a solution to an application with a particular length scale will also apply to a different system with a different length scale but whose other values are adjusted so that the dimensionless groups are the same.

A similar analysis can be undertaken for the case of the Navier-Stokes equation in a rotating frame of reference. Equation 2.92 involves six parameters, density, gravity, viscosity, angular velocity, length and velocity. Dimensionless analysis of this would normally lead to 6-3=3 dimensionless parameters. An alternative form however more relevant and useful to rotating flows involving only two dimensionless parameters can be developed by using $1/\Omega$ as the relevant characteristic time and assuming that the body force $\boldsymbol{F}$ is a conservative force, such as gravity, so that it can expressed as the gradient of a potential function.

It is sometimes convenient to modify the definition of pressure to account for the effects of body forces. The resulting pressure is known as the 'reduced pressure' or 'modified pressure'. Unfortunately although it may be stated explicitly within a proof that the reduced pressure is being employed it is often difficult to distinguish the difference between static pressure and reduced pressure in an equation as the symbol p may be used for both. The development of the reduced pressure to account for the effect of body force and simplification of the Navier-Stokes equation is developed in this text following the procedure outlined by Bachelor (1967). This is subsequently extended to account the effects of the centrifugal acceleration. The Navier-Stokes equation of motion for a viscous fluid is given by

$$\rho\frac{D\boldsymbol{u}}{Dt} = -\nabla p + \mu\nabla^2\boldsymbol{u} + \boldsymbol{F} \tag{2.148}$$

In the case of fluid in the Earth's atmosphere or oceans the body force can be represented by the Earth's gravitational acceleration, such that

$$\rho\frac{D\boldsymbol{u}}{Dt} = -\nabla p + \mu\nabla^2\boldsymbol{u} + \rho\boldsymbol{g} \tag{2.149}$$

In Equation 2.149 when the density is uniform it can be seen that the force per unit volume due to gravity is balanced by a pressure equal to $\rho\boldsymbol{g}.\boldsymbol{x}$. The pressure could be defined as the following sum of terms

$$p = p_o + \rho\boldsymbol{g}.\boldsymbol{x} + p_{reduced} \tag{2.150}$$

where p_o is a constant, and the sum $p_o + \rho\boldsymbol{g}.\boldsymbol{x}$ is the pressure that would exist in the same body at rest and which would in a moving body of fluid give rise to a pressure gradient in balance with gravity. $p_{reduced}$ is the remaining part of the pressure and arises from the effect of motion of the fluid. If the reduced pressure is used in the Navier-Stokes equations the motion of the fluid is defined by

$$\rho\frac{Du}{Dt} = -\nabla p_{reduced} + \mu\nabla^2 u \qquad (2.151)$$

The reduced pressure can only be introduced if the density in the fluid is uniform and the gravitational force per unit volume being the gradient of a scalar quantity. The acceleration due to gravity does not now explicitly appear in Equation 2.151. Provided gravity does not appear in the boundary conditions it can be inferred that gravity has no effect on the velocity distribution in the fluid. If the absolute pressure occurs in the boundary conditions then Equation 2.150 must be used and in this way the effect of gravity is incorporated in the model. The use of reduced pressure is principally suitable when the boundary conditions involve only velocity (Bachelor (1967)).

In a similar fashion to the use of reduced pressure for subtracting out hydrostatic pressure to eliminate the non-dynamical effects of gravitational forces in a uniform fluid, the terms associated with centrifugal force can be combined with the static pressure to form a further form of reduced pressure.

The momentum equation in a reference frame rotating at constant angular velocity is given by Equation 2.92, but repeated here.

$$\rho\left(\frac{Dw}{Dt}\right)_{rotating\ frame} = \rho\left[\frac{\partial w}{\partial t} + (w\cdot\nabla)w + 2\Omega \times w + \Omega \times (\Omega \times r)\right]$$

$$= -\nabla p + \mu\nabla^2 w + F \qquad (2.152)$$

The term $\Omega\times(\Omega\times r)$ can be written as $-\nabla(\Omega^2 r^2/2)$ where r is the distance from the axis of rotation. For a fluid of uniform density the term $-\nabla(\Omega^2 r^2/2)$ can be combined with the static pressure to define a reduced static pressure.

$$p_{reduced} = p - \frac{1}{2}\rho\Omega^2 r^2 \qquad (2.153)$$

Alternatively (see Greenspan (1968)), noting that

$$\Omega \times (\Omega \times r) = -\frac{1}{2}\nabla[(\Omega \times r)\cdot(\Omega \times r)] \qquad (2.154)$$

the reduced pressure incorporating body forces due to, for example gravity, can be expressed by

$$p_{reduced} = p + \rho\Phi - \frac{1}{2}\rho(\Omega \times r)\cdot(\Omega \times r) \qquad (2.155)$$

Here it has been assumed that the body force F is a conservative force, such as gravity, so that it can expressed as the gradient of a potential function, Φ (Vanyo (2001)). Hence

$$\nabla p_{reduced} = \nabla\left[p + \rho\Phi - \frac{1}{2}\rho(\Omega \times r)\cdot(\Omega \times r)\right] \qquad (2.156)$$

In the relative system it is gradients of reduced static pressure that cause accelerations. For example (see Greitzer, Tan and Graf (2004)), for a fluid in solid body rotation, the pressure field is given by

$$p - p_{axis} = \frac{1}{2}\rho\Omega^2 r^2 \tag{2.157}$$

The pressure gradient is given by

$$\nabla p = \rho\Omega^2 \mathbf{r} \tag{2.158}$$

The reduced static pressure is therefore constant throughout the fluid.

The measurement of static pressure in rotating machinery can be used to illustrate the interpretation of reduced static pressure (Greitzer, Tan and Graf (2004) and Moore (1973)). Suppose a static pressure tap is located on the blades of a turbomachine at a radial location r, and the pressure recorded by a transducer at the axis of the turbomachine. The fluid in the tube between the tapping and the transducer will be in hydrostatic equilibrium due to the pressure gradient dp/dr and the centrifugal force $\rho\Omega^2 r$. The pressure difference between the tap and the transducer can be found by integration giving $0.5\rho\Omega^2 r^2$. The reduced static pressure in this case can be viewed as the pressure felt by a measuring device or observer located on the axis of rotation.

For the case of a uniform density fluid it is sometimes useful to work in terms of reduced static pressure, provided none of the boundary conditions involve static pressure. An application where it is useful to use the reduced pressure is for steady flow when both the Rossby and Ekman number are small.

Employing Equations 2.154 and 2.156 with the pressure being made dimensionless using $\rho L\Omega U$, Equation 2.92 can be stated in dimensionless form, as

$$\frac{\partial w_*}{\partial t_*} + \frac{U_o}{\Omega L}(w_* \cdot \nabla)w_* + 2k \times w_* = -\nabla p_* + \frac{\nu}{\Omega L^2}\nabla^2 w_* \tag{2.159}$$

where k is a unit vector.

The groups $U_o/\Omega L$ and $\nu/\Omega L^2$ are known as the Rossby, Ro, and Ekman, Ek, numbers respectively as given in Equations 2.160 and 2.161.

$$Ro = \frac{U_o}{\Omega L} \tag{2.160}$$

$$Ek = \frac{\nu}{\Omega L^2} \tag{2.161}$$

The momentum equation can be stated using these as

$$\frac{\partial w_*}{\partial t} + Ro(w_* \cdot \nabla)w_* + 2k \times w_* = -\nabla p_* + Ek\nabla^2 w_* \tag{2.162}$$

The Rossby number provides an indication of the basic characteristic of fluid motion in a rotating system and indicates the ratio of inertial forces (or momentum advection) and Coriolis forces. If the Rossby number is large in comparison to unity, then the influence of rotation is small. If $Ro<<1$ then rotation is decisive in determining the flow regime and its characteristic flow pattern. The relative strength of viscous forces to Coriolis forces is described by the Ekman number.

When both the Rossby and the Ekman numbers are small this indicates that the viscous forces are negligible and this condition is referred to as the geostrophic approximation. Some atmospheric flows can be modelled using the geostrophic approximation.

For steady rotating flow when both the Rossby number and Ekman number are small, indicating that the Coriolis force is significant in comparison to inertial forces and viscous forces respectively, then the equation of motion for a rotating flow can be stated as

$$2\boldsymbol{\Omega} \times \boldsymbol{u} = -\frac{1}{\rho}\nabla p_{reduced} \qquad (2.163)$$

where the pressure has been transformed by removing the centrifugal forces. This equation defines geostrophic balance where the pressure gradient is balanced by the Coriolis term.

Just as for non-rotating fluid applications, viscous effects can be important in the boundary layer close to a surface. Here, however, flow away from the surfaces will be considered. In a rotating flow the Coriolis force is always perpendicular to the flow direction. As a result the pressure gradient is also perpendicular to the flow direction. This means that the pressure is constant along a streamline. Such behaviour is markedly different to that for a non-rotating application, where pressure variation along a streamline is common. For example considering the flow along a stationary converging or diverging duct modelled by Bernoulli's equation there will be a pressure velocity exchange along a streamline with duct area. A comparison between rotating and non-rotating duct flow is illustrated in Figure 2.9 for a converging duct. As

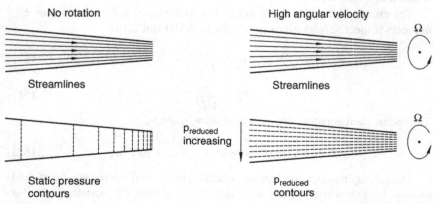

FIGURE 2.9 Variation of pressure in a stationary converging duct and in a rotating converging duct.

can be seen although the streamlines are comparable, the pressure field for the non-rotating and rotating applications are completely different.

The phenomenon of the pressure gradient being perpendicular to the flow direction in geostrophic flows can be exploited in interpreting weather patterns. From a knowledge of the pressure distribution, lines of constant pressure known as isobars can be constructed. The Earth's rotation causes the Coriolis force to dominate over both inertial and viscous forces away from the Earth's surface and the isobars can be taken to be the lines along which the wind is blowing. This topic is explored further in Section 8.3.1.

Rotation and the distribution of energy sources are responsible for the general circulation of the Earth's atmosphere and its oceans. For many geophysical and astrophysical objects rotation, together with spatial inhomegenity of energy sources allow the origin and characteristics of fluid motion to be determined. In the northern hemisphere, the Coriolis force is responsible for a tendency for the wind to turn to the right of its direction of motion. In the southern hemisphere, the Coriolis force is responsible for a tendency for the wind to turn to the left of its direction of motion. The Coriolis force is zero at the equator. Although the Coriolis force is as real as any other force in that it causes a mass to be accelerated, it is sometimes referred to as a fictitious force because it only arises as a result of an arbitrary definition of the reference frame.

The complexity of modelling required for a fluid system can often be reduced by neglecting insignificant terms and factors. It is therefore important to be able to determine whether the rotation of a convective system is negligible and when it must be taken into account. The relative magnitude of centrifugal and Coriolis forces can be considered for the case of an arbitrary mass. Here a mass of 0.1 kg has been chosen which is rotating about an axis as shown in Figure 2.10.

In addition to rotating the axis, the mass also is moving with a velocity v, relative to a rotating frame of reference rotating at an angular velocity of Ω. The corresponding relative magnitude for the centrifugal force and the Coriolis force on a body are considered in Tables 2.1 and 2.2 and Figure 2.11 for a

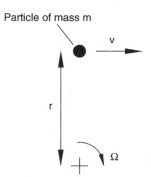

FIGURE 2.10   Mass moving at velocity v, relative to a rotating frame of reference rotating at an angular velocity Ω.

TABLE 2.1 Variation of the centrifugal force, for a body with an arbitrary mass of 0.1 kg rotating about an axis for a range of angular velocities and radii

Ω (rad/s)	r=0.1 m	r=10 m	r=1000 m	r=100000 m	r=10000000 m
0.00001	1×10^{-12}	1×10^{-10}	0.00000001	0.000001	0.0001
0.0001	1×10^{-10}	1×10^{-8}	0.000001	0.0001	0.01
0.001	0.00000001	0.000001	0.0001	0.01	1
0.01	0.000001	0.0001	0.01	1	100
0.1	0.0001	0.01	1	100	10000
1	0.01	1	100	10000	1000000
10	1	100	10000	1000000	100000000
100	100	10000	1000000	100000000	10000000000
1000	10000	1000000	100000000	10000000000	1×10^{12}
10000	1000000	1×10^{8}	1×10^{10}	1×10^{12}	1×10^{14}

TABLE 2.2 Variation of the Coriolis force for a body with an arbitrary mass of 0.1 kg rotating about an axis for a range of angular velocities and velocities relative to the rotating coordinate frame

Ω (rad/s)	v=0.1 m/s	v=1 m/s	v=10 m/s	v=100 m/s	v=1000 m/s
0.00001	0.0000002	0.000002	0.00002	0.0002	0.002
0.0001	0.000002	0.00002	0.0002	0.002	0.02
0.001	0.00002	0.0002	0.002	0.02	0.2
0.01	0.0002	0.002	0.02	0.2	2
0.1	0.002	0.02	0.2	2	20
1	0.02	0.2	2	20	200
10	0.2	2	20	200	2000
100	2	20	200	2000	20000
1000	20	200	2000	20000	200000
10000	200	2000	20000	200000	2000000

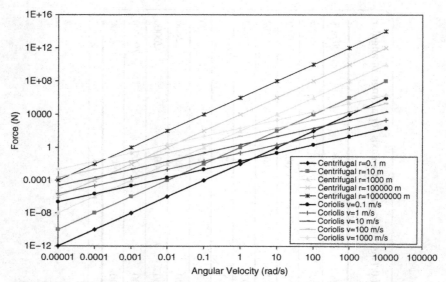

FIGURE 2.11 Comparison of the magnitude of centrifugal force and Coriolis force for a body of mass 0.1 kg for a range of angular velocities, radii, and velocities relative to the rotating coordinate frame.

body with an a mass of 0.1 kg. The mass has been selected to be representative of many components used in machinery. The centrifugal force, calculated from $m\Omega^2 r$, and the Coriolis force calculated from $2m\Omega v$, are presented for a range of angular velocities, radii and velocities relative to the rotating coordinate frame. The range for the angular velocity has been chosen to be representative of values for the Earth's rotation (7.27×10^{-5} rad/s) to those associated with some turbomachines. Similarly the radius variation has been selected to cover rotating machinery to the Earth's radius (polar radius ca. 6357 km, equatorial radius ca. 6378 km). The range taken for the velocity relative to the rotating coordinate frame is meant to span common engineering applications.

The relative magnitude of the centrifugal force to the Coriolis force can be expressed as a function of a dimensionless ratio. The ratio of the centrifugal force to the Coriolis force is given by

$$\frac{F_{centrifugal}}{F_{Coriolis}} = \frac{m\Omega^2 r}{2\Omega vm} = \frac{1}{2}\left(\frac{\Omega r}{v}\right) \tag{2.164}$$

This ratio is plotted as a function of the dimensionless function $\Omega r/v$ in Figure 2.12, again allowing comparison of the relative magnitude of centrifugal to Coriolis forces for a given application to be determined.

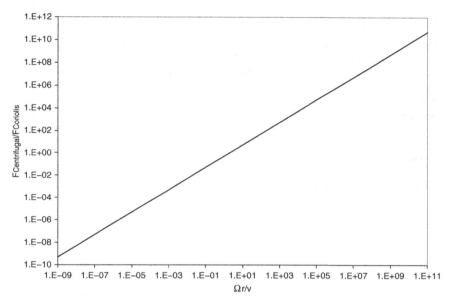

FIGURE 2.12   Variation of the ratio of centrifugal to Coriolis force as a function of $\Omega r/\nu$.

2.7. CONCLUSIONS

The Navier-Stokes and continuity equations provide the foundations for modeling fluid motion. These equations will be extensively used and referred to within this text. For certain applications they can be simplified to enable the analytical modeling of, for example, laminar and turbulent flow over a rotating disc, see Chapter 4, and laminar flow in a concentric annulus with cylinder rotation, see Chapter 6. Vorticity, rotation and vortex flow are important subjects in fluid mechanics and are considered in the next chapter and Chapter 7 for the case of flow in the cavity formed between two co-rotating discs and in Chapter 8 for the case of geophysical flows.

REFERENCES

Bachelor, G.K. *An introduction to fluid mechanics*. Cambridge University Press, 1967.

Douglas, J.F., Gasiorek, J.M., Swaffield, J.A., and Jack, L.B. Fluid mechanics. 5th Edition. Pearson, 2005.

Euler, L. Principes généraux du movement des fluids. *Hist. de l'Acad. de Berlin*, 1755.

Euler, L. De principiis motus fluidorum. *Novi. Comm. Acad. Petrop.*, xiv, 1, 1759.

Greenspan, H.P. *The theory of rotating fluids*. Cambridge, 1968.

Greitzer, E.M., Tan, C.S., and Graf, M.B. *Internal flow*. Cambridge University Press, 2004.

Jones, G.W., Jr. Unsteady lift forces generated by vortex shedding about a large, stationary oscillating cylinder at high Reynolds numbers. *ASME Symp. Usteady Flow*, (1968).

Kármán, Th., von. Aerodynamics: *Selected topics in the light of their historical development.* Dover edition, 2004.

Moore, J.A wake and an eddy in a rotating, radial flow passage. Part 1: Experimental observations. *ASME Journal of Engineering for Power*, 95, pp. 205–212, 1973.

Navier, C.L.M.H. Memoire sur les lois du mouvement des fluids. *Mem. Acad. Sci, Inst. Fr.*, Vol. 6 (1823), pp. 389–416.

Roshko, A. On the development of turbulent wakes from vortex streets. *NACA Report* 1191, (1954).

Schlichting, H. *Boundary layer theory.* 7th Edition. McGraw-Hill, 1979.

Stokes, G.G. On the theories of the internal friction of fluids in motion and of the equilibrium and motion of elastic solids. *Trans. Cambridge Philos. Soc.*, 8 (1845), pp. 287–341.

Stroud, K.A. *Engineering mathematics.* Palgrave Macmillan, 2001.

Vanyo, J.P. *Rotating fluids in engineering and science.* Dover edition, 2001.

White, F.M. *Fluid mechanics*, 5th Edition. McGraw-Hill, 2003.

Chapter 3

Vorticity and Rotation

This chapter explores the mechanisms responsible for the interactions between fluids and associated structures and boundaries that produce vorticity in a flow. In addition, modeling approaches for some of the principal types of vortex formations are introduced.

3.1. INTRODUCTION

The term *vortex* is commonly used to designate a region of concentrated rotation in a flow, such as an eddy, a whirlpool, or the depression at the center of a whirling body of air or water. Naturally occurring vortices include hurricanes, tornadoes, waterspouts, and dust devils. Vortices also occur in the wake of an object, such as, for example, when a paddle is dragged through water or a spoon is moved through a cup of tea.

Vorticity and rotation are almost always present in a moving fluid, even though a vortex may not be evident. One of the most important mechanisms in the generation of vorticity is the no-slip condition. An understanding of vorticity in flows where rotation is not evident can be developed by consideration of laminar developing uniform flow of a viscous fluid in a duct (Figure 3.1). When the flow is fully developed, it will have a parabolic velocity distribution. Ink or dye that is slowly injected into the flow would move in straight lines, indicating straight streamlines. However, if small objects, say small corks, were placed in the flow, they would also rotate as they translate along a straight path, as indicated by the half-filled circles in Figure 3.1. This signifies that the flow has vorticity and that the physical mechanism responsible is viscosity. At the wall the fluid particles will have zero velocity due to the no-slip condition. A small distance away from the wall the fluid has a higher axial velocity component. Through the action of viscosity, this difference results in a difference in shear stress across a fluid element; it is this difference in shear that causes the local rotation of the fluid element.

Another extreme in which flow is moving in circular paths but has no vorticity is the inviscid vortex. A practical embodiment of this is illustrated in Figure 3.2. Small objects placed in the fluid would not rotate, indicating that the

Rotating Flow, DOI: 10.1016/B978-0-12-382098-3.00003-2

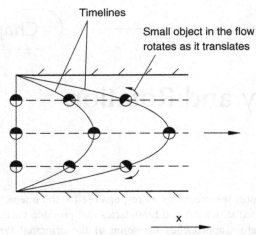

FIGURE 3.1 Laminar developing flow along in a straight horizontal channel. Vorticity is evident by the rotation of the small objects placed in the flow.

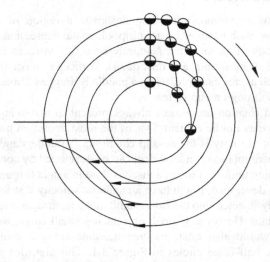

FIGURE 3.2 The translation of small objects in an inviscid vortex.

fluid is not rotating but is instead translating in circular paths. In this flow an infinitesimal fluid element does not experience different shear stresses, and therefore no rotation or vorticity is generated. In reality, the experiment illustrated in Figure 3.2 does actually produce some vorticity because of the effects of fluid viscosity. The fluid surface will not be flat, and a secondary flow will pump fluid radially inward.

There is therefore a distinction between vorticity and curved or circular translation of fluid elements. If the motion between particles is purely translational, and the distortion of the fluid element concerned is symmetrical, the flow is defined as irrotational.

The modeling of vorticity can be developed by consideration of the motion and deformation of an infinitesimal fluid element. In a similar fashion to a solid, a fluid element can undergo four different types of motion or deformation.

- Translation
- Rotation
- Dilation or extensional strain
- Shear strain

An infinitesimal fluid element, as illustrated in Figure 3.3, bounded by slower and faster moving fluid, will experience different shear stresses on each edge generating rotation of the fluid element. At time t the fluid element is taken to be initially square, and at time $t + \delta t$ it has adopted a new shape.

There has been:

- translation of the reference corner B to B′
- anticlockwise rotation of the diagonal BD to B′D′
- a dilation and the area of the fluid element has changed
- shear strain and the fluid element shape has changed from square to rhombic

The translation can be defined by the displacements $u_x dt$ and $u_y dt$ of the point B. The rates of translation are u_x and u_y. In three-dimensional motion, the rate of translation is defined by the velocity vector $\boldsymbol{u}$ with velocity components u_x, u_y, u_z.

The rotation of the diagonal BD is given by $\Omega_z = \phi + d\alpha - 45°$. Here the subscript z denotes rotation about an axis parallel to the z-axis. The angles ϕ and $45°$ can be eliminated by noting that $2\phi + d\alpha + d\beta = 90°$. Hence

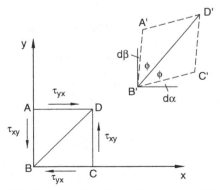

FIGURE 3.3   A square fluid element in shear flow and its resultant deformation.

$$\Omega_z = \frac{1}{2}(d\alpha - d\beta) \tag{3.1}$$

The anticlockwise direction has been deduced from the relative magnitudes of $d\alpha$ and $d\beta$.

The angles $d\alpha$ and $d\beta$ are both directly related to velocity derivatives:

$$d\alpha = \lim_{dt\to 0}\left(\tan^{-1}\frac{\dfrac{\partial u_y}{\partial x}dxdt}{dx + \dfrac{\partial u_x}{\partial x}dxdt}\right) = \frac{\partial u_y}{\partial x}dt \tag{3.2}$$

$$d\beta = \lim_{dt\to 0}\left(\tan^{-1}\frac{\dfrac{\partial u_x}{\partial x}dydt}{dy + \dfrac{\partial u_y}{\partial x}dydt}\right) = \frac{\partial u_x}{\partial y}dt \tag{3.3}$$

Substituting for $d\alpha$ and $d\beta$ using Equations 3.2 and 3.3 in 3.1 gives the rate of rotation, the angular velocity, about the z-axis.

$$\Omega_z = \frac{1}{2}\left(\frac{\partial u_y}{\partial x} - \frac{\partial u_x}{\partial y}\right) \tag{3.4}$$

In a similar fashion, the rates of rotation about the x- and y-axes are

$$\Omega_x = \frac{1}{2}\left(\frac{\partial u_z}{\partial y} - \frac{\partial u_y}{\partial z}\right) \tag{3.5}$$

$$\Omega_y = \frac{1}{2}\left(\frac{\partial u_x}{\partial z} - \frac{\partial u_z}{\partial x}\right) \tag{3.6}$$

These three components define the velocity vector Ω.

For convenience a quantity called the vorticity, ω, is defined, which is twice the local angular velocity and provides a measure of the local rate of rotation of a fluid particle or element.

$$\omega = 2\Omega \tag{3.7}$$

The rotation convention is represented by the right-hand rule between velocity and vorticity directions.

The components of vorticity in Cartesian coordinates are

$$\omega_x = \left(\frac{\partial u_z}{\partial y} - \frac{\partial u_y}{\partial z}\right) \tag{3.8}$$

$$\omega_y = \left(\frac{\partial u_x}{\partial z} - \frac{\partial u_z}{\partial x}\right) \tag{3.9}$$

$$\omega_z = \left(\frac{\partial u_y}{\partial x} - \frac{\partial u_x}{\partial y}\right) \tag{3.10}$$

The physical significance of vorticity can be developed by imagining a small spherical element of fluid in a flow to be suddenly frozen. If the solid sphere has any rotation associated with it as it moves in the flow, then the fluid has vorticity at the point concerned. The numerical magnitude of the vorticity will be twice the angular velocity of the solid sphere.

It is often convenient in rotating flow applications to use cylindrical polar coordinates, and the components of vorticity in cylindrical coordinates (r, ϕ, z) are:

$$\omega_r = \left(\frac{1}{r}\frac{\partial u_z}{\partial \phi} - \frac{\partial u_\phi}{\partial z} \right) \tag{3.11}$$

$$\omega_\phi = \left(\frac{\partial u_r}{\partial z} - \frac{\partial u_z}{\partial r} \right) \tag{3.12}$$

$$\omega_z = \left(\frac{u_\phi}{r} + \frac{\partial u_\phi}{\partial r} - \frac{1}{r}\frac{\partial u_r}{\partial \phi} \right) \tag{3.13}$$

Inspection of Equations 3.4-3.7 shows that vorticity and velocity are related by vector calculus by

$$\boldsymbol{\omega} = curl\ \boldsymbol{u} = \nabla \times \boldsymbol{u} \tag{3.14}$$

where $\boldsymbol{u}$ is the velocity vector, the operator $\times$ denotes the cross product, and ∇ is the del operator.

$$\nabla = \frac{\partial}{\partial x}\boldsymbol{i} + \frac{\partial}{\partial y}\boldsymbol{j} + \frac{\partial}{\partial z}\boldsymbol{k} \tag{3.15}$$

The rotation of particles is also expressed by a quantity known as the circulation, Γ. The circulation is defined as the integral of the tangential velocity component along a closed curve fixed in the flow.

$$\Gamma = \oint_s \boldsymbol{u}{\cdot}\boldsymbol{t}ds \tag{3.16}$$

In SI units the units of circulation are m^2/s. Stokes's theorem states that the circulation around a contour is equal to the sum of the total vorticity enclosed within the area of the contour. From Stokes' theorem the line integral can be converted to a surface integral over the area A enclosed by s.

$$\Gamma = \oint_s \boldsymbol{u}{\cdot}\boldsymbol{t}ds = \int_A (\nabla \times \boldsymbol{u}){\cdot}\boldsymbol{n}dA \tag{3.17}$$

where **n** is the outward pointing unit normal vector to the surface. Since $\boldsymbol{\omega} = curl\boldsymbol{u} = \nabla \times \boldsymbol{u}$, the circulation can also be expressed in terms of the vorticity vector by

$$\Gamma = \int_A \boldsymbol{\omega}{\cdot}\boldsymbol{n}dA \tag{3.18}$$

If the flow is irrotational, $\omega = 0$, then the circulation, Γ, will also be zero.

In the case of two-dimensional flow, the vorticity at a point is defined as the ratio of the circulation around an infinitesimal circuit to the area of that circuit; that is,

$$\omega = \frac{\Gamma}{A} \tag{3.19}$$

3.2. VORTEX FLOW

As described in Section 3.1, the term *vortex* is used to describe a region of concentrated rotation in a flow, such as an eddy, a whirlpool, or the depression at the center of a whirling body fluid. Vortices and their formation and decay are very important in fluid mechanics because of their common occurrence in nature and machinery.

A flow pattern in which the streamlines are concentric circles is called a plane circular vortex. There are two fundamental types of vortex flow, and they are distinguished by whether the flow is rotational or irrotational. If the particles of fluid moving in these concentric circles do not rotate on their own axes, the flow is irrotational and the vortex is known as an irrotational flow vortex or as a free vortex. If the particles of fluid moving in these concentric circles rotate on their own axes, the flow is rotational and the vortex is known as a rotational flow vortex or as a forced vortex. In a forced vortex the fluid rotates as a solid body with a constant rotational velocity. A comparison of the rotation of "floats" placed in a free vortex and a forced vortex is shown in Figure 3.4. In addition, combinations of irrotational and rotational flow are possible, and these are referred to as compound vortices or Rankine vortices.

Free, forced and compound vortices are introduced in Sections 3.2.1, 3.2.2, and 3.2.3, respectively. The models developed are based on simplifications and idealizations. Potential flow relationships can be exploited to enable the modeling of a free spiral vortex flow where fluid is supplied or extracted at the axis; this method is introduced in Section 3.2.4. The fundamental theorems developed by Helmholtz for vortex nature are given in Section 3.2.5. In general, vortex flows are complex and difficult to model with accuracy owing to the

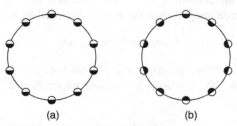

(a)   (b)

FIGURE 3.4   (a) Floats placed in a free vortex. (b) Floats placed in a forced vortex.

interplay between fast tangential motion and intense localized vorticity, which leads to physical modeling outside the scope of potential flow and standard boundary layer theory. Examples of the complex nature of vortex flow are the smoke ring, which is described in Section 3.2.6, and the bathtub vortex, which is described in Section 3.2.7.

3.2.1 Free Vortex

An irrotational flow vortex is also known as a free vortex, a line vortex, or a potential vortex. Here the term *free vortex* will principally be used. The fundamental relationships governing the flow associated with a free vortex can be developed by considering the small fluid element illustrated in Figure 3.5 where the fluid element in the reference plane is bounded by two streamlines and two radii.

The radius r here represents the radius of curvature and is used to ensure the generality of the results. The circulation around the element is given by

$$\Gamma = (u_\phi + \delta u_\phi)(r + \delta r)\delta\phi - u_\phi r\delta\phi \tag{3.20}$$

Neglecting high orders of small magnitude, Equation 3.20 becomes

$$\Gamma = (r\delta u_\phi + u_\phi\delta r)\delta\phi \tag{3.21}$$

The vorticity is given by the ratio of circulation and area

$$\omega = \frac{circulation}{area} = \frac{(r\delta u_\phi + u_\phi\delta r)\delta\phi}{r\delta r\delta\phi} = \frac{u_\phi}{r} + \frac{\delta u_\phi}{\delta r} \tag{3.22}$$

In the limit as $\delta r \rightarrow 0$

$$\omega = \frac{u_\phi}{r} + \frac{\partial u_\phi}{\partial r} \tag{3.23}$$

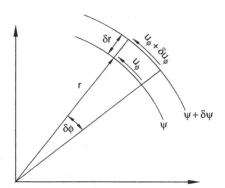

FIGURE 3.5 Fluid element.

The relationship for vorticity given in Equation 3.23 could be obtained directly from Equation 3.13, recognizing that u_r and u_z are both zero, and hence in cylindrical polar coordinates,

$$\omega_z = \left(\frac{u_\phi}{r} + \frac{\partial u_\phi}{\partial r}\right) \tag{3.24}$$

For irrotational flow, $\omega = 0$, so from Equation 3.23,

$$\frac{u_\phi}{r} + \frac{\partial u_\phi}{\partial r} = 0 \tag{3.25}$$

Since the streamlines are concentric circles in a circular vortex, the cross-sectional area of a streamtube will be constant along its length. Continuity of matter means that the velocity will be constant along each streamline. The velocity therefore only varies as a function of the radius, and Equation 3.25 can be rewritten as an ordinary differential equation.

$$\frac{u_\phi}{r} + \frac{du_\phi}{dr} = 0 \tag{3.26}$$

This can be stated as

$$\frac{du_\phi}{dr} = -\frac{u_\phi}{r} \tag{3.27}$$

which can be integrated to give

$$u_\phi r = \text{constant} = C \tag{3.29}$$

The constant C is known as the vortex strength at any radius r, and the angular momentum in a free vortex is constant.

Equation 3.29 states that, as the radius tends to zero, the velocity tends to infinity. This in practice is of course impossible. In a real fluid the effects of friction become significant as the radius tends to zero, and in a real fluid behaving as a free vortex, the central region tends to rotate like a solid body. The analysis presented for a free vortex is usually only used for regions away from the central core.

The circulation around a circuit corresponding to a streamline of a free vortex is defined by the line integral of the tangential velocity component taken once around a closed circuit in the fluid. The circulation at a radius r is therefore given for anticlockwise rotation by

$$\Gamma = 2\pi r u_\phi \tag{3.30}$$

For clockwise rotation the circulation is given by $-2\pi r u_\phi$.

Since $u_\phi r = \text{constant}$, the circulation is also constant for the entire vortex. This theoretically includes the center of the vortex. The circulation around an infinitesimally small circuit at the center has a value of the same nonzero constant and is therefore not zero. The free vortex, though irrotational everywhere else, has a rotational core at the center. The center is a "singular" point at which the velocity is theoretically infinite. As this is practically impossible, the equations do not necessarily apply here, and the results obtained with them for this region need to be viewed with appropriate caution.

For a vortex centered at the origin of the coordinates, the stream function, ψ, is given by

$$\psi = \int \frac{\partial \psi}{\partial r} dr + \int \frac{\partial \psi}{\partial \phi} d\phi = \int u_\phi dr + 0 = \int \frac{\Gamma}{2\pi r} dr = \frac{\Gamma}{2\pi} \ln\left(\frac{r}{r_o}\right) \tag{3.31}$$

where r_o is radius at which the stream function is zero.

The velocity potential, Φ, is given by

$$d\Phi = u_r dr + r u_\phi d\phi \tag{3.32}$$

$$\Phi = \int \frac{\Gamma}{2\pi} d\phi = \frac{\Gamma}{2\pi} \phi \tag{3.33}$$

From Equation 2.58 with $F_r = 0$, $\mu = 0$, and $u_r = 0$,

$$-\frac{\partial p}{\partial r} = -\rho \frac{u_\phi^2}{r} \tag{3.34}$$

or

$$\frac{\partial p}{\partial r} = \rho \frac{u_\phi^2}{r} \tag{3.35}$$

From Equation 2.60 with $u_z = 0$, and the body force $F_z = -\rho g$.

$$-\frac{\partial p}{\partial z} - \rho g = 0 \tag{3.36}$$

The pressure, p is varying in both r and z, so $p = f(r,z)$,

$$\frac{dp}{dr} = \frac{\partial p}{\partial r} + \frac{\partial p}{\partial z} \frac{dz}{dr} \tag{3.37}$$

$$dp = \frac{\partial p}{\partial r} dr + \frac{\partial p}{\partial z} dz \tag{3.38}$$

Hence substituting for $\partial p/\partial r$ and $\partial p/\partial z$

$$dp = \rho \frac{u_\phi^2}{r} dr - \rho g dz \tag{3.39}$$

Rearranging and dividing through by ρg gives

$$\frac{dp}{\rho g} + dz = \frac{u_\phi^2}{rg} dr \tag{3.40}$$

For horizontal flow from Bernoulli's equation with $z = 0$ and for vortex flow $u = u_\phi$, the total head, H, is given by

$$H = \frac{p}{\rho g} + \frac{u_\phi^2}{2g} + z \qquad (3.41)$$

Differentiating with respect to radius gives

$$\frac{dH}{dr} = \frac{1}{\rho g}\frac{dp}{dr} + \frac{u_\phi}{g}\frac{du_\phi}{dr} + \frac{dz}{dr} \qquad (3.42)$$

or

$$dH = \frac{dp}{\rho g} + \frac{u_\phi}{g}du_\phi + dz \qquad (3.43)$$

Substituting for $dp/\rho g + dz$ in Equation 3.43 using Equation 3.40 gives

$$dH = \frac{u_\phi}{g}\left(\frac{du_\phi}{dr} + \frac{u_\phi}{r}\right)dr \qquad (3.44)$$

In general $\delta\psi = u_\phi\delta n$, so

$$\delta\psi = u_\phi\delta r \qquad (3.45)$$

The vorticity for a free vortex is given in Equation 3.24, and comparison between Equations 3.44 and 3.45 yields the relationship

$$dH = \omega\frac{\delta\psi}{g} \qquad (3.46)$$

Both the total head and the stream function are constant along any stream-line, and as a result the space between two adjacent streamlines corresponds to fixed values of δH and $\delta\psi$. From Equation 3.46 the vorticity must also have a fixed value between two streamlines, and the limit is constant along a streamline. Particles undergoing steady motion therefore have constant vorticity as they move along a streamline. If the flow is irrotational, as in a free vortex, then $\omega = 0$ and hence $\delta H = 0$. This means that the total head is constant not only along a streamline but across the entire region of the flow.

Substituting for the tangential velocity in Equation 3.39 using Equation 3.29 and integrating gives

$$\frac{p - p_o}{\rho} = -\frac{C^2}{2r^2} + g(z_o - z) \qquad (3.47)$$

For the case of a liquid free vortex, the profile of the free surface of the liquid can be obtained by setting $p = p_o =$ atmospheric pressure at the free surface. This gives

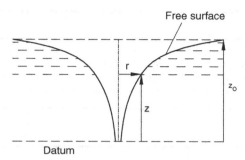

FIGURE 3.6 The free surface between a liquid free vortex and an atmosphere.

$$z = z_o - \frac{C^2}{2gr^2} \tag{3.48}$$

Examination of Equation 3.48 implies that the height of the free surface would be $-\infty$ at the axis, $r = 0$, along with an infinite velocity. The profile indicated by Equation 3.48 for the free surface between a liquid free vortex and an atmosphere is illustrated in Figure 3.6.

The analysis and resulting insight given here for a free vortex is based on the assumption of an ideal fluid with no viscosity. In general, the flow in a free vortex is characterized by the pressure decreasing and tangential velocity increasing as the radius decreases. The motion of many real fluid applications does, however, approach that of a free vortex. Examples include the flow around a bend in a pipe, the vortex formed at the outlet of a bath or sink, whirlpools, the vortices shed from aircraft wing tips, and tornadoes.

In an ideal fluid, a free vortex is permanent and indestructible. Once set in motion, the velocity at the core is infinite, and the vortex can only be stopped by the application of a tangential force, which in an ideal fluid, where the viscosity is zero, is impossible. In a real fluid, however, vortices are formed as a consequence of viscosity, and they are dissipated also as a result of viscosity. In an ideal fluid, a vortex cannot have a free end because that would involve a discontinuity of pressure. It must instead terminate at a solid boundary or a free surface or form a closed loop.

An indication of the manner in which the behavior of vortices in a real fluid approaches that of an ideal fluid is given by the examples of trailing wing tip vortices and the vortices generated by paddles in water. Both of these take a substantial length of time to decay in comparison to the time-scale of their motions. It is standard operational practice in air transport to ensure set distances between aircraft in order to mitigate against an aircraft flying into the vortical flow structures generated in the wake of a preceding aircraft.

Example 3.1

For a free vortex in air, calculate the pressure and tangential velocity components at a radius of 0.05 m and at a radius of 0.03 m, if the tangential velocity component and static pressure at a radius of 0.06 m are 15 m/s and 100000 Pa, respectively. The density can be assumed constant with a value of 1.1 kg/m^3.

Solution

The local static pressure can be determined from Equation 3.47, neglecting differences in elevation

$$\frac{p - p_o}{\rho} = -\frac{C^2}{2r^2}$$

From Equation 3.29, the vortex strength is given by

$$C = u_\varphi r = 15 \times 0.06 = 0.9 \text{ m}^2/\text{s}$$

At a radius of 0.05 m, the pressure is given by

$$p = -\rho\frac{C^2}{2r^2} + p_o = -1.1\frac{0.9^2}{2 \times 0.05^2} + 10^5 = 99820 \text{ Pa}$$

The tangential velocity is given by

$$u_\varphi = \frac{C}{r} = \frac{0.9}{0.05} = 18 \text{ m/s}$$

At a radius of 0.03 m, the pressure is given by

$$p_o = -1.1\frac{0.9^2}{2 \times 0.03^2} + 10^5 = 99510 \text{ Pa}$$

The tangential velocity is given by

$$u_\varphi = \frac{C}{r} = \frac{0.9}{0.03} = 30 \text{ m/s}$$

As can be seen, the static pressure decreases, and the tangential velocity component increases as the radius decreases.

Example 3.2

The free vortex that tends to form above an open drain in a relatively shallow tank is characterized by a tangential velocity that varies inversely with radial distance from a vertical axis through the drain. If the tangential velocity is 0.1 m/s at a radius of 0.3 m, what is the decrease in surface elevation at this radius, and what is the decrease in elevation at a radius of 0.03 m and at a radius of 0.01 m?

Solution

From Equation 3.29, the vortex strength is given by

$$C = u_\varphi r = 0.1 \times 0.3 = 0.03 \text{ m}^2/\text{s}$$

From Equation 3.48,

$$z - z_o = -\frac{C^2}{2gr^2} = -\frac{0.03^2}{2 \times 9.81 \times 0.3^2} = -0.5097 \times 10^{-3} \text{ m}$$

At $r = 0.03$ m,

$$z - z_o = -\frac{0.03^2}{2 \times 9.81 \times 0.03^2} = -0.05097 \text{ m}$$

At $r = 0.01$ m,

$$z - z_o = -\frac{0.03^2}{2 \times 9.81 \times 0.01^2} = -0.4587 \text{ m}$$

This example illustrates the substantial decrease in surface elevation with decreasing radius.

3.2.2 Forced Vortex

A relatively simple example of rotational flow is the case of the forced vortex, which is also known as a flywheel vortex. A fluid may be subjected to translation or rotation at constant accelerations without relative motion between particles. If there is no relative motion between particles, then the fluid is free from shear stresses and is in a state of relative equilibrium. For example, a body of fluid contained in a vessel that is rotating about a vertical axis with constant angular velocity will eventually reach a state of relative equilibrium and rotate with the same angular velocity as the vessel, forming a forced vortex. Three body forces arise relative to a spinning coordinate system: the Coriolis body force, the body force from angular acceleration, and the centrifugal body force. If the fluid is assumed to be at rest in the coordinate frame, the Coriolis force will be zero. If the angular velocity is constant, then the only rotational body force remaining is the centrifugal body force. The acceleration of any particle of fluid at radius r due to rotation will be $-\Omega^2 r$ perpendicular to the axis of rotation.

In a forced vortex, all the fluid rotates with uniform angular velocity Ω about an axis, which for convenience here will be taken to be the z-axis. The fluid essentially rotates as a solid body, and the tangential velocity at a radius r will be given by Ωr. If the angular velocity is constant, there will be no tangential component of acceleration in the ϕ direction. With no relative movement between fluid elements, the flow can be modeled as inviscid, and the Navier-Stokes equation can be reduced to Euler's equation.

$$\rho\frac{Du}{Dt} = -\nabla p + F \tag{3.49}$$

As the acceleration on an element of fluid at radius r must be u_ϕ^2/r, for radial equilibrium from Equation 2.58,

$$\frac{\partial p}{\partial r} = \rho\frac{u_\phi^2}{r} \tag{3.50}$$

For vertical equilibrium from Equation 2.60,

$$\frac{\partial p}{\partial z} = -\rho g \tag{3.51}$$

Substituting for the tangential velocity in terms of Ω and r gives

$$\frac{\partial p}{\partial r} = \rho\Omega^2 r \tag{3.52}$$

Equations 3.51 and 3.52 can be integrated to give the pressure distribution. The static pressure, p, is varying in both r and z, so $p = f(r,z)$,

$$\frac{dp}{dr} = \frac{\partial p}{\partial r} + \frac{\partial p}{\partial z}\frac{dz}{dr} \tag{3.53}$$

$$dp = \frac{\partial p}{\partial r}dr + \frac{\partial p}{\partial z}dz \tag{3.54}$$

Hence substituting for $\partial p/\partial r$ and $\partial p/\partial z$

$$dp = \rho\Omega^2 r\,dr - \rho g\,dz \tag{3.55}$$

Integration of Equation 3.55, gives

$$p - p_o = \frac{1}{2}\rho\Omega^2 r^2 - \rho g(z - z_o) \tag{3.56}$$

If a forced vortex is produced in a liquid in a container open to an atmosphere, the pressure at the free surface of the liquid will be the same as that of the atmosphere and therefore constant. The pressure at the free surface can be taken as p_o, and the vertical distance from the datum to the free surface minimum taken as z_o (Figure 3.7). If $z = z_o$ when $r = 0$,

$$z - z_o = \frac{\Omega^2 r^2}{2g} \tag{3.57}$$

Equation 3.56 gives the free surface in the form of a parabolic distribution. Surfaces of constant pressure are therefore paraboloids of revolution. If the fluid is a liquid, the free surface will also be a paraboloid, as illustrated in Figure 3.7.

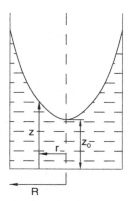

FIGURE 3.7 Forced vortex.

Rearranging Equation 3.55 in a form analogous to Bernoulli's equation gives

$$p + \frac{1}{2}\rho r^2 \Omega^2 + \rho g z = p_o + \rho r^2 \Omega^2 + \rho g z_o \tag{3.58}$$

It can be seen from Equation 3.58 that the energy associated with the streamlines in a forced vortex increases with increasing radius. The flow in a forced vortex can therefore only be maintained by the addition of energy from some external source. In the case of a rotating vessel containing a liquid, the energy to sustain the flow comes from the motor driving the vessel. Examples of a forced vortex occurring in practice include samples in a centrifuge, the central core of a stirred mixing vessel, and the motion in the impeller of a centrifugal pump under shut-off conditions when the delivery valve is closed.

The vorticity associated with a forced vortex can be evaluated from Equations 3.11–3.13. u_r and u_z are both zero, and $u_\phi = \Omega r$, so the vorticity in a forced vortex has only one component in the z direction

$$\omega_z = \left(\frac{u_\phi}{r} + \frac{\partial u_\phi}{\partial r} \right) = 2\Omega \tag{3.59}$$

Vorticity in a forced vortex is constant, and its amplitude is twice the angular velocity.

The stream function for a forced vortex can be obtained as for the free vortex. For a forced vortex $u_\phi = \Omega r$. For a vortex centered at the origin of the coordinates, the stream function is given by

$$\psi = \int \frac{\partial \psi}{\partial r} dr + \int \frac{\partial \psi}{\partial \phi} d\phi = \int \Omega r \, dr = -\frac{1}{2}\Omega r^2 + \text{constant} \tag{3.60}$$

For $\psi = 0$ at $r = 0$,

$$\psi = -\frac{1}{2}\Omega r^2 \tag{3.61}$$

for anticlockwise rotation.

Since a forced vortex is rotational there is no velocity potential corresponding to it.

Example 3.3

A 1 m tall vertical cylinder with an internal diameter of 500 mm is filled with oil with a density of 900 kg/m³ to a depth of 0.3 m. The cylinder is open to the atmosphere. If the cylinder is rotated about its vertical axis so that the oil just begins to uncover the base of the vessel, determine the angular velocity of the cylinder at the height to which the oil rises up the walls of the cylinder.

Solution

As the liquid is contained within a rotating cylinder, it is assumed that a forced vortex will be formed.

The volume of oil is

$$V_{oil} = \pi R^2 z_2 = \pi \times 0.25^2 \times 0.3 = 0.05890 \text{ m}^3$$

As the cylinder is open to the atmosphere, the free surface of the oil will form a paraboloid.

The height to which the oil will rise can be determined from the volume of the paraboloid. This is given by half the volume of the circumscribing cylinder, that is,

$$V_{paraboloid} = \frac{1}{2}\pi R^2 z_1$$

Provided that no oil is lost, an assumption that will need to be verified later, then the volume of the oil will be equal to the volume of the cylindrical section to which the oil rises minus the volume of the paraboloid of oil.

$$V_{oil} = \pi R^2 z_1 - V_{paraboloid} = \frac{1}{2}\pi R^2 z_1$$

This gives height to which the oil will rise up the cylinder walls as

$$z_1 = 2z_2 = 0.6 \text{ m}$$

As this is less than the height of the vessel, it can be assumed that no oil is lost.

The rotational speed at which the base of the drum just becomes uncovered can be determined from

$$z - z_o = \frac{\Omega^2 r^2}{2g}$$

giving

$$\Omega = \sqrt{\frac{2gz_1}{R^2}} = \sqrt{\frac{2 \times 9.81 \times 0.6}{0.25^2}} = 13.72 \text{ rad/s} \approx 131 \text{ rpm}$$

3.2.3 Compound Vortex

In a free vortex the tangential velocity is given by $u_\varphi = C/r$, and theoretically the velocity becomes infinite at the center. Near the axis, in a real fluid, velocities would indeed be high, but as frictional losses tend to vary as the square of the velocity, they would not be negligible and the assumption that the total head remains constant will not be valid. The central part of a vortex in a real fluid tends to rotate as a solid body. This can be modeled as a forced vortex surrounded by a free vortex as illustrated in Figure 3.8.

The velocity at the common radius between the forced vortex and free vortex must be equal in order to avoid a nonphysical discontinuity.

For the free vortex if Δz_1 is equal to the depression of the free surface at a radius R below the level of the free surface at $r = \infty$, then from Equations 3.48 and 3.29,

$$\Delta z_1 = \frac{C^2}{2gR^2} = \frac{u_\phi^2}{2g} = \frac{\Omega^2 R^2}{2g} \qquad (3.62)$$

For the forced vortex if Δz_2 is equal to the height of the free surface at radius R above the center of the depression, then from Equation 3.57

$$\Delta z_2 = \frac{u_\phi^2}{2g} = \frac{C^2}{2gR^2} \qquad (3.63)$$

The total depression of the free surface at a radius of $r = 0$ below the level of the free surface at $r = \infty$ is given, using Equations 3.62 and 3.63, by

$$\Delta z = \Delta z_1 + \Delta z_2 = \frac{C^2}{2gR^2} + \frac{C^2}{2gR^2} = \frac{C^2}{gR^2} \qquad (3.64)$$

For the forced vortex, the velocity at radius R is ΩR, and for the free vortex, the velocity at R is C/R. The common radius at which these two velocities is the same is given by

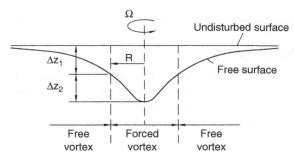

FIGURE 3.8   Compound vortex.

$$\Omega R = \frac{C}{R} \tag{3.65}$$

that is,

$$R = \sqrt{\frac{C}{\Omega}} \tag{3.66}$$

Example 3.4

An impeller of diameter 500 mm rotating at 75 rpm about the vertical axis inside a large vessel containing water produces a circular vortex motion with a local depression of the free surface above the impeller, the depth of which reduces with radius as indicated in Figure 3.8. Determine the level of the free surface at a radius equal to that of the impeller and also the total depression of the free surface below that of the free surface at a considerable distance away from the impeller.

Solution

Within the cylindrical region of the impeller diameter, the motion produced can be assumed to be that of a forced vortex and at a diameter greater than this the motion can be modeled as a free vortex.

The depression in the free surface level at the radius of the impeller can be determined from Equation 3.57 or Equation 3.62.

$$z - z_o = \frac{\Omega^2 r^2}{2g} = \frac{(75 \times 2\pi/60)^2 \times 0.25^2}{2 \times 9.81} = 0.1964 \text{ m}$$

The total depression can be determined from Equation 3.64. The vortex strength is given by

$$C = u_\varphi r = \Omega r^2 = (75 \times 2\pi/60) \times 0.25^2 = 0.4908 \text{ m}^2/\text{s}$$

Hence from Equation 3.64, the total depression of the free surface level below that of the surface at a considerable radius from the impeller is given by

$$\Delta z = \frac{C^2}{gR^2} = \frac{0.4908^2}{9.81 \times 0.25^2} = 0.3929 \text{ m}$$

3.2.4 Free Spiral Vortex

A free spiral vortex is the combination of a free vortex and radial flow. It can be modeled by the superposition of the stream functions of a free vortex and either a sink or source, depending on the direction of the radial flow.

For a radially outward flow with a clockwise free vortex,

$$\psi_{spiral\ vortex} = \psi_{source} + \psi_{free\ vortex} = \frac{\dot{q}\theta}{2\pi} + \frac{\Gamma}{2\pi}\ln r = \frac{1}{2\pi}(\dot{q}\theta + \Gamma\ln r) \tag{3.67}$$

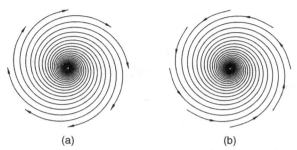

(a)   (b)

FIGURE 3.9   (a) Radially outward free spiral vortex. (b) Radially inward free spiral vortex.

$$\Phi_{spiral\ vortex} = \Phi_{source} + \Phi_{free\ vortex} = \frac{\dot{q}}{2\pi}\ln r + \frac{\Gamma\theta}{2\pi} = \frac{1}{2\pi}(\dot{q}\ln r + \Gamma\theta) \quad (3.68)$$

For a radially inward flow with a clockwise free vortex,

$$\psi_{spiral\ vortex} = \psi_{sink} + \psi_{free\ vortex} = -\frac{\dot{q}\theta}{2\pi} + \frac{\Gamma}{2\pi}\ln r = \frac{1}{2\pi}(\Gamma\ln r - \dot{q}\theta) \quad (3.69)$$

$$\Phi_{spiral\ vortex} = \Phi_{sink} + \Phi_{free\ vortex} = -\frac{\dot{q}}{2\pi}\ln r + \frac{\Gamma\theta}{2\pi} = \frac{1}{2\pi}(\Gamma\theta - \dot{q}\ln r) \quad (3.70)$$

The resulting flows for a radially outward and a radially inward free spiral vortex are illustrated in Figure 3.9.

3.2.5 Helmholtz's Theorems

Helmholtz's theorems provide useful guidance concerning the nature of a vortex. The theorems are as follows.

1. A vortex tube cannot end in a fluid. It must extend to infinity, end at a solid boundary or form a closed loop within the fluid (for example, the torus formed by a smoke ring).
2. Vortices remain attached to the same fluid particles as the fluid moves. The vortex strength of a flow, like velocity, can be specified relative to a fixed Eulerian grid.

Helmholtz's theorem 1 holds strictly for an inviscid, incompressible fluid. In a viscous fluid an open vortex can be initiated, but cannot persist, unless maintained by an external source of energy and angular momentum.

3.2.6 Toroidal Vortex

For the case of vortex formed by, for example, the expulsion of fluid from a cylinder through a body of more dense fluid, the no-slip condition at the

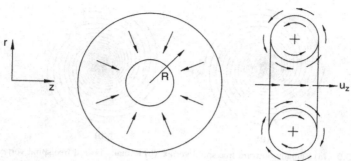

FIGURE 3.10 Toroidal vortex.

cylinder walls means that there is a velocity gradient across the fluid, and this results in vorticity in the fluid and a toroidal vortex exiting from the cylinder. From Helmholtz's first theorem, a vortex tube cannot end in a fluid, and the highly rotational flow exiting from the cylinder forms a closed loop or circular vortex ring (Figure 3.10).

Integration of the net effect of each element, dL of the circular vortex line of strength C, shows that a circular vortex ring, with radius R and its axis coincident with the z-axis will propel itself forward along the axis. The induced velocity of an arbitrary point on the z-axis is given by

$$u_z = \frac{C}{2} \frac{R^2}{(R^2 + z^2)^{1.5}} \tag{3.71}$$

At the center of the ring, the induced velocity, u_z, is given by

$$u_z = \frac{C}{2R} \tag{3.72}$$

The ring does not induce any component of velocity that could change R. As a result, a toroidal vortex, undisturbed by other influences, retains a constant radius in an inviscid fluid. For the case of a toroidal vortex ring in a viscid fluid, the action of viscosity results in the ring growing in radius. This can be observed for the case of smoke rings, which as they propel themselves forward grow in radius, until viscous dissipation of their vorticity results in the breakup of the toroidal vortex.

3.2.7 The Bathtub Vortex

For the case of flow down a sink or bathtub drain, the exiting flow will be due to the velocity gradients between the surface, where the flow is stationary owing to the no-slip condition, and the flow in the middle of the drain will generate vorticity. Leftover spin from filling the sink (even when the water looks still, it may be rotating slowly for a long time after it seems to stop), irregularities in the construction of the basin and convection currents if the water is warmer or colder than the basin, retraction of any plug, and to a very weak extent Coriolis effects—all result in rotation of the fluid in the sink about a vertical axis. Whether the rotation is clockwise or anticlockwise will depend on the relative magnitude of these various factors. The direction of rotation is only marginally affected by the Coriolis force. The influence of Coriolis effects can only be observed as a significant factor under laboratory conditions where exact symmetry is maintained and extensive time allowed for the liquid to settle prior to drainage of the sink (see, for example, Shapiro, 1962 and Trefethen et al., 1965). Coriolis accelerations are significant when the Rossby number is small, normally much less than unity, but for the purpose of this discussion say 0.1. Taking the length scale for a bathtub vortex as 0.1 m, the angular velocity of the Earth as 7.272×10^{-5} rad/s, Equation 2.160 gives the required velocity for Coriolis effects to be significant at approximately 7×10^{-5} m/s. This is such a small velocity that it is next to impossible to realize within a body of liquid and therefore can normally be discounted as a significant factor. The flow exiting a drain hole, once a vortex has formed, exhibits a fall in the liquid surface that corresponds to that shown in Figure 3.6.

While the bathtub vortex can crudely be modeled by a free vortex, the detailed flow phenomena are more complex, involving interactions between the axial and radial flow. Andersen et al. (2003), using the example of a cylindrical container rotating about its axis with a central drain at the base, show that the shape adopted by the free surface depends on the outflow rate and also, but to a lesser extent, on the angular velocity.

The development of a bathtub vortex is illustrated in Figure 3.11. At low values of angular velocity, a small depression is observed, but this grows at higher angular velocities until it forms a needlelike shape that rotates very rapidly. At very high rotation rates, the core becomes unstable and air bubbles detach and get dragged down by the surrounding flow. If the angular velocity is increased further, the frequency of the bubble shedding increases until the air filled core extends through the drain hole.

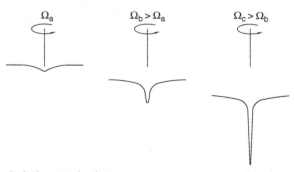

FIGURE 3.11 Bathtub vortex development.

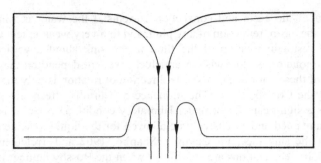

FIGURE 3.12 Flow structure of a bathtub vortex.

As indicated in the flow structure sketch shown in Figure 3.12, nearly all of the exiting flow is supplied by a thin boundary layer, known as an Ekman layer at the base of the container. In this boundary layer the flow spirals inward, and close to the drain a significant proportion is diverted upward and the rest goes down the drain.

In the bulk of the vessel, the flow is essentially two-dimensional and well described by a free vortex superimposed on the rigid motion of the container with

$$u_r = 0 \tag{3.73}$$

$$u_z = 0 \tag{3.74}$$

$$u_\phi \approx \Omega r + C/r \tag{3.75}$$

In an extension of Lundgren's (1985) model, Andersen et al. (2003) provide the following model for a bathtub vortex.

$$\frac{1}{r}\frac{d(rhu_r)}{dr} = u_{z,z=0} \tag{3.76}$$

Inside the downflow region, a downward plug flow is assumed, Equation 3.77. In the upflow region an exponential profile is used to match experimental measurements, Equation 3.78.

$$u_{z,z=0} = -\frac{Q}{\pi R^2} \text{ for } r \leq R \tag{3.77}$$

$$u_{z,z=0} = -\frac{Qe^{-(r-R)/d}}{2\pi d(R+d)} \text{ for } r > R \tag{3.78}$$

The length scales R and d are determined from experiments.

The Navier-Stokes equations for $Ro \ll 1$ and $u_z \ll 2gH$ reduce to

$$\frac{u_\varphi^2}{r} = g\frac{dh}{dr} - \frac{\alpha_{fs}}{\rho}\frac{ds}{dr} \qquad (3.79)$$

where α_{fs} is the free surface tension of the air–water interface. The curvature of the free surface, s_{fs}, is given by

$$s_{fs} = \frac{h'}{r\sqrt{1+h'^2}} + \frac{h''}{[1+h'^2]^{1.5}} \qquad (3.80)$$

The tangential Navier-Stokes equation reduces to

$$u_r\left(\frac{du_\phi}{dr} + \frac{u_\phi}{r}\right) = \frac{\mu}{\rho}\frac{d^2u_\phi}{dr^2} \qquad (3.81)$$

Equations 3.79 and 3.81 can be solved to give h and u_ϕ and Equation 3.76 and the velocity profiles in Equation 3.77 and 3.78 to eliminate u_r.

3.3. TAYLOR-PROUDMAN THEOREM

The Taylor-Proudman theorem states that there is no variation of the velocity field in the direction parallel to the axis of rotation. This is a physically significant theorem for rotating fluids and can be proved readily from manipulation of the Navier-Stokes and continuity equations. Here proofs are presented based on the Navier-Stokes equations defined in terms of vorticity and velocity and also using vector analysis of the equations of motion. Only one proof is necessary, but all are presented here to show the equivalence of the analytical techniques and to illustrate this important result.

The Navier-Stokes equations can be presented using vorticity as the variable by a number of vector manipulations. The Navier-Stokes equation for unsteady flow of a constant density and constant viscosity fluid where body forces are insignificant, from Equation 2.52, is

$$\frac{\partial \mathbf{u}}{\partial t} + (\mathbf{u}\cdot\nabla)\mathbf{u} = -\frac{1}{\rho}\nabla p + \nu\nabla^2\mathbf{u} \qquad (3.82)$$

Applying $\nabla\times$ to each term and use of vector identities gives

$$\frac{\partial \boldsymbol{\omega}}{\partial t} + (\mathbf{u}\cdot\nabla)\boldsymbol{\omega} - (\boldsymbol{\omega}\cdot\nabla)\mathbf{u} = \nu\nabla^2\boldsymbol{\omega} \qquad (3.83)$$

From Equation 3.14, $\boldsymbol{\omega} = \nabla\times\boldsymbol{u}$, and from the vector identity the divergence of a curl is equal to zero, $\nabla\cdot\boldsymbol{\omega} = 0$. The pressure term has disappeared because the curl of gradient is equal to zero, so $\nabla \times \nabla p = 0$. Using the scaling factors given in Section 2.6, Equation 3.83 in dimensionless form is given by

$$\frac{\partial \boldsymbol{\omega}_*}{\partial t_*} + (\mathbf{u}_* \cdot \nabla)\boldsymbol{\omega}_* - (\boldsymbol{\omega}_* \cdot \nabla)\mathbf{u}_* = \frac{1}{Re}\nabla^2 \boldsymbol{\omega}_* \qquad (3.84)$$

In a rotating frame of reference, Equation 3.83 becomes

$$\frac{\partial \boldsymbol{\omega}}{\partial t} + (\mathbf{u} \cdot \nabla)\boldsymbol{\omega} - (\boldsymbol{\omega} \cdot \nabla)\mathbf{u} - 2\Omega\frac{\partial \mathbf{u}}{\partial z} = \nu\nabla^2 \boldsymbol{\omega} \qquad (3.85)$$

In a rotating frame of reference, Equation 3.84 becomes

$$\frac{\partial \boldsymbol{\omega}_*}{\partial t_*} + Ro(\mathbf{u}_* \cdot \nabla)\boldsymbol{\omega}_* - (\boldsymbol{\omega}_* \cdot \nabla)\mathbf{u}_* - 2\frac{\partial \mathbf{u}_*}{\partial z_*} = Ek\nabla^2 \boldsymbol{\omega}_* \qquad (3.86)$$

For steady flow $\partial \boldsymbol{\omega}_*/\partial t_* \approx 0$. If the convective velocities in the flow are negligible relative to rotational speeds, then the Rossby number, $Ro \approx 0$. If the flow is inviscid then the Ekman number, $Ek \approx 0$. Substitution of $\partial \boldsymbol{\omega}_*/\partial t_* = 0$, $Ro = 0$ and $Ek = 0$ in Equation 3.86 gives

$$\frac{\partial \mathbf{u}_*}{\partial z_*} = 0 \qquad (3.87)$$

or in dimensional form

$$\frac{\partial u}{\partial z} = 0 \qquad (3.88)$$

The implication of Equation 3.88 is that for a system with solid boundaries perpendicular to the axis of rotation the flow is restricted to two-dimensional motion in planes perpendicular to the axis of rotation. Conditions where $\partial \boldsymbol{\omega}_*/\partial t_* \approx 0$, $Ro \approx 0$, and $Ek \approx 0$ exist in some oceanic flows and can also be reproduced in laboratory experiments. Examples of the physical manifestation of the Taylor-Proudman theorem are transverse and longitudinal Taylor columns.

For steady rotating flow when both the Rossby number and Ekman number are small, indicating that the Coriolis force is significant in comparison to inertial forces and viscous forces, respectively, then the equation of motion for a rotating flow can be stated, from Equation 2.92, as

$$2\boldsymbol{\Omega} \times \mathbf{w} = -\frac{1}{\rho}\nabla p \qquad (3.89)$$

where the pressure has been transformed by removing the centrifugal forces. The Taylor-Proudman theorem can also be developed by taking the curl of Equation 3.89.

$$\nabla \times (\boldsymbol{\Omega} \times \mathbf{w}) = 0 \qquad (3.90)$$

$$\boldsymbol{\Omega}\cdot\nabla\mathbf{w} - \mathbf{w}\cdot\nabla\boldsymbol{\Omega} + \mathbf{w}(\nabla\cdot\boldsymbol{\Omega}) - \boldsymbol{\Omega}(\nabla\cdot\mathbf{w}) = 0 \qquad (3.91)$$

The angular velocity, $\mathbf{\Omega}$, is not a function of position so $\mathbf{w} \cdot \nabla \mathbf{\Omega} = 0$ and $\mathbf{w}(\nabla \cdot \mathbf{\Omega}) = 0$. The continuity equation in a rotating from of reference for steady flow, from Equation 2.109, is $\nabla \cdot \mathbf{w} = 0$, so

$$\mathbf{\Omega} \cdot \nabla \mathbf{w} = 0 \tag{3.92}$$

If the axis is selected so that the angular velocity is in the z direction,

$$\Omega \frac{\partial \mathbf{w}}{\partial z} = 0 \tag{3.93}$$

that is,

$$\frac{\partial \mathbf{w}}{\partial z} = 0 \tag{3.94}$$

This states that there is no variation of the velocity field in the direction parallel to the axis of rotation.

In component form, Equation 3.94 states

$$\frac{\partial u}{\partial z} = \frac{\partial v}{\partial z} = \frac{\partial w}{\partial z} \tag{3.95}$$

For the case of a system with solid boundaries perpendicular to the axis of rotation, w will be zero at some value of z. So

$$\frac{\partial u}{\partial z} = \frac{\partial v}{\partial z} = 0 \tag{3.96}$$

The Taylor-Proudman theorem stating that for a system with solid boundaries perpendicular to the axis of rotation the flow is restricted to two-dimensional motion in planes perpendicular to the axis of rotation can be demonstrated in the laboratory using a turntable, light source, and transparent cylindrical container (Greenspan, 1968). The closed container is completely filled with water into which a small quantity of aluminium powder has been suspended by mixing it with detergent. The container is illuminated with a vertical slit beam, or laser sheet, from the side. If a small hemispherical or cylindrical object is stuck to the base of the cylinder and the entire system rotated at a constant speed, say, for example, 20 rpm, then after a few minutes constant body rotation will be attained with all of the particles rotating with the same angular velocity. Once steady motion has been attained, the speed of the turntable is deliberately reduced by a small quantity, say 0.5% so that the container and the fluid no longer have the same angular velocity. The focus of the experiment is the fluid motion relative to the obstacle. What is observed in this experiment is a column of fluid rigidly attached to the protuberance as illustrated in Figure 3.13. The bulk of the fluid, with the exception of the fluid immediately above the protuberance, flows around the column formed above the protuberance as if the column of fluid was an impermeable solid cylinder. As indicated by the analysis

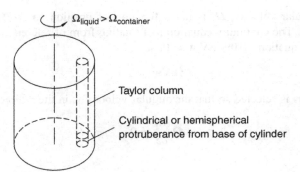

$\Omega_{liquid} > \Omega_{container}$

Taylor column

Cylindrical or hemispherical
protruberance from base of cylinder

FIGURE 3.13   A Taylor-Proudman column.

for a rotating fluid, when the Rossby and Ekman numbers approach zero, the
fluid motion is restricted to two-dimensional motion in planes perpendicular to
the axis of rotation, and the fluid cannot flow over the protuberance but must
instead circumscribe it. This was first predicted by Proudman (1916) and
subsequently confirmed by Taylor (1921a, 1921b, 1923). The resulting fluid
columns are known as Taylor-Proudman columns.

The Taylor-Proudman theorem provides insight into why vortices, when
frictional effects are small, are columnar. Changes in the velocity parallel to the
axis tend to be suppressed. This results in particle paths that are approximately
circular or helical (Lugt, 1995).

3.4. CONCLUSIONS

This chapter has introduced the concept of vorticity and a number of types of
vortices, a vortex being a region of concentrated rotation in a flow. Vorticity
involves the rotation of a flow element, representing a chunk of fluid particles, as
the flow element moves through a flow field. There is a fundamental distinction
between vorticity and curved or circular translation of fluid elements. If the
motion between particles is purely translational, and the distortion of the fluid
element concerned is symmetrical, the flow is defined as irrotational and vorticity
is not present. A number of analytical ideals representing different types of
vortices have been introduced. Some of these such as free and Rankine vortices
involve vorticity. By comparison, the forced vortex is irrotational and vorticity is
not present. Vortices are common in rotating flow applications and the techniques
introduced in this chapter will be developed and applied in Chapters 4, 5, 6, 7 and 8.

REFERENCES

Andersen, A., Bohr, T., Stenum, B., Juul Rasmussen, J., and Lautrup, B. Anatomy of a bathtub
    vortex. *Physical Review Letters*, 91 (2003).
Greenspan, H.P. *The theory of rotating fluids*. Cambridge, 1968.

Lugt, H.J. *Vortex flow in nature and technology.* Kreiger Publishing Company, 1995.

Proudman, J. On the motion of solids in liquids possessing vorticity. *Proc. Roy. Soc. A*, 92 (1916), pp. 408–424.

Shapiro, A.H. Bath-tub vortex. *Nature*, 196 (December 15, 1962), pp. 1080–1081.

Taylor, G.I. Experiments with rotating fluids. *Proc. Cambridge Phil. Soc.*, 20 (1921a), pp. 326–329,

Taylor, G.I. Experiments with rotating fluids. *Proc. Roy. Soc. A*, 100 (1921b), pp. 114–121.

Taylor, G.I. Experiments on the motion of solid bodies in rotating fluids. *Proc. Roy. Soc. A*, 104 (1923), pp. 213–218.

Trefethen, L.M., Bilger, R.W., Fink, P.T., Luxton, R.E., and Tanner, R.I. The bath-tub vortex in the Southern Hemisphere. *Nature*, 207 (September 4, 1965), pp. 1084–1085.

Chapter 4

Introduction to Rotating Disc Systems

Rotating discs are found in a range of engineering applications such as gas turbine engines, flywheels, gears, and brakes. This chapter considers the subject of a plain rotating disc in detail, thereby providing an in-depth development of understanding of the flow physics and modeling approach for both laminar and turbulent flow.

4.1. INTRODUCTION

The flow associated with discs is important in a number of applications ranging from disc drives for the memories of computers and electronic products to automotive discs brakes, flywheels, cutting discs, gear wheels, and discs to support turbomachinery blades. Shear stresses between the disc and the fluid in which it is rotating dictate the power required to drive the disc to overcome frictional drag, and the local flow field will influence the heat transfer. Unfortunately, a number of factors combine to frustrate any universal analysis, and the flow conditions and the proximity of local geometry need to be taken into account.

The schematics in Figure 4.1 illustrate some of the geometrical configurations of interest in disc flows. A large range of options are possible. In considering a simple disc, the flow characteristics are considerably different, depending, for example, on whether the disc is rotating in an initially static atmosphere or whether the disc is stationary and the surrounding fluid is rotating (Figures 4.1a and 4.1b, respectively). A common configuration in engineering is the rotor-stator wheel-space formed between a rotating disc and a stationary disc (e.g., Figure 4.1e–4.1j).

In order to develop an understanding of the flow associated with rotating discs, the case of a single disc rotating in an initially stationary fluid will be considered first. This is further developed to model a rotating fluid adjacent to a stationary disc, a case of relevance to some atmospheric flows such as torna-does. Subsequently, additional factors and geometrical complexities will be introduced, along with correlations for quantifying mass flows and windage. A theoretical treatment based on consideration of the momentum of the flow in

Rotating Flow, DOI: 10.1016/B978-0-12-382098-3.00004-4
81

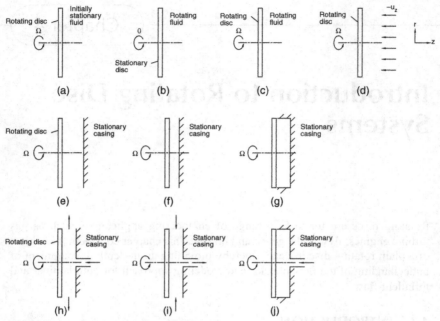

FIGURE 4.1 Selected disc flow configurations (a) The free disc. A rotating disc in an initially static flow. (b) Stationary disc in a rotating flow, (c) Rotating disc in a rotating flow. (d) Rotating disc with an impinging flow. (e) Rotor-stator disc wheelspace with a large gap. (f) Rotor-stator disc wheelspace with a narrow gap. (g) Enclosed rotor-stator disc wheelspace. (h) Rotor-stator disc wheelspace with radial outflow. (i) Rotor-stator disc wheelspace with radial inflow. (j) Rotor-stator disc wheelspace with stationary shroud and radial outflow. (ESDU, 2007).

the boundary layer is developed in detail in Section 4.4.1 dealing with turbulent flow for a rotating disc. This gives confirmation of results presented previously de facto in Section 4.4 and provides the foundation for subsequent analysis of the flow in the rotor and stator boundary layers for the rotor-stator wheelspace configuration, which is considered in Chapter 5.

A further set of configurations relevant to disc flows is the cavity formed between two coaxially located co-rotating or contra-rotating discs. Here the flow characteristics are dependent on a function of both rotational and convective effects in a fashion comparable with that of Rayleigh-Bénard flow for a simple stationary rectangular cavity. Rotating cavity flows are discussed in Chapter 7.

It is now standard practice for the fluid flow modeling for many applications to be undertaken using computational fluid dynamics (CFD). Use of CFD, as with any modeling method, requires appropriate attention to the model physics and boundary conditions. There is likely to be a trade-off between the accuracy of the model to the true conditions and the length of time it takes to develop and obtain answers from the model. The computational cost of a CFD model is highly dependent on how many dimensions the flow is modeled in and on whether it is modeled as steady or unsteady. Further factors include the flow characteristics, whether laminar, transitional, or turbulent. If the flow is turbulent, then it is necessary to select an appropriate turbulence model for the flow concerned, with options including the one equation turbulence model, the two equation turbulence model, direct numerical solution, large eddy simulation, and a choice of standard or near wall functions. Further considerations include whether it is necessary to also solve the energy equation and whether to model across fluid-solid boundaries to obtain, for instance, structure and thermal information.

If the flow concerned is likely to be unsteady with three-dimensional features, then careful consideration is required to determine whether it is appropriate to also model the flow as unsteady in three dimensions. If, however, the unsteady phenomena are not significant in determining the solution, it may be possible to obtain acceptable results with a comparatively simpler steady calculation. It is quite possible to adopt a modeling strategy, obtain solutions, and then be faced with the question of whether the data is valid. In making this judgement, prior experience is useful, combined with appropriate model validation against experimental data and theoretical solutions. Modeling a flow with CFD therefore requires application of careful judgements.

Prior to the advent of CFD, many of the advances in understanding associated with rotating disc flows were achieved through a combination of theoretical, numerical, and experimental approaches. Many of the theoretical approaches have now been superseded by CFD negating their detailed application. Here, however, detailed consideration will be given to the theoretical development for the case of laminar flow over a rotating disc. This is used to form the basis for developing an understanding of rotating disc flows for other important applications, in which relatively simple modeling and correlations can be used to provide, for example, estimation of flow rates and windage.

4.2. THE FREE DISC

The case of a single disc rotating in an initially stationary fluid is known as the free disc. In this analysis, a disc of radius b rotating with an angular velocity of Ω about the z-axis is considered (Figure 4.2). A boundary layer will develop on each side of the disc through which the tangential velocity of the fluid u_ϕ is sheared from the disc speed, Ωr, by the no-slip condition, to zero in the "free-stream" outside the boundary layer. Because of symmetry, only one side of the disc needs to be considered in developing the analysis.

The "centrifugal forces" generated by the shear between the rotating disc and the fluid cause a radial flow of fluid in the boundary layer. As the radial component of velocity is zero at the disc surface and is also zero in the free-stream, in order to satisfy conservation of mass and provide the radial outflow,

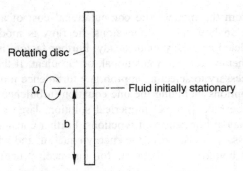

FIGURE 4.2 A single disc rotating in an initially stationary fluid.

fluid must be entrained axially into the boundary layer. The radial outflow resulting from a disc rotating in a fluid is sometimes referred to as the pumped flow and in the case of a free disc as the free-disc pumping effect. The principal flows and associated boundary layers for this case are illustrated in Figure 4.3.

Inside the disc boundary layer, the flow may be laminar or turbulent. The flow regime is governed by the relative magnitude of inertia to viscous effects as defined by the local rotational Reynolds number, which is defined in Equation 4.1.

$$\mathrm{Re}_{\phi,\,local} = x^2 \mathrm{Re}_{\phi} = \left(\frac{r}{b}\right)^2 \frac{\rho \Omega b^2}{\mu} = \frac{\rho \Omega r^2}{\mu} \tag{4.1}$$

where x is the ratio of the local radius, r, and the disc outer radius, b.

The flow near the axis of rotation tends to be laminar. For large values of angular velocity, the flow can become turbulent even at a low radius, and there

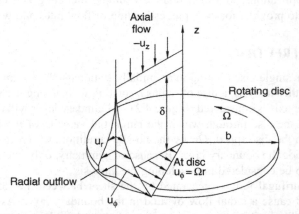

FIGURE 4.3 The radial and tangential boundary layers and associated inflow and outflow induced by a rotating disc.

will be a transition region between the laminar flow and turbulent flow zones. Transitional flow is considered further at the end of Section 4.2.1.

4.2.1 Laminar Flow over the Free Disc

As introduced in Chapter 2, using cylindrical polar coordinates, defining r as the distance from the axis, ϕ as the azimuthal angle, and z as the distance along the axis, the continuity equation in a cylindrical coordinate system is given by

$$\frac{\partial u_r}{\partial r} + \frac{u_r}{r} + \frac{1}{r}\frac{\partial u_\phi}{\partial \phi} + \frac{\partial u_z}{\partial z} = 0 \tag{4.2}$$

The Navier-Stokes equations in cylindrical polar coordinates are given by

$$\rho\left(\frac{\partial u_r}{\partial t} + u_r\frac{\partial u_r}{\partial r} + \frac{u_\phi}{r}\frac{\partial u_r}{\partial \phi} + u_z\frac{\partial u_r}{\partial z} - \frac{u_\phi^2}{r}\right) = -\frac{\partial p}{\partial r}$$

$$+ \mu\left(\frac{\partial^2 u_r}{\partial r^2} + \frac{1}{r}\frac{\partial u_r}{\partial r} - \frac{u_r}{r^2} + \frac{1}{r^2}\frac{\partial^2 u_r}{\partial \phi^2} + \frac{\partial^2 u_r}{\partial z^2} - \frac{2}{r^2}\frac{\partial u_\phi}{\partial \phi}\right) + F_r \tag{4.3}$$

$$\rho\left(\frac{\partial u_\phi}{\partial t} + u_r\frac{\partial u_\phi}{\partial r} + \frac{u_r u_\phi}{r} + \frac{u_\phi}{r}\frac{\partial u_\phi}{\partial \phi} + u_z\frac{\partial u_\phi}{\partial z}\right) = -\frac{1}{r}\frac{\partial p}{\partial \phi}$$

$$+ \mu\left(\frac{\partial^2 u_\phi}{\partial r^2} + \frac{1}{r}\frac{\partial u_\phi}{\partial r} - \frac{u_\phi}{r^2} + \frac{1}{r^2}\frac{\partial^2 u_\phi}{\partial \phi^2} + \frac{\partial^2 u_\phi}{\partial z^2} + \frac{2}{r^2}\frac{\partial u_r}{\partial \phi}\right) + F_\phi \tag{4.4}$$

$$\rho\left(\frac{\partial u_z}{\partial t} + u_r\frac{\partial u_z}{\partial r} + \frac{u_\phi}{r}\frac{\partial u_z}{\partial \phi} + u_z\frac{\partial u_z}{\partial z}\right) = -\frac{\partial p}{\partial z}$$

$$+ \mu\left(\frac{\partial^2 u_z}{\partial r^2} + \frac{1}{r}\frac{\partial u_z}{\partial r} + \frac{1}{r^2}\frac{\partial^2 u_z}{\partial \phi^2} + \frac{\partial^2 u_z}{\partial z^2}\right) + F_z \tag{4.5}$$

If rotational symmetry and steady flow are assumed, then the $\partial/\partial\phi$ and $\partial/\partial t$ terms can be neglected. This assumption is acceptable for flows that remain laminar, but for higher angular velocities flows that have periodic features in the circumferential direction or are turbulent, then the assumption that the $\partial/\partial\phi$ terms can be neglected is invalid. In addition, the terms u_r/r^2 and u_ϕ/r^2 can be ignored because they are insignificant in magnitude relative to the other viscous terms. If we further assume that body forces are insignificant, the continuity and Navier-Stokes equations in an inertial reference system reduce to

$$\frac{\partial u_r}{\partial r} + \frac{u_r}{r} + \frac{\partial u_z}{\partial z} = 0 \tag{4.6}$$

$$\rho\left(u_r\frac{\partial u_r}{\partial r}+u_z\frac{\partial u_r}{\partial z}-\frac{u_\phi^2}{r}\right)=-\frac{\partial p}{\partial r}+\mu\left(\frac{\partial^2 u_r}{\partial r^2}+\frac{1}{r}\frac{\partial u_r}{\partial r}+\frac{\partial^2 u_r}{\partial z^2}\right)\qquad(4.7)$$

$$\rho\left(u_r\frac{\partial u_\phi}{\partial r}+\frac{u_r u_\phi}{r}+u_z\frac{\partial u_\phi}{\partial z}\right)=\mu\left(\frac{\partial^2 u_\phi}{\partial r^2}+\frac{1}{r}\frac{\partial u_\phi}{\partial r}+\frac{\partial^2 u_\phi}{\partial z^2}\right)\qquad(4.8)$$

$$\rho\left(u_r\frac{\partial u_z}{\partial r}+u_z\frac{\partial u_z}{\partial z}\right)=-\frac{\partial p}{\partial z}+\mu\left(\frac{\partial^2 u_z}{\partial r^2}+\frac{1}{r}\frac{\partial u_z}{\partial r}+\frac{\partial^2 u_z}{\partial z^2}\right)\qquad(4.9)$$

For the region $z > 0$ and $r < b$ the boundary conditions, consistent with the no-slip condition at the rotating surface are

$$u_r = 0,\ u_\phi = r\Omega,\ u_z = 0 \text{ at } z = 0 \qquad(4.10)$$

and

$$u_r = 0, u_\phi = 0 \text{ as } z\rightarrow\infty \qquad(4.11)$$

The value of the axial velocity, u_z, as z tends to infinity is not specified. It adjusts to a nonzero negative value in order to provide the necessary flow for the disc pumping effect.

The analysis strategy is to reduce the Navier-Stokes equations by transformation to an exact set of ordinary differential equations that can be solved by numerical or other means. This problem was originally investigated by von Kármán (1921) and subsequently by Cochran (1934). It is relevant to note that the growth of a viscous boundary layer for flow over a flat stationary plate is proportional to $\sqrt{\nu x/u_x}$, giving in the case of an infinite plate an infinite boundary layer thickness. If an infinite plate is, however, oscillated in its plane with a velocity $u_x = U\cos(\omega t)$, where U is the amplitude of the velocity, ω is the circular frequency of oscillation, t is time, and the boundary layer does not grow to infinity but instead reaches a finite limit known as the depth of penetration, with the boundary layer thickness proportional to $\sqrt{\nu/\omega}$ (see, for example, Schlicting, 1979, for a more detailed treatment of the boundary layer growth on an oscillating plate). Flows induced in opposite directions as the plate oscillates diffuse together to cancel out of the order of 99% of each other in terms of the boundary layer thickness. Boundary layers in rotating flows are found to also reach finite thicknesses that tend to be proportional to $\sqrt{\nu/\Omega}$, where Ω is the angular velocity.

Equations 4.6 to 4.9 can be made dimensionless by dividing through by appropriate parameters using the transformations listed in Equations 4.12 to 4.16. Selection of the dimensions requires consideration of the physical process concerned, and obviously foreknowledge of previous experience is useful. Here the * subscript has been used following the variable to signify that the variable is dimensionless. For example, z_* is therefore a dimensionless axial location, and u_{r*} is a dimensionless radial velocity component. In Equation 4.13, the notation $u_{r*}(z_*)$ is used to signify that the dimensionless radial velocity u_{r*} is a function of the dimensionless axial location z_*.

$$z_* = z\sqrt{\frac{\Omega}{\nu}} \qquad (4.12)$$

$$u_{r*}(z_*) = \frac{u_r}{r\Omega} \qquad (4.13)$$

$$u_{\phi*}(z_*) = \frac{u_\phi}{r\Omega} \qquad (4.14)$$

$$u_{z*}(z_*) = \frac{u_z}{\sqrt{\nu\Omega}} \qquad (4.15)$$

$$p_*(z_*) = -\frac{p}{\mu\Omega} = -\frac{p}{\rho\nu\Omega} \qquad (4.16)$$

Substitution of Equations 4.12 to 4.16 in Equations 4.6 to 4.9 gives a set of four coupled ordinary differential equations.

$$\frac{du_{z*}}{dz_*} + 2u_{r*} = 0 \qquad (4.17)$$

$$\frac{d^2u_{r*}}{dz_*^2} - u_{z*}\frac{du_{r*}}{dz_*} - u_{r*}^2 + u_{\phi*}^2 = 0 \qquad (4.18)$$

$$\frac{d^2u_{\phi*}}{dz_*^2} - u_{z*}\frac{du_{\phi*}}{dz_*} - 2u_{r*}u_{\phi*} = 0 \qquad (4.19)$$

$$\frac{d^2u_{z*}}{dz_*^2} - u_{z*}\frac{du_{z*}}{dz_*} - \frac{dp_*}{dz_*} = 0 \qquad (4.20)$$

For the region $z > 0$ and $r < b$ the boundary conditions, consistent with the no-slip condition at the rotating surface, are

$$u_{r*} = 0, u_{\phi*} = 1, \ u_{z*} = 0, p_* = 0 \ at \ z_* = 0 \qquad (4.21)$$

and

$$u_{r*} = 0, \ u_{\phi*} = 0 \ as \ z_* \rightarrow \infty \qquad (4.22)$$

Equations 4.17–4.20 can now be solved using numerical techniques. As they are dimensionless, any solution is not dependent on specific values of rotational speed, density and viscosity. The solution for the dimensionless variables is tabulated in Table 4.1 and presented graphically in Figure 4.4 using data from Owen and Rogers (1989).

Examination of Figure 4.4 reveals the relative magnitude of $u_{\phi*}$, u_{z*}, and u_{r*}. Fluid in contact with the surface rotates at the same angular velocity as the surface, in accordance with the no-slip condition, and experiences the same

TABLE 4.1 Numerical solution of von Kármán's equations for the free disc (Owen and Rogers, 1989)

z_*	u_{r*}	$u_{\phi*}$	u_{z*}	p_*	$\partial u_{r*}/\partial z_*$	$\partial u_{*}/\partial z_*$
0.0	0.0000	1.0000	0.0000	0.0000	0.5102	−0.6159
0.1	0.0462	0.9386	−0.0048	0.0925	0.4163	−0.6112
0.2	0.0836	0.8780	−0.0179	0.1674	0.3338	−0.5987
0.3	0.1133	0.8190	−0.0377	0.2274	0.2620	−0.5803
0.4	0.1364	0.7621	−0.0628	0.2747	0.1999	−0.5577
0.5	0.1536	0.7076	−0.0919	0.3115	0.1467	−0.5321
0.6	0.1660	0.6557	−0.1239	0.3396	0.1015	−0.5047
0.7	0.1742	0.6067	−0.1580	0.3608	0.0635	−0.4764
0.8	0.1789	0.5605	−0.1934	0.3764	0.0317	−0.4476
0.9	0.1807	0.5171	−0.2294	0.3877	0.0056	−0.4191
1.0	0.1802	0.4766	−0.2655	0.3955	−0.0157	−0.3911
1.1	0.1777	0.4389	−0.3013	0.4008	−0.0327	−0.3641
1.2	0.1737	0.4038	−0.3365	0.4041	−0.0461	−0.3381
1.3	0.1686	0.3712	−0.3707	0.4059	−0.0564	−0.3133
1.4	0.1625	0.3411	−0.4038	0.4066	−0.0640	−0.2898
1.5	0.1559	0.3132	−0.4357	0.4066	−0.0693	−0.2677
1.6	0.1487	0.2875	−0.4661	0.4061	−0.0728	−0.2470
1.7	0.1414	0.2638	−0.4952	0.4053	−0.0747	−0.2276
1.8	0.1338	0.2419	−0.5227	0.4043	−0.0754	−0.2095
1.9	0.1263	0.2218	−0.5487	0.4031	−0.0751	−0.1927
2.0	0.1189	0.2033	−0.5732	0.4020	−0.0739	−0.1771
2.1	0.1115	0.1864	−0.5962	0.4008	−0.0721	−0.1627
2.2	0.1045	0.1708	−0.6178	0.3998	−0.0698	−0.1494
2.3	0.0976	0.1565	−0.6380	0.3987	−0.0671	−0.1371
2.4	0.0910	0.1433	−0.6569	0.3978	−0.0643	−0.1258
2.5	0.0848	0.1313	−0.6745	0.3970	−0.0612	−0.1153
2.6	0.0788	0.1202	−0.6908	0.3962	−0.0580	−0.1057

2.7	0.0732	0.1101	−0.7060	0.3955	−0.0548	−0.0969
2.8	0.0678	0.1008	−0.7201	0.3949	−0.0517	−0.0888
2.9	0.0628	0.0923	−0.7332	0.3944	−0.0485	−0.0814
3.0	0.0581	0.0845	−0.7452	0.3939	−0.0455	−0.0745
3.2	0.0496	0.0708	−0.7668	0.3932	−0.0397	−0.0625
3.4	0.0422	0.0594	−0.7851	0.3926	−0.0343	−0.0524
3.6	0.0358	0.0498	−0.8007	0.3922	−0.0296	−0.0440
3.8	0.0304	0.0417	−0.8139	0.3919	−0.0253	−0.0369
4.0	0.0257	0.0349	−0.8251	0.3917	−0.0216	−0.0309
4.5	0.0168	0.0225	−0.8460	0.3914	−0.0144	−0.0199
5.0	0.0109	0.0144	−0.8596	0.3912	−0.0094	−0.0128
5.5	0.0070	0.0093	−0.8684	0.3912	−0.0062	−0.0082
6.0	0.0045	0.0060	−0.8741	0.3912	−0.0040	−0.0053
7.0	0.0019	0.0025	−0.8802	0.3912	−0.0017	−0.0022
8.0	0.0008	0.0010	−0.8827	0.3911	−0.0007	−0.0009
9.0	0.0003	0.0004	−0.8837	0.3911	−0.0003	−0.0004
10.0	0.0001	0.0002	−0.8842	0.3911	−0.0001	−0.0002
∞	0.0000	0.0000	−0.8845	0.3911	0.0000	0.0000

centripetal acceleration. As the disc begins to rotate, a boundary layer will form in the circumferential direction. Fluid in the boundary layer just above the surface also begins to rotate but cannot maintain the same centripetal acceleration as the surface and acquires an outward radial component. As the radial component of velocity increases in magnitude, a secondary boundary layer develops in the radial direction. The stresses in this secondary boundary layer do provide a central force and centripetal acceleration but less than that at the surface. At distances beyond the boundary layer thickness, the fluid is not rotating, and there is no mechanism available to sustain radial flow. Conservation of mass demands a flow to match the outward radial flow. This results in zero momentum fluid being drawn in axially, being given momentum in the boundary layer and then being pumped radially outward as high angular momentum fluid. The rotating boundary layer associated with a free disc is commonly referred to as an Ekman layer.

If the boundary layer thickness is defined to be the axial distance away from the rotating disc at which the tangential velocity is 1% of the disc speed, that is,

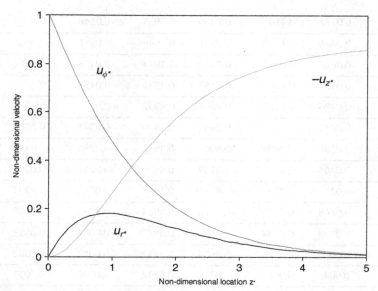

FIGURE 4.4 Dimensionless laminar flow velocity profiles for the free disc.

$u_\phi = 0.01\Omega r$, then Table 4.1 can be examined to give the boundary layer thickness for laminar flow over a free disc.

$u_{\phi*} = u_\phi/\Omega r = 0.01$ corresponds to a value for z_* of about 5.5, which on substitution in Equation 4.12 gives the following relationship for a laminar boundary layer on a free disc.

$$\delta \approx 5.5\sqrt{\frac{\nu}{\Omega}} \tag{4.23}$$

The boundary layer thickness on the disc, from Equation 4.23, is independent of the radius. In the boundary layer, the radial component of stress is balanced by the centrifugal force per unit area $(\tau_r = \rho 2\pi r\delta r\Omega^2 r\delta/2\pi r\delta r = \rho\Omega^2 r\delta)$. The tangential component of stress is proportional to the axial shear of the tangential component of velocity $(\rho\Omega r/\delta)$. Hence

$$\frac{\tau_r}{\tau_\phi} \propto \frac{\rho\Omega^2 r\delta}{\rho\Omega r/\delta} = \frac{\Omega\delta^2}{\nu} \tag{4.24}$$

The ratio of the components of radial and tangential stress in Equation 4.24 is independent of the radius. If it is accepted that the boundary layer thickness is independent of radius then

$$\delta \approx K_{bl}\sqrt{\frac{\nu}{\Omega}} \tag{4.25}$$

The value of K_{bl} will depend on the definition used for the boundary layer. If the 99% convention is applied to the tangential velocity component, then $K_{bl} = 5.5$ as given in Equation 4.23.

Examination of Equations 4.13 and 4.14 shows that the radial and tangential velocity components for free disc flows increase linearly with radius. As the radius increases, the area of the surface in contact with the fluid also increases with the square of the radius. Thus although the volume available to the flow in the boundary layer is increasing with radius, implying a decelerating flow, this is countered by the surface shear that causes the pumping effect with entrainment of flow axially. The net result is radial and tangential velocities in the boundary layer that increase as a linear function of radius.

For low values of the local rotational Reynolds number, the flow over a rotating disc will be laminar. As the rotational Reynolds number rises, the flow will become unstable with either small patches of turbulence or a more regular oscillating vortex disturbance. This disturbed flow will then break down into fully turbulent flow at a higher critical value of the local rotational Reynolds number. Experimental observations by Gregory, Stuart, and Walker (1955) for a rotating perspex disc showed that the flow became unstable at local rotational Reynolds numbers between 1.78×10^5 and 2.12×10^5, breaking down into a system of spiral vortices, illustrated schematically in Figure 4.5.

The transition to complete turbulence occurred for values of local rotational Reynolds numbers between 2.7×10^5 and 2.99×10^5. The experiments by Theodorsen and Regier (1944) for a polished rotating disc found that laminar flow began to become unstable for a local rotational Reynolds numbers of about 3.1×10^5 and for a roughened disc that this value could be as low as 2.2×10^5. They found that fully turbulent flow was established for a local rotational Reynolds number of about 7×10^5. These data combined with theoretical investigations provide evidence for the initial breakdown of laminar flow for values of the local rotational Reynolds number of about 2×10^5.

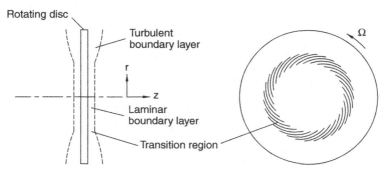

FIGURE 4.5   Schematic diagram of the boundary layers on a rotating disc.

$$\text{Re}_{\phi,critical} = x^2 \text{Re}_\phi = \left(\frac{r}{b}\right)^2 \frac{\rho\Omega b^2}{\mu} = \frac{\rho\Omega r^2}{\mu} \approx 2 \times 10^5 \qquad (4.26)$$

$\text{Re}_{\phi,critical}$ represents the value of the local rotational Reynolds number at the flow conditions become "critical" with breakdown of laminar flow and the onset of transitional flow. The actual value will depend on the specific local condition of the flow and surface roughness.

Example 4.1.

For a disc of radius 0.15 m spinning in atmospheric air at 950 rpm, determine the radial, tangential, and axial velocity components for the following locations.

i) r = 0.05 m, 0.1 m and 0.15 m for a plane 1 mm from the disc
ii) r = 0.05 m, 0.1 m and 0.15 m for a plane 1.6 mm from the disc
iii) r = 0.05 m, 0.1 m and 0.15 m for a plane distant from the disc

Take the density and viscosity of air as 1.226 kg/m³ and 1.795 × 10⁻⁵ Pa s, respectively.

Solution

The rotational Reynolds number for these conditions is given by

$$Re_\phi = \frac{1.226 \times (950 \times 2\pi/60) \times 0.15^2}{1.795 \times 10^{-5}} = 1.529 \times 10^5$$

As the value of the rotational Reynolds number is below 200000, the flow can be taken as laminar across the whole disc.

i) The nondimensional location, z_*, is given by

$$z_* = 0.001\sqrt{\frac{99.48}{1.795 \times 10^{-5}/1.226}} = 2.607$$

From Table 4.1, for $z_* = 2.6$, $u_{r*} = 0.0788$, $u_{\phi*} = 0.1202$, and $u_{z*} = -0.6908$
The radial, tangential, and axial velocities can now be determined for $r = 0.05$ m.

$$u_r = r\Omega u_{r*} = 0.05 \times 99.48 \times 0.0788 = 0.3920 \ m/s$$

$$u_\phi = r\Omega u_{\phi*} = 0.05 \times 99.48 \times 0.1202 = 0.5979 \ m/s$$

$$u_z = \sqrt{(\Omega\mu/\rho)}u_{z*} = \sqrt{(99.48 \times 1.795 \times 10^{-5}/1.226)} \times -0.6908 = -0.02636 \ m/s$$

The radial, tangential, and axial velocities can now be determined for $r = 0.1$ m.

$$u_r = r\Omega u_{r*} = 0.1 \times 99.48 \times 0.0788 = 0.7839 \ m/s$$

$$u_\phi = r\Omega u_{\phi*} = 0.1 \times 99.48 \times 0.1202 = 1.196 \ m/s$$

$$u_z = \sqrt{(\Omega\mu/\rho)}u_{z*} = \sqrt{(99.48 \times 1.795 \times 10^{-5}/1.226)} \times -0.6908 = -0.02636 \; m/s$$

The radial, tangential, and axial velocities can now be determined for $r = 0.15$ m.

$$u_r = r\Omega u_{r*} = 0.15 \times 99.48 \times 0.0788 = 1.176 \; m/s$$

$$u_\phi = r\Omega u_{\phi*} = 0.15 \times 99.48 \times 0.1202 = 1.794 \; m/s$$

$$u_z = \sqrt{(\Omega\mu/\rho)}u_{z*} = \sqrt{(99.48 \times 1.795 \times 10^{-5}/1.226)} \times -0.6908 = -0.02636 \; m/s$$

These calculations show the radial velocity increasing linearly with radius.
ii) The nondimensional location, z_*, is given by

$$z_* = 0.0016\sqrt{\frac{99.48}{1.795 \times 10^{-5}/1.226}} = 4.171$$

From Table 4.1, for $z_* = 4.171$, $u_{r*} = 0.0227$, $u_{\phi*} = 0.0307$ and $u_{z*} = -0.8323$
The radial, tangential, and axial velocities can now be determined for $r = 0.05$ m.

$$u_r = r\Omega u_{r*} = 0.05 \times 99.48 \times 0.0227 = 0.1129 \; m/s$$

$$u_\phi = r\Omega u_{\phi*} = 0.05 \times 99.48 \times 0.0307 = 0.1527 \; m/s$$

$$u_z = \sqrt{(\Omega\mu/\rho)}u_{z*} = \sqrt{(99.48 \times 1.795 \times 10^{-5}/1.226)} \times -0.8323 = -0.03176 \; m/s$$

The radial, tangential, and axial velocities can now be determined for $r = 0.1$ m.

$$u_r = r\Omega u_{r*} = 0.1 \times 99.48 \times 0.0227 = 0.2258 \; m/s$$

$$u_\phi = r\Omega u_{\phi*} = 0.1 \times 99.48 \times 0.0307 = 0.3054 \; m/s$$

$$u_z = \sqrt{(\Omega\mu/\rho)}u_{z*} = \sqrt{(99.48 \times 1.795 \times 10^{-5}/1.226)} \times -0.8323 = -0.03176 \; m/s$$

The radial, tangential, and axial velocities can now be determined for $r = 0.15$ m.

$$u_r = r\Omega u_{r*} = 0.15 \times 99.48 \times 0.0227 = 0.3387 \; m/s$$

$$u_\phi = r\Omega u_{\phi*} = 0.15 \times 99.48 \times 0.0307 = 0.4581 \; m/s$$

$$u_z = \sqrt{(\Omega\mu/\rho)}u_{z*} = \sqrt{(99.48 \times 1.795 \times 10^{-5}/1.226)} \times -0.8323 = -0.03176 \; m/s$$

These calculations also show the radial velocity increasing linearly with radius, but the magnitude of the radial velocity component has decreased with the increased distance from the disc for the equivalent radius.

iii) The nondimensional location distant from the disc is given by $z_* = \infty$.
From Table 4.1, for $z_* = \infty$, $u_{r*} = 0$, $u_{\phi*} = 0$, and $u_{z*} = -0.8845$
The radial, tangential, and axial velocities can now be determined for $r = 0.05$ m, $r = 0.1$ m, and $r = 0.15$ m.
$u_r = 0$
$u_\phi = 0$

$$u_z = \sqrt{(\Omega\mu/\rho)}u_{z*} = \sqrt{(99.48 \times 1.795 \times 10^{-5}/1.226)} \times -0.8845 = -0.03375 \; m/s$$

Distant from the disc the only component of velocity is the axial component.

The mass flow entrained by the boundary layer on one side of the rotating disc will be equal to the axial flow rate into the boundary layer. This can be found by integration.

$$\dot{m}_o = -2\pi\rho \int_0^r r[u_z]_{z\to\infty} \; dr = -\pi\rho r^2 \sqrt{(\Omega\nu)}u_{z*=\infty} \qquad (4.27)$$

The total mass flow for a disc of radius b, is found by substituting $u_{z*} = -0.8845$, the corresponding value of u_{z*} for $z_* = \infty$ from Table 4.1, which gives

$$\dot{m}_o = 2.779\rho b^2 \sqrt{\Omega\nu} \qquad (4.28)$$

It is often convenient to use a nondimensional mass flow rate, Cw, defined by

$$Cw = \frac{\dot{m}_o}{\mu b} \qquad (4.29)$$

Restating Equations 4.27 and 4.28 using 4.29 gives the nondimensional mass flow rate entrained by a rotating disc of radius b.

$$Cw = -\pi u_{z*=\infty}\text{Re}_\phi^{0.5} = 2.779\text{Re}_\phi^{0.5} \qquad (4.30)$$

A frequent requirement in rotating machinery is the evaluation of the power required to overcome viscous drag due to rotation of both discs and cylinders. The frictional moment on one side of the disc can be obtained by integrating the tangential component of the fluid stress at the surface over the area concerned.

The total moment, or torque, T_q, can be evaluated from

$$T_q = \tau A r \qquad (4.31)$$

The tangential component of shear stress is given by

$$\tau_\phi = \mu \left[\frac{\partial u_\phi}{\partial z}\right]_{z=0} \qquad (4.32)$$

From Equation 4.14, $u_\phi = r\Omega u_{\phi*}$, so

$$\frac{\partial u_\phi}{\partial z} = r\Omega\frac{\partial u_{\phi*}}{\partial z} \qquad (4.33)$$

From Equation 4.12, $z_* = z\sqrt{\Omega/\nu}$ which on differentiation gives

$$\frac{dz_*}{dz} = \sqrt{\frac{\Omega}{\nu}} \qquad (4.34)$$

Substitution in Equation 4.33 gives

$$\frac{\partial u_\phi}{\partial z} = \frac{r\Omega^{1.5}}{\sqrt{\nu}}\frac{\partial u_{\phi*}}{\partial z_*} \tag{4.35}$$

Considering an elemental ring of disc with area $\delta A = 2\pi r \delta r$ and integrating the tangential component of the fluid stress at the surface over the area concerned gives the frictional moment, T_q, on one side of the disc.

$$T_q = -\int_0^b 2\pi r \times \mu \left[\frac{\partial u_\phi}{\partial z}\right]_{z=0} \times r \, dr$$

$$= -\int_0^b 2\pi r^2 \mu \frac{r\Omega^{1.5}}{\sqrt{\nu}}\left[\frac{\partial u_{\phi*}}{\partial z_*}\right]_{z*=0} dr = -\int_0^b 2\pi r^3 \mu \frac{\Omega^{1.5}}{\sqrt{\nu}}\left[\frac{\partial u_{\phi*}}{\partial z_*}\right]_{z*=0} dr \tag{4.36}$$

This gives

$$T_q = -\frac{1}{2}\pi b^4 \rho \nu^{0.5}\Omega^{1.5}\left[\frac{\partial u_{\phi*}}{\partial z_*}\right]_{z_*=0} \tag{4.37}$$

It is convenient to use the moment coefficient, which is derived from the moment definition for drag on a rotating disc. The moment coefficient for one side of a disc of radius b is given by

$$C_m = \frac{T_q}{0.5\rho\Omega^2 b^5} \tag{4.38}$$

For laminar flow, for $z_* = 0$, $\partial u_{\phi*}/\partial z_* = -0.6159$ from Table 4.1. Substitution in Equation 4.37 for $\partial u_{\phi*}/\partial z_*$ and the rotational Reynolds number gives the moment coefficient for one side of a disc of radius b

$$C_m = -\pi \left[\frac{\partial u_{\phi*}}{\partial z_*}\right]_{z_*=0} \mathrm{Re}_\phi^{-0.5} = -\pi \times -0.6159\mathrm{Re}_\phi^{-0.5} = 1.935\mathrm{Re}_\phi^{-0.5} \tag{4.39}$$

It should be noted that the solution approach, and hence results obtained using Equations 4.27–4.31 and 4.37–4.39, assume that edge effects, that is, the windage contributions due to the thickness of the disc, are negligible.

Example 4.2.

For the disc of the previous example, with a radius of 150 mm and rotating at 950 rpm, in air with density and viscosity of 1.226 kg/m³ and 1.795 × 10⁻⁵ Pa s, respectively, determine:

i) the boundary layer thickness
ii) the nondimensional and the total mass flow entrained in the boundary layer due to its rotation
iii) the moment coefficient
iv) the total moment
v) the power required to drive the disc to overcome frictional drag

Solution

i) The rotational Reynolds number at the outer radius for these conditions is given by

$$Re_\phi = \frac{1.226 \times (950 \times 2\pi/60) \times 0.15^2}{1.795 \times 10^{-5}} = 1.529 \times 10^5$$

As the rotational Reynolds number is below 200000, the flow can be taken as laminar.

The boundary thickness can be determined from Equation 4.23.

$$\delta \approx 5.5\sqrt{\frac{\nu}{\Omega}} = 5.5\sqrt{\frac{\mu/\rho}{\Omega}} = 5.5\sqrt{\frac{1.795 \times 10^{-5}/1.226}{950 \times 2\pi/60}} = 2.110 \times 10^{-3} \ m \approx 2.1 \ mm$$

ii) The nondimensional mass flow rate of air entrained in the disc boundary layer can be determined from Equation 4.30.

$$Cw = 2.779Re_\phi^{0.5} = 2.779(1.529 \times 10^5)^{0.5} = 1087$$

The mass flow entrained is

$$\dot{m}_o = \mu bCw = 1.795 \times 10^{-5} \times 0.15 \times 1087 = 2.927 \times 10^{-3} \ kg/s$$

iii) The moment coefficient for one side of the disc can be determined from Equation 4.39

$$C_m = 1.935Re_\phi^{-0.5} = 1.935(1.529 \times 10^5)^{-0.5} = 4.949 \times 10^{-3}$$

iv) The moment on the disc face can be determined from Equation 4.38

$$T_q = 0.5\rho\Omega^2 b^5 C_m = 0.5 \times 1.226 \times 99.48^2 \times 0.15^5 \times 4.949 \times 10^{-3}$$
$$= 2.280 \times 10^{-3} \ N \, m$$

v) The power required to spin this disc face at 950 rpm to overcome frictional drag can be determined by

$$Power = T_q\Omega = 0.2268 \ W \approx 0.23 \ W$$

As there are two sides to the disc, the total power would be 0.46 W.

Comment

In this example, the power required to drive the disc is relatively small, a fraction of a Watt. It should, however, be noted that in laminar flow the moment, $T_q = 0.5 \times 1.935\rho\Omega^2 b^5(\rho\Omega b^2/\mu)^{-0.5} = 0.8675\rho^{0.5}\Omega^{1.5}b^4\mu^{0.5}$, is a function of the 1.5 power of the angular velocity and the fourth power of the disc radius. The power, $T_q\Omega$, is therefore a function of the 2.5 power of the angular velocity. Doubling the speed results in increasing the power by a factor of 6.25, and doubling the radius results in increasing the power required by a factor of 16, assuming no flow transition.

4.3. A ROTATING FLUID ABOVE A STATIONARY DISC

The case of a rotating fluid above a stationary disc (Figure 4.6) is of interest particularly for providing insight to atmospheric flows. The same equations as for the rotating free disc, Equations 4.6 to 4.9, provide the basis for a model and has been considered by Bödewadt (1940), Rogers and Lance (1960), Nydahl (1971), and Schlicting (1979).

The boundary conditions for an infinite mass of rotating fluid at a uniform angular velocity Ω in contact with a nonrotating surface are

$$u_r = 0, u_\phi = 0, u_z = 0 \; at \, z = 0 \tag{4.40}$$

and

$$u_r = 0, u_\phi = r\Omega \, at \, z = \infty \tag{4.41}$$

In this case it can be assumed that the pressure above the boundary layer is constant across the boundary layer thickness. The pressure distribution in the fluid rotating at a constant angular velocity Ω can be obtained using

$$\frac{\partial p}{\partial r} = \rho r \Omega^2 \tag{4.42}$$

On substitution a set of three coupled ordinary differential equations result.

$$\frac{d^2 u_{r*}}{dz_*^2} - u_{z*} \frac{du_{r*}}{dz_*} - u_{r*}^2 + u_{\phi*}^2 - 1 = 0 \tag{4.43}$$

$$\frac{d^2 u_{\phi*}}{dz_*^2} - u_{z*} \frac{du_{\phi*}}{dz_*} - 2u_{r*} u_{\phi*} = 0 \tag{4.44}$$

$$\frac{du_{z*}}{dz_*} + 2u_{r*} = 0 \tag{4.45}$$

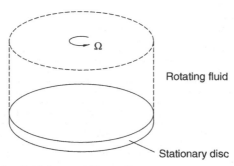

FIGURE 4.6 A rotating fluid above a stationary disc.

Pressure does not appear in these equations as the pressure in the main body of rotating fluid is known and the boundary layer approximation assumes the same value.

The dimensionless boundary conditions are

$$u_{r*} = 0, \ u_{\phi*} = 0, \ u_{z*} = 0, \ \text{at } z = 0 \quad (4.46)$$

and

$$u_{r*} = 0, \ u_{\phi*} = 1, \ \text{at } z = \infty \quad (4.47)$$

Solutions for the dimensionless velocity components have been performed by Nydahl (1971) and more recently updated by Owen and Rogers (1989); they are presented in Table 4.2 and plotted in Figure 4.7. A schematic of the

TABLE 4.2 Dimensionless velocity profiles for a stationary disc in a rotating fluid, (Owen and Rogers, 1989)

z_*	u_{r*}	$u_{\phi*}$	u_{z*}
0	0	0	0
0.5	−0.3487	0.3834	0.1944
1.0	−0.4788	0.7354	0.6241
1.5	−0.4496	1.0134	1.0987
2.0	−0.3287	1.1924	1.4929
2.5	−0.1762	1.2721	1.7459
3.0	−0.0361	1.2714	1.8496
3.5	0.0663	1.2182	1.8308
4.0	0.1227	1.1413	1.7325
4.5	0.1371	1.0640	1.5995
5.0	0.1210	1.0016	1.4685
5.5	0.0878	0.9611	1.3632
6.0	0.0499	0.9427	1.2944
6.5	0.0162	0.9421	1.2620
7.0	−0.0084	0.9530	1.2589
7.5	−0.0223	0.9693	1.2751
8.0	−0.0268	0.9857	1.3004
8.5	−0.0243	0.9990	1.3264

9.0	−0.0179	1.0078	1.3476
9.5	−0.0102	1.0118	1.3617
10.0	−0.0033	1.0121	1.3683
10.5	0.0018	1.0099	1.3689
11.0	0.0047	1.0065	1.3654
11.5	0.0057	1.0031	1.3601
12.0	0.0052	1.0003	1.3545
12.5	0.0038	0.9984	1.3500
∞	0	1	1.3494

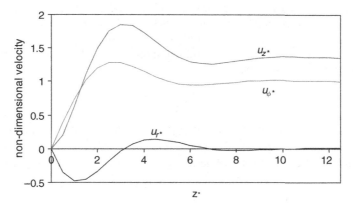

FIGURE 4.7   Dimensionless laminar flow velocity profiles for a stationary disc in a rotating fluid.

boundary layer velocity profiles is given in Figure 4.8. Fluid in the boundary layer near the stationary surface has less centripetal pressure differential than the rotating fluid. Peripheral fluid flows radially inward through the thin boundary layer toward the axis. Conservation of angular momentum of the inward flow causes the tangential velocity to be slightly elevated above the free-stream rotational speed at approximately $z_* = 1.5\delta$ that leads to the flow reversals. The radial inward flow is the mechanism responsible for the accumulation of sugar and tea leaf debris at the center of a stirred cup of tea. Velocity profiles for conditions comparable to that of a stirred cup of tea are illustrated in Figure 4.9, and those for an arbitrary set of conditions relevant to air flow are shown in Figure 4.10.

If the boundary layer thickness is defined to be the axial distance away from the disc at which the tangential velocity is 99% of the free-stream

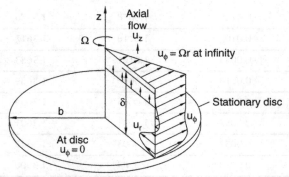

FIGURE 4.8 The radial and tangential boundary layers and associated inflow and outflow induced by a rotating fluid adjacent to a stationary disc.

rotational speed, that is, $u_\phi = 0.99\Omega r$, then Table 4.2 can be examined to give the boundary layer thickness for laminar flow over a stationary disc. $u_{\phi*} = u_\phi/\Omega r = 0.99$ corresponds to a value for z_* of about 8, which on substitution in Equation 4.12 gives the following relationship for a rotating laminar boundary layer on a stationary disc.

$$\delta \approx 8\sqrt{\frac{\nu}{\Omega}} \qquad (4.48)$$

The boundary layer thickness is thus almost 50% more than that of the free disc. In the case of the free disc, the axial flow toward the disc tends to suppress the boundary layer thickness, while in this case the axial flow away from the

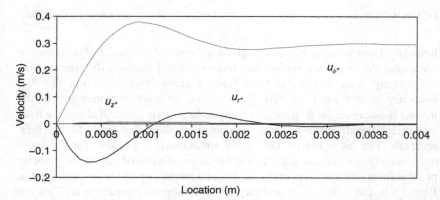

FIGURE 4.9 Example velocity profiles for a stationary disc in a rotating fluid. $b = 0.035$ m, $\Omega = 60$ rpm $= 8.442$ rad/s, $\rho = 1000$ kg/m^3, $\mu = 0.001$ Pa s. Conditions equivalent to those for a stirred cup of tea.

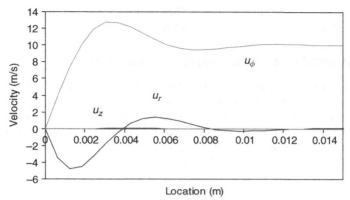

FIGURE 4.10   Example velocity profiles for a stationary disc in a rotating fluid. $b = 1$ m, $\Omega = 10$ rad/s, $\rho = 1.2047$ kg/m^3, $\mu = 1.82 \times 10^{-5}$ Pa s.

surface and conservation of angular momentum tend to move the boundary layer outward.

The net radial flow through a cylinder of radius b can be obtained by integrating the radial velocity over the cylinder. From Equation 4.30 and Table 4.2 with $u_{z_*=\infty} = 1.3494$, the nondimensional mass flow rate for the entrained flow is

$$Cw = -4.239 \mathrm{Re}_\phi^{0.5} \tag{4.49}$$

The velocity vector can be plotted for a horizontal plane through the rotating fluid. The envelope of the horizontal velocity, $\sqrt{u_{r*}^2 + u_{\phi*}^2}$, as a function of z_* is known as an Ekman spiral. This has been plotted in Figure 4.11. The curve traced out by the endpoints of the vector form a logarithmic spiral. An Ekman spiral illustrates the relationship of the radial velocity to the tangential velocity

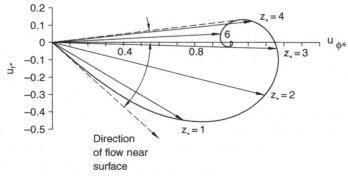

FIGURE 4.11   An Ekman spiral for the flow induced by a viscous fluid rotating with constant angular velocity in the vicinity of a flat surface.

as a function of distance from the surface in a boundary layer influenced by rotation. In an Ekman spiral, z_* becomes a polar angle.

4.4. TURBULENT FLOW OVER A SINGLE DISC

Von Kármán (1921) used an integral momentum technique applied to turbulent flow on a free disc. From this, incorporating calculation corrections identified by Dorfman (1963), a relationship for the moment on one side of the disc was determined, Equation 4.50, along with relationships for the boundary layer thickness, Equation 4.51, and dimensionless entrained mass flow, Equation 4.52 using a 1/7th power law assumption for the velocity profile.

$$C_m = 0.073 \text{Re}_\phi^{-0.2} \tag{4.50}$$

$$\delta = 0.526r \left[\left(\frac{r}{b} \right)^2 \text{Re}_\phi \right]^{-0.2} \tag{4.51}$$

$$Cw = 0.219 \text{Re}_\phi^{0.8} \tag{4.52}$$

The von Kármán relationship for the moment coefficient for turbulent flow over a free disc, Equation 4.50, is in close agreement with data from Theodorsen and Regier (1944) for the range of rotational Reynolds numbers from 5×10^5 to 2×10^6. At higher rotational Reynolds numbers, however, Equation 4.50 significantly underpredicts the measured data and the data obtained later by Bayley and Owen (1969). The detailed derivation of Equations 4.50 to 4.52 is considered in Section 4.4.1.

Dorfman (1963), assuming logarithmic profiles for the flow near the rotating disc, obtained an alternative solution for moment coefficient for one side of a disc, given in Equation 4.54, for rotational Reynolds numbers ranging from

$$1 \times 10^4 \le Re_\phi \le 1 \times 10^9 \tag{4.53}$$

$$C_m = 0.491 [\log_{10}(\text{Re}_\phi)]^{-2.58} \tag{4.54}$$

For a disc rotating in air at rotational Reynolds numbers up to approximately 4.2×10^6, Bayley and Owen (1969) give the correlation presented in Equation 4.55.

$$C_m = 0.0655 \text{Re}_\phi^{-0.186} \tag{4.55}$$

For rotational Reynolds numbers in the range $1 \times 10^5 < Re_\phi < 1 \times 10^7$, Equations 4.55 and 4.54 differ by less than 5%. Although Equation 4.50 has been relatively widely used for the prediction of the moment coefficient on a rotating disc, as stated previously it provides a reasonable agreement with the data from Theodorsen and Regier (1944) in the range $5 \times 10^5 < Re_\phi < 2 \times 10^6$ and at higher rotational Reynolds numbers significantly underpredicts the measured data and the data obtained later by Bayley and Owen (1969). In modeling it is therefore necessary to determine the rotational Reynolds number range for the application concerned and then use Equation 4.50, 4.54, or 4.55 as appropriate.

In general, roughness has little effect on shear stress, provided the flow is laminar or the roughness elements do not exceed the laminar sublayer thickness and protrude through to disrupt the flow. The width of the viscous sublayer is conventionally taken as corresponding to a universal coordinate of $y_+ \approx 5$. Other values are sometimes quoted, but this represents the approximate dimensionless wall distance at which the viscous shear stress and the Reynolds shear stress become equal.

$$y_+ = yu_\tau/\nu \tag{4.56}$$

u_τ is the friction velocity given by

$$u_\tau = \sqrt{\tau_s/\rho} \tag{4.57}$$

For $y_+ = 5$, the width of the viscous sublayer is

$$y_{sublayer} = \frac{5\nu}{\sqrt{\tau_s/\rho}} \tag{4.58}$$

A surface can be regarded as hydraulically smooth if the sand-grain roughness of the surface, $k_s < y_{sublayer}$. Turbulent flow can be classified according to the roughness parameter, taken as $k_s u_\tau/\nu$, and three regimes can be defined.

1. Hydraulically smooth; $k_s u_\tau/\nu < 5$
2. Transition; $5 < k_s u_\tau/\nu < 70$
3. Completely rough; $k_s u_\tau/\nu > 70$.

For the completely rough regime, Dorfman (1958) developed the solution given in Equation 4.59 for the momentum integral equations in terms of k_s', taken to be the size of the individual grains used. The sand-grain roughness, k_s, might typically be 60% of this value ($k_s = 0.6k_s'$) owing to the proximity of grains and the center line average roughness $R_a \approx k_s/2 = 0.3k_s'$ (see Childs and Noronha, 1997), allowing the substitutions given below.

$$C_m = 0.054\left(\frac{k_s'}{b}\right)^{0.272} = 0.062\left(\frac{k_s}{b}\right)^{0.272} = 0.075\left(\frac{R_a}{b}\right)^{0.272} \tag{4.59}$$

Equation 4.59 is in reasonable agreement with Kanaev's data (1953) for a steel disc rotating in water with $k_s/b = 3.33 \times 10^{-3}$ but underestimates Theodorsen and Regier's data (1944) for $k_s/b = 8.3 \times 10^{-4}$ by approximately 15%.

4.4.1 Momentum Integral Equations for Turbulent Flow over a Rotating Disc

The sum of the momentum of a flow across a boundary layer thickness can be integrated to give relationships for the boundary layer thickness, total mass flow, and frictional moment on the disc. The technique involves a number of simplifications of the flow physics and therefore limitations in its accuracy and scope of application. Nevertheless, the models provide valuable correlations and relationships that are valid under limited conditions.

The momentum integral method can be applied to almost any boundary layer flow. As developed here, it is applied to the boundary layer on a rotating disc for turbulent flow. The analysis is extended in Section 5.4 for the rotor-stator wheelspace configuration to provide relationships for the mass flow associated with the rotor boundary layer and also for the stator boundary layer. As the analytical treatment here results in some relationships previously presented in Section 4.4, the reader may choose to skip over the extended analysis, which does cover several pages of involved mathematical development, and proceed directly to Section 4.4.2.

The starting point for this analysis is the statement of the relevant boundary layer equations. In the development of these equations, various assumptions are made. If the distance between the discs is small in comparison to their radius then:

- the axial velocity component, u_z, can be assumed to be small in comparison with the other two components of velocity, u_r and u_ϕ
- the rate of change of any variable, other than pressure, in the direction normal to the disc can be assumed to be much greater than its rate of change in the radial or tangential direction
- the pressure, p, only depends on the distance from the axis of rotation.

It should be noted that the resulting boundary layer equations will not be valid in a region near a shroud or axis of rotation.

The resulting boundary layer equations in a stationary frame of reference, for steady axisymmetric flow, are given in Equations 4.60 to 4.63. The continuity equation remains unchanged, albeit with the time and tangential derivatives removed, reflecting the steady-state axisymmetric assumptions, respectively, Equation 4.60. The turbulent Navier-Stokes equations (see Equations 2.78 and 2.62 to 2.64) reduce to Equations 4.61 to 4.63.

$$\frac{\partial u_r}{\partial r} + \frac{u_r}{r} + \frac{\partial u_z}{\partial z} = 0 \tag{4.60}$$

$$u_r \frac{\partial u_r}{\partial r} + u_z \frac{\partial u_r}{\partial z} - \frac{u_\phi^2}{r} = -\frac{1}{\rho}\frac{\partial p}{\partial r} + \frac{1}{\rho}\frac{\partial \tau_r}{\partial z} \tag{4.61}$$

$$u_r \frac{\partial u_\phi}{\partial r} + \frac{u_r u_\phi}{r} + u_z \frac{\partial u_\phi}{\partial z} = \frac{1}{\rho} \frac{\partial \tau_\phi}{\partial z} \qquad (4.62)$$

$$0 = -\frac{\partial p}{\partial z} \qquad (4.63)$$

where

$$\tau_r = \mu \frac{\partial u_r}{\partial z} - \rho \overline{u'_r u'_z} \qquad (4.64)$$

$$\tau_\phi = \mu \frac{\partial u_\phi}{\partial z} - \rho \overline{u'_\phi u'_z} \qquad (4.65)$$

One of the assumptions of the boundary layer analysis outlined here is that the pressure depends only on the distance from the axis of rotation and is therefore independent of the axial location throughout the boundary layer. For a single disc the pressure, p, within the boundary layer can therefore be taken as $p = p_\infty$ where p_∞ is the pressure at a location far from the disc. From Equation 4.7 if viscous effects are negligible and radial components of velocity are very small and therefore also negligible (see also Section 5.3), then,

$$\frac{1}{\rho} \frac{dp}{dr} = \frac{u_\phi^2}{r} \qquad (4.66)$$

The term involving $\partial p/\partial r$ in Equation 4.61 can therefore be replaced by $-u_\phi^2/r$. Using this substitution and rearrangement of terms, Equations 4.60–4.62 become

$$\frac{\partial}{\partial r}(r u_r) + \frac{\partial}{\partial z}(r u_z) = 0 \qquad (4.67)$$

$$\frac{\partial}{\partial r}(r u_r^2) + \frac{\partial}{\partial z}(r u_r u_z) - u_\phi^2 = -u_{\phi,\infty}^2 + \frac{r}{\rho} \frac{\partial \tau_r}{\partial z} \qquad (4.68)$$

$$\frac{\partial}{\partial r}(r^2 u_r u_\phi) + \frac{\partial}{\partial z}(r^2 u_\phi u_z) = \frac{r^2}{\rho} \frac{\partial \tau_\phi}{\partial z} \qquad (4.69)$$

Integration of Equations 4.67 to 4.69 gives Equations 4.70 to 4.72. It is now possible to state these as ordinary differential equations, for each of the equations has reduced to a relationship between an independent variable and a dependent variable.

$$\frac{d}{dr}\left(r \int_0^\delta u_r dz\right) + r u_{z,\infty} = 0 \qquad (4.70)$$

$$\frac{d}{dr}\left(r \int_0^\delta u_r^2 dz\right) - \int_0^\delta (u_\phi^2 - u_{\phi,\infty}^2) dz = -\frac{r}{\rho} \tau_{r,o} \qquad (4.71)$$

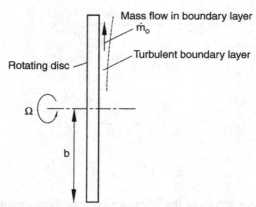

FIGURE 4.12   Rotating disc boundary layer.

$$\frac{d}{dr}\left(r^2\int_0^\delta u_r u_\phi dz\right) + r^2 u_{\phi,\infty} u_{z,\infty} = -\frac{r^2}{\rho}\tau_{\phi,o} \tag{4.72}$$

The mass flow in the boundary layer on a rotating disc (see Figure 4.12) can be determined by the integral given in Equation 4.73.

$$\dot{m}_o = 2\pi r\rho\int_0^\delta u_r dz \tag{4.73}$$

From Equation 4.73,

$$\int_0^\delta u_r dz = \frac{\dot{m}_o}{2\pi r\rho} \tag{4.74}$$

Substitution of Equation 4.74 in Equation 4.70 gives

$$\frac{d}{dr}\left(\frac{r\dot{m}_o}{2\pi r\rho}\right) + ru_{z,\infty} = 0 \tag{4.75}$$

$$\frac{1}{2\pi\rho}\frac{d\dot{m}_o}{dr} + ru_{z,\infty} = 0 \tag{4.76}$$

which on rearrangement gives

$$u_{z,\infty} = -\frac{1}{2\pi\rho r}\frac{d\dot{m}_o}{dr} \tag{4.77}$$

Substitution for $u_{z,\infty}$ in Equation 4.72 gives

$$\frac{d}{dr}\left(r^2\int_0^\delta u_r(u_\phi - u_{\phi,\infty})dz\right) + \frac{\dot{m}_o}{2\pi\rho}\frac{d}{dr}(ru_{\phi,\infty}) = -\frac{r^2}{\rho}\tau_{\phi,o} \tag{4.78}$$

In order to determine a relationship for the shear stress for a rotating disc, von Kármán (1921) assumed distributions for the radial and tangential velocity

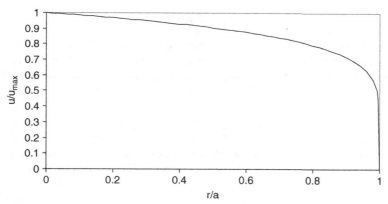

FIGURE 4.13   1/7th power law velocity profile for turbulent pipe flow. It is assumed that the boundary layer profile for the radial velocity component on a rotating disc for turbulent flow has a similar profile.

components analogous to the 1/7th velocity profiles for pipe flow. For turbulent pipe flow, the local velocity u is found to be related to the maximum velocity at the pipe axis by the empirical formula:

$$u = u_{max}\left(1-\frac{r}{a}\right)^{1/7} \tag{4.79}$$

where r = radial distance from the pipe axis, a = radius of the pipe. The profile is illustrated in Figure 4.13.

The skin friction coefficient, C_f, and the friction factor, f, are defined here by Equations 4.80 and 4.81, respectively

$$C_f = \frac{\tau_s}{\frac{1}{2}\rho u^2} \tag{4.80}$$

$$f = 4C_f = \frac{8\tau_s}{\rho u^2} \tag{4.81}$$

where τ_s is the shear stress at the wall.

The shear stress can be determined using the Blasius equation (1913) for the friction factor for turbulent flow in smooth pipes, Equation 4.82.

$$f = 0.316Re_d^{-0.25} = 0.316\left(\frac{\nu}{\bar{u}d}\right)^{0.25} = 0.316\left(\frac{\nu}{2\bar{u}a}\right)^{0.25} \tag{4.82}$$

where Re_d is the pipe Reynolds number, $Re_d = \rho\bar{u}d/\mu$, $\bar{u}$ is the average pipe velocity, and d is the pipe diameter. The frictional loss in a pipe flow, over a pipe length L, can be determined using

$$\Delta p = f \frac{L}{d} 0.5 \rho \bar{u}^2 \tag{4.83}$$

Substitution of the Blasius relationship, Equation 4.82, for the friction factor for a smooth pipe in Equation 4.83 reveals that the pressure drop, and therefore by inference the surface shear stress, is proportional to the fluid speed to the power 7/4; see Equation 4.84.

$$\Delta p = 0.316 \times 0.5 \left(\frac{\nu}{\bar{u}d} \right)^{0.25} \frac{L}{d} \rho \bar{u}^2 = 0.158 \frac{\rho L \nu^{1/4} \bar{u}^{7/4}}{d^{5/4}} \tag{4.84}$$

By integration of Equation 4.79, a relationship between the average pipe velocity and the maximum pipe velocity for turbulent flow can be found.

$$u_{max} = \frac{60}{49} \bar{u} = 1.224 \bar{u} = \frac{\bar{u}}{0.816} \tag{4.85}$$

It is assumed that the velocity distribution in the vicinity of the wall depends only on the viscosity, the density, the shear stress at the wall, and the distance from the wall, $y = a\text{-}r$.

$$u = f(\mu, \rho, \tau_s, y) \tag{4.86}$$

Assuming u is independent of pipe dimensions and expressing Equation 4.86 by increasing powers of y,

$$u = f_1(\mu, \rho, \tau_s) y^x \tag{4.87}$$

Dimensional analysis of Equation 4.87 leads to

$$u = B \left(\frac{\tau_s}{\rho} \right)^{\left(\frac{1+x}{2} \right)} \left(\frac{y}{\nu} \right)^x \tag{4.88}$$

where B = a nondimensional constant.

Since u increases in proportion to the throughflow volume, that is, doubling the maximum velocity in the center of the pipe results in doubling of all the velocities, see Equation 4.79, whereas τ_o increases with the 7/4 power of the throughflow volume,

$$\frac{1+x}{2} = \frac{4}{7} \tag{4.89}$$

Hence, $x = 1/7$, so from Equation 4.88,

$$u = B \left(\frac{\tau_s}{\rho} \right)^{4/7} \left(\frac{y}{\nu} \right)^{1/7} \tag{4.90}$$

From Equation 4.79, substituting $y = r - a$,

$$u = u_{max} \left(1 - \frac{r}{a} \right)^{1/7} = u_{max} \left(\frac{y}{a} \right)^{1/7} \tag{4.91}$$

We now have two equations for u, Equations 4.90 and 4.91, so

$$B\left(\frac{\tau_s}{\rho}\right)^{4/7}\left(\frac{y}{\nu}\right)^{1/7} = u_{max}\left(\frac{y}{a}\right)^{1/7} \tag{4.92}$$

Substituting for u_{max} and the shear stress at the wall gives

$$B = \frac{\bar{u}\nu^{1/7}}{0.816a^{1/7}}\left(\frac{8}{0.316\left(\frac{\nu}{2\bar{u}a}\right)^{1/4}\bar{u}^2}\right)^{4/7} = 2^{1/7}\left(\frac{8}{0.316}\right)^{4/7}/0.816 = 8.57 \tag{4.93}$$

From Equation 4.90,

$$\tau_s = \frac{\rho u^2}{B^{7/4}}\left(\frac{\nu}{u(a-r)}\right)^{1/4} \tag{4.94}$$

The best value however for the constant B characterizing the turbulent flow data available was found to be $B \approx 8.74$, giving von Kármán's (1921) relationship for the shear stress at the wall, τ_s.

$$\tau_s = 0.0225\rho u^2\left(\frac{\nu}{u(a-r)}\right)^{1/4} \tag{4.95}$$

For the case of a rotating disc, the radial velocity in the boundary layer is assumed to be zero at the disc, $z = 0$, and zero beyond the boundary layer, $z > \delta$, and approximated by

$$u_r = ar\Omega\left(\frac{z}{\delta}\right)^{1/7}\left(1-\frac{z}{\delta}\right) \quad \text{for } z \leq \delta \tag{4.96}$$

and $u_r = 0$ for $z > \delta$, where u_r is the radial component of velocity.

The tangential velocity is approximated by

$$u_\phi = r\Omega\left[1-\left(\frac{z}{\delta}\right)^{1/7}\right] \quad \text{for } z \leq \delta \tag{4.97}$$

α is the ratio of the radial and tangential components of relative velocity in the boundary layer and therefore also the ratio of the radial and tangential components of stress.

$$\alpha = -\frac{\tau_r}{\tau_\phi} \tag{4.98}$$

The resultant velocity near the surface of the disc is given by

$$u_{resultant} = \sqrt{u_r^2 + (u_\phi-r\Omega)^2} \tag{4.99}$$

Substitution for u_r and u_ϕ using Equations 4.96 and 4.97 gives

$$u_{resultant} = \left(\alpha^2 r^2 \Omega^2 \left(\frac{z}{\delta}\right)^{2/7} \left(1-\frac{z}{\delta}\right)^2 + \left(-r\Omega\left(\frac{z}{\delta}\right)^{1/7}\right)^2\right)^{0.5} \approx (\alpha^2+1)^{0.5} r\Omega\left(\frac{z}{\delta}\right)^{1/7}$$

(4.100)

From the empirical equation developed for pipe flow, Equation 4.95, substituting $a - r = \delta$

$$\tau_o = 0.0225\rho u_{resultant}^2 \left(\frac{\nu}{u_{resultant}\delta}\right)^{1/4}$$

(4.101)

Or, on rearranging

$$u_{resultant} = 8.74\left(\frac{\delta}{\nu}\right)^{1/7}\left(\frac{\tau_o}{\rho}\right)^{4/7}$$

(4.102)

From Equations 4.100 and 4.102

$$\frac{1}{8.74}(\alpha^2+1)^{0.5} r\Omega\left(\frac{z}{\delta}\right)^{1/7} = \left(\frac{\delta}{\nu}\right)^{1/7}\left(\frac{\tau_o}{\rho}\right)^{4/7}$$

(4.103)

For $z = \delta$

$$\frac{1}{8.74}(\alpha^2+1)^{0.5} r\Omega\left(\frac{z}{\delta}\right)^{1/7}\left(\frac{\nu}{\delta}\right)^{1/7} = \left(\frac{\tau_o}{\rho}\right)^{4/7}$$

(4.104)

$$\frac{\tau_o}{\rho} = \frac{1}{(8.74)^{7/4}}\left((\alpha^2+1)^{0.5}\right)^{7/4}(r\Omega)^{7/4}\left(\frac{\nu}{\delta}\right)^{1/4}$$

(4.105)

$$\tau_o = 0.0225\rho\left((\alpha^2+1)^{0.5}\right)^{7/4}(r\Omega)^{7/4}\left(\frac{\nu}{\delta}\right)^{1/4}$$

(4.106)

$$\tau_o = 0.0225\rho(\alpha^2+1)^{7/8}(r\Omega)^{7/4}\left(\frac{\nu}{\delta}\right)^{1/4}$$

(4.107)

$$\tau_o = 0.0225\rho\alpha^{7/4}\left(1+\frac{1}{\alpha^2}\right)^{7/8}(r\Omega)^{7/4}\left(\frac{\nu}{\delta}\right)^{1/4}$$

(4.108)

$$\tau_o = \sqrt{\tau_r^2 + \tau_\phi^2}$$

(4.109)

From Equation 4.98, $\tau_\phi = -\tau_r/\alpha$, so

$$\tau_o = \sqrt{\tau_r^2 + \frac{\tau_r^2}{\alpha^2}} = \tau_r\sqrt{1+\frac{1}{\alpha^2}}$$

(4.110)

$$\tau_r = \tau_o \Big/ \sqrt{1 + \frac{1}{\alpha^2}} \qquad (4.111)$$

So from Equation 4.108, substituting for τ_o in Equation 4.111, the radial stress component, τ_r, is given by

$$\tau_r = 0.0225\rho\alpha^{7/4}(r\Omega)^{7/4}\left(\frac{\nu}{\delta}\right)^{1/4}\left(1 + \frac{1}{\alpha^2}\right)^{3/8} \qquad (4.112)$$

Equation 4.112 can be restated in the form

$$\tau_r = 0.0225\rho\alpha(r\Omega)^{7/4}\left(\frac{\nu}{\delta}\right)^{1/4}(1 + \alpha^2)^{3/8} \qquad (4.113)$$

From Equation 4.98, $\tau_\phi = -\tau_r/\alpha$, the tangential stress component, τ_ϕ, is given by

$$\tau_\phi = -0.0225\rho(r\Omega)^{7/4}\left(\frac{\nu}{\delta}\right)^{1/4}(1 + \alpha^2)^{3/8} \qquad (4.114)$$

Equations 4.113 and 4.114 can be restated to give the radial and tangential shear stress at the rotor, $\tau_{r,o}$ and $\tau_{\phi,o}$, respectively, by substituting for the velocity using $\Omega r = u_{\phi,o} - u_{\phi,\infty}$ where $u_{\phi,o}$ = the tangential velocity at the rotor and $u_{\phi,\infty}$ = the tangential velocity outside the boundary layer. $u_{\phi,\infty}$ may be zero, as would be the case for a disc rotating in a stationary fluid, or it could be nonzero as for a disc in a rotor-stator wheelspace with an inviscid rotating core region in which case $u_{\phi,\infty} = u_{\phi,c}$.

$$\tau_{r,o} = 0.0225\rho\left(\frac{\mu}{\rho\delta}\right)^{1/4}(u_{\phi,o}-u_{\phi,\infty})^{7/4}\alpha[1 + \alpha^2]^{3/8} \qquad (4.115)$$

$$\tau_{\phi,o} = -0.0225\rho\left(\frac{\mu}{\rho\delta}\right)^{1/4}(u_{\phi,o}-u_{\phi,\infty})^{7/4}[1 + \alpha^2]^{3/8} \qquad (4.116)$$

Conveniently, Equations 4.113 and 4.114 or Equations 4.115 and 4.116 involve no differential terms.

Solution of the momentum integral equations can be implemented using similarity solutions of the form

$$u_r = \alpha(r)\,[u_{\phi,o}(r)-u_{\phi,\infty}(r)]\,f(z_*) \quad \text{for } 0 \le z* \le 1 \qquad (4.117)$$

$$u_\phi = u_{\phi,o}(r)- [u_{\phi,o}(r)-u_{\phi,\infty}(r)]\,g(z_*) \quad \text{for } 0 \le z* \le 1 \qquad (4.118)$$

where $\alpha(r)$ is a function to be determined. z_* is the dimensionless axial location, given by $z_* = z/\delta$. The functions $f(z_*)$ and $g(z_*)$ are selected to give good approximations to the expected or known profiles for the radial and tangential velocity components, u_r and u_ϕ.

Assuming 1/7th power law profiles for $f(z_*)$ and $g(z_*)$:

$$f(z_*) = z_*^{1/7}(1-z_*) \qquad (4.119)$$

$$g(z_*) = z_*^{1/7} \tag{4.120}$$

Taking $z_* = z/\delta$,

$$z = z_*\delta \tag{4.121}$$

$$\frac{dz}{dz_*} = \delta \tag{4.122}$$

so

$$dz = \delta dz_* \tag{4.123}$$

If $z = \delta$, then from Equation 4.121, $z_* = 1$.

From Equations 4.73, 4.117, and 4.119

$$\dot{m}_o = 2\pi r\rho\int_0^\delta \alpha(r)[u_{\phi,o}(r)-u_{\phi,\infty}(r)]z_*^{1/7}(1-z_*)dz \tag{4.124}$$

Changing variables using Equations 4.123 and substituting for the limits of integration,

$$\dot{m}_o = 2\pi r\rho\int_0^1 \alpha(r)[u_{\phi,o}(r)-u_{\phi,\infty}(r)]z_*^{1/7}(1-z_*)\delta dz_* \tag{4.125}$$

$$= 2\pi r\rho\alpha\delta(u_{\phi,o}-u_{\phi,\infty}) \int_0^1(z_*^{1/7}-z_*^{8/7})dz_* \tag{4.126}$$

$$= 2\pi r\rho\alpha\delta(u_{\phi,o}-u_{\phi,\infty}) \left[\frac{7}{8}z_*^{8/7}-\frac{7}{15}z_*^{15/7}\right]_0^1 \tag{4.127}$$

$$= 2\pi r\rho\alpha\delta(u_{\phi,o}-u_{\phi,\infty}) \left(\frac{7}{8}-\frac{7}{15}\right) = 2\pi r\rho\alpha\delta(u_{\phi,o}-u_{\phi,\infty})\frac{(15 \times 7)-(8 \times 7)}{120} \tag{4.128}$$

So the mass flow in the boundary layer, for turbulent flow, is given by

$$\dot{m}_o = \frac{49\pi}{60}r\rho\alpha\delta(u_{\phi,o}-u_{\phi,\infty}) \tag{4.129}$$

The integrated forms of the Navier-Stokes equations for disc boundary flow, Equations 4.71 and 4.72, can be restated as ordinary differential equations (see Equations 4.135 and 4.142).

Substituting for u_r and u_ϕ in Equation 4.71, using Equations 4.117 and 4.118,

$$\frac{d}{dr}\left[r\int_0^\delta \alpha^2(u_{\phi,o}-u_{\phi,\infty})^2\left(z_*^{1/7}(1-z_*)\right)^2 dz\right]-\int_0^\delta \left(\left(u_{\phi,o}-(u_{\phi,o}-u_{\phi,\infty})z_*^{1/7}\right)^2-u_{\phi,\infty}^2\right)dz$$

$$= -\frac{r}{\rho}\tau_{r,o}$$

$$\tag{4.130}$$

Changing variables using $z_* = z/\delta$, so $dz = \delta dz_*$ and $z_* = 0$ when $z = 0$, $z_* = 1$ when $z = \delta$ and expanding terms gives

$$\frac{d}{dr}\left[r\int_0^1 \alpha^2\delta(u_{\phi,o}-u_{\phi,\infty})^2(z_*^{2/7}-2z_*^{9/7}+z_*^{16/7})dz_*\right]$$
$$-\int_0^\delta\left(\left(u_{\phi,o}-(u_{\phi,o}-u_{\phi,\infty})z_*^{1/7}\right)^2-u_{\phi,\infty}^2\right)dz = -\frac{r}{\rho}\tau_{r,o}$$

(4.131)

$$\frac{d}{dr}\left[r\int_0^1 \alpha^2\delta(u_{\phi,o}-u_{\phi,\infty})^2(z_*^{2/7}-2z_*^{9/7}+z_*^{16/7})dz_*\right]-$$
$$\delta\int_0^1\left(u_{\phi,o}^2-u_{\phi,\infty}^2-2u_{\phi,o}^2z_*^{1/7}+2u_{\phi,o}u_{\phi,\infty}z_*^{1/7}+(u_{\phi,o}-u_{\phi,\infty})^2z_*^{2/7}\right)dz_* = -\frac{r}{\rho}\tau_{r,o}$$

(4.132)

$$\frac{d}{dr}r\alpha^2\delta(u_{\phi,o}-u_{\phi,\infty})^2\left[\frac{7}{9}z_*^{9/7}-\frac{7}{8}z_*^{16/7}+\frac{7}{23}z_*^{23/7}\right]_0^1-$$

$$\delta\left[u_{\phi,o}^2z_*-u_{\phi,\infty}^2z_*-\frac{7}{4}u_{\phi,o}^2z_*^{8/7}+\frac{7}{4}u_{\phi,o}u_{\phi,\infty}z_*^{8/7}+\frac{7}{9}(u_{\phi,o}-u_{\phi,\infty})^2z_*^{9/7}\right]_0^1 = -\frac{r}{\rho}\tau_{r,o}$$

(4.133)

$$\frac{d}{dr}\left(r\alpha^2\delta(u_{\phi,o}-u_{\phi,\infty})^2\frac{7\times184+7\times207+7\times72}{1656}\right)-$$

$$\delta\left(u_{\phi,o}^2-u_{\phi,\infty}^2-\frac{7}{4}u_{\phi,o}^2+\frac{7}{4}u_{\phi,o}u_{\phi,\infty}+\frac{7}{9}u_{\phi,o}^2-\frac{14}{9}u_{\phi,o}u_{\phi,\infty}+\frac{7}{9}u_{\phi,\infty}^2\right) = -\frac{r}{\rho}\tau_{r,o}$$

(4.134)

which finally gives

$$\frac{343}{1656}\frac{d}{dr}\left(r\alpha^2\delta(u_{\phi,o}-u_{\phi,\infty})^2\right)-\delta\left(\frac{1}{8}(u_{\phi,o}^2-u_{\phi,\infty}^2)-\frac{7}{72}(u_{\phi,o}-u_{\phi,\infty})^2\right) = -\frac{r}{\rho}\tau_{r,o}$$

(4.135)

with $\tau_{r,o}$ given by Equation 4.115 or 4.113.

Similarly, substituting for u_r and u_ϕ in Equation 4.78, using Equations 4.117, 4.118, 4.119, 4.120, and 4.123,

$$\frac{d}{dr}\left(r^2 \int_0^1 \alpha\delta(u_{\phi,o}-u_{\phi,\infty})(z_*^{1/7}-z_*^{8/7})(u_{\phi,o}-u_{\phi,\infty}z_*^{1/7}+u_{\phi,\infty}z_*^{1/7}-u_{\phi,\infty})dz_*\right)$$

$$+\frac{\dot{m}_o}{2\pi\rho}\frac{d}{dr}(ru_{\phi,\infty})=-\frac{r^2}{\rho}\tau_{\phi,o} \tag{4.136}$$

$$\frac{d}{dr}\left(r^2\alpha\delta\int_0^1 (u_{\phi,o}-u_{\phi,\infty})(u_{\phi,o}z_*^{1/7}-u_{\phi,o}z_*^{8/7}-u_{\phi,o}z_*^{2/7}+u_{\phi,o}z_*^{9/7}+u_{\phi,\infty}z_*^{2/7}\right.$$

$$\left.-u_{\phi,\infty}z_*^{9/7}-u_{\phi,\infty}z_*^{1/7}+u_{\phi,\infty}z_*^{8/7})dz_*\right)+\frac{\dot{m}_o}{2\pi\rho}\frac{d}{dr}(ru_{\phi,\infty})=-\frac{r^2}{\rho}\tau_{\phi,o}$$

$$\tag{4.137}$$

$$\frac{d}{dr}\left(r^2\alpha\delta(u_{\phi,o}-u_{\phi,\infty})\left[\frac{7}{8}u_{\phi,o}z_*^{8/7}-\frac{7}{15}u_{\phi,o}z_*^{15/7}-\frac{7}{9}u_{\phi,o}z_*^{9/7}+\frac{7}{16}u_{\phi,o}z_*^{16/7}+\frac{7}{9}u_{\phi,\infty}z_*^{9/7}\right.\right.$$

$$\left.\left.-\frac{7}{16}u_{\phi,\infty}z_*^{16/7}-\frac{7}{8}u_{\phi,\infty}z_*^{8/7}+\frac{7}{15}u_{\phi,\infty}z_*^{15/7}\right]_0^1\right)+\frac{\dot{m}_o}{2\pi\rho}\frac{d}{dr}(ru_{\phi,\infty})=-\frac{r^2}{\rho}\tau_{\phi,o}$$

$$\tag{4.138}$$

$$\frac{d}{dr}\left(r^2\alpha\delta(u_{\phi,o}-u_{\phi,\infty})\left(\frac{7}{8}u_{\phi,o}-\frac{7}{15}u_{\phi,o}-\frac{7}{9}u_{\phi,o}+\frac{7}{16}u_{\phi,o}+\frac{7}{9}u_{\phi,\infty}-\frac{7}{16}u_{\phi,\infty}\right.\right.$$

$$\left.\left.-\frac{7}{8}u_{\phi,\infty}+\frac{7}{15}u_{\phi,\infty}\right)\right)+\frac{\dot{m}_o}{2\pi\rho}\frac{d}{dr}(ru_{\phi,\infty})=-\frac{r^2}{\rho}\tau_{\phi,o} \tag{4.139}$$

$$\frac{d}{dr}\left(r^2\alpha\delta(u_{\phi,o}-u_{\phi,\infty})\left(\frac{630-336-560+315}{720}u_{\phi,o}+\frac{560-315-630+336}{720}u_{\phi,\infty}\right)\right)$$

$$+\frac{\dot{m}_o}{2\pi\rho}\frac{d}{dr}(ru_{\phi,\infty})=-\frac{r^2}{\rho}\tau_{\phi,o} \tag{4.140}$$

$$\frac{d}{dr}\left(r^2\alpha\delta(u_{\phi,o}-u_{\phi,\infty})\left(\frac{49}{720}u_{\phi,o}-\frac{49}{720}u_{\phi,\infty}\right)\right)+\frac{\dot{m}_o}{2\pi\rho}\frac{d}{dr}(ru_{\phi,\infty})=-\frac{r^2}{\rho}\tau_{\phi,o}$$

$$\tag{4.141}$$

$$\frac{49}{720}\frac{d}{dr}\left(r^2\alpha\delta(u_{\phi,o}-u_{\phi,\infty})^2\right)+\frac{\dot{m}_o}{2\pi\rho}\frac{d}{dr}(ru_{\phi,\infty})=-\frac{r^2}{\rho}\tau_{\phi,o} \tag{4.142}$$

with $\tau_{\phi,o}$ given by Equation 4.116.

For the case of a rotating disc, the power law velocity profiles developed for pipe flow can be used to approximate the radial and tangential velocity profiles for the rotating disc flow. Assuming a 1/7th power law:

$$u_r = a(r)\Omega r f(z_*) = a(r)\Omega r z_*^{1/7}(1-z_*) \tag{4.143}$$

$$u_\phi = \Omega r[1-g(z_*)] = \Omega r(1-z_*^{1/7}) \tag{4.144}$$

For the case of a rotating disc in a quiescent environment, $u_{\phi,o} = \Omega r$ and $u_{\phi,\infty} = 0$, then from Equation 4.129, the mass flow in the rotor boundary layer is given by

$$\dot{m}_o = \frac{49\pi}{60}\rho\Omega r^2 a\delta \tag{4.145}$$

From Equation 4.135 with $u_{\phi,o} = \Omega r$ and $u_{\phi,\infty} = 0$,

$$\frac{343}{1656}\frac{d}{dr}(ra^2\Omega^2 r^2\delta)-\delta\left(\frac{1}{8}\Omega^2 r^2-\frac{7}{72}\Omega^2 r^2\right) = -\frac{r}{\rho}\tau_{r,o} \tag{4.146}$$

$$\frac{343}{1656}\frac{d}{dr}(r^3 a^2\Omega^2\delta)-\delta\Omega^2 r^2\left(\frac{1}{8}-\frac{7}{72}\right) = -\frac{r}{\rho}\tau_{r,o} \tag{4.147}$$

$$\frac{343}{1656}\left(3r^2 a^2\Omega^2\delta + 2ar^3\Omega^2\delta\frac{da}{dr} + r^3 a^2\Omega^2\frac{d\delta}{dr}\right)-\frac{\delta\Omega^2 r^2}{36} = -\frac{r}{\rho}\tau_{r,o} \tag{4.148}$$

Dividing by $\Omega^2 r^2 a^2\delta$

$$\frac{343}{1656}\left(3 + 2\frac{r}{a}\frac{da}{dr} + \frac{r}{\delta}\frac{d\delta}{dr}\right)-\frac{1}{36a^2} = -\frac{\tau_{r,o}}{\Omega^2 r\rho a^2\delta} \tag{4.149}$$

Substituting for $\tau_{r,o}$ from Equation 4.113,

$$\frac{343}{1656}\left(3 + 2\frac{r}{a}\frac{da}{dr} + \frac{r}{\delta}\frac{d\delta}{dr}\right)-\frac{1}{36a^2} = -\frac{0.0225\rho}{\Omega^2 r\rho a^2\delta}\left(\frac{\mu}{\rho\delta}\right)^{1/4}(\Omega r)^{7/4}a(1 + a^2)^{3/8} \tag{4.150}$$

Rearranging to give $\mu/\rho\Omega r^2$, the inverse of the rotational Reynolds number gives

$$\frac{343}{1656}\left(3 + 2\frac{r}{a}\frac{da}{dr} + \frac{r}{\delta}\frac{d\delta}{dr}\right)-\frac{1}{36a^2} = -0.0225\left(\frac{\mu}{\rho\Omega r^2}\right)^{1/4}\frac{(1 + a^2)^{3/8}}{a}\left(\frac{\delta}{r}\right)^{-5/4} \tag{4.151}$$

$$\frac{343}{1656}\left(3 + 2\frac{r}{a}\frac{da}{dr} + \frac{r}{\delta}\frac{d\delta}{dr}\right)-\frac{1}{36a^2} = -0.0225(\mathrm{Re}_\phi)^{-0.25}\frac{(1 + a^2)^{3/8}}{a}\left(\frac{\delta}{r}\right)^{-5/4} \tag{4.152}$$

From Equation 4.142 with $u_{\phi,o} = \Omega r$ and $u_{\phi,\infty} = 0$,

$$\frac{49}{720}\frac{d}{dr}(r^2 a\Omega^2 r^2 \delta) = \frac{49}{720}\frac{d}{dr}(r^4 a\Omega^2 \delta) = -\frac{r^2}{\rho}\tau_{\phi,o} \tag{4.153}$$

$$\frac{49}{720}\left(4r^3 a\Omega^2 \delta + r^4\Omega^2\delta\frac{da}{dr} + r^4 a\Omega^2\frac{d\delta}{dr}\right) = -\frac{r^2}{\rho}\tau_{\phi,o} \tag{4.154}$$

Dividing by $\Omega^2 r^3 a\delta$

$$\frac{49}{720}\left(4 + \frac{r}{a}\frac{da}{dr} + \frac{r}{\delta}\frac{d\delta}{dr}\right) = -\frac{\tau_{\phi,o}}{\rho r\Omega^2 a\delta} \tag{4.155}$$

Substituting for $\tau_{\phi,o}$ from Equation 4.114

$$\tau_{\phi,o} = -0.0225\rho(r\Omega)^{7/4}\left(\frac{\nu}{\delta}\right)^{1/4}(1+a^2)^{3/8} \tag{4.156}$$

$$\frac{49}{720}\left(4 + \frac{r}{a}\frac{da}{dr} + \frac{r}{\delta}\frac{d\delta}{dr}\right) = -\frac{0.0225\rho}{\rho r\Omega^2 a\delta}(r\Omega)^{7/4}\left(\frac{\nu}{\delta}\right)^{1/4}(1+a^2)^{3/8} \tag{4.157}$$

The right-hand side of this equation can be rearranged in a similar fashion to Equation 4.152 as

$$\frac{49}{720}\left(4 + \frac{r}{a}\frac{da}{dr} + \frac{r}{\delta}\frac{d\delta}{dr}\right) = -0.0225(\mathrm{Re}_\phi)^{-0.25}\frac{(1+a^2)^{3/8}}{a}\left(\frac{\delta}{r}\right)^{-5/4} \tag{4.158}$$

In order to determine quantitative values for the boundary layer thickness, moment, and mass flow, it is necessary to determine a numerical value for a. From Equations 4.71 and 4.72

$$\rho\frac{d}{dr}(r\int_0^1 u_r^2 dz) - \rho\int_0^1 (u_\phi^2 - u_{\phi,\infty}^2)dz = -r\tau_{r,o} \tag{4.159}$$

$$\rho\frac{d}{dr}(r^2\int_0^\delta u_r u_\phi dz) = -r^2\tau_{\phi,o} \tag{4.160}$$

The left-hand side of Equation 4.159 is known from Equation 4.135, and the right-hand side from Equation 4.113, giving for $u_{\phi,o} = \Omega r$ and $u_{\phi,\infty} = 0$

$$\frac{343}{1656}\frac{d}{dr}(r^3 a^2\delta\Omega^2) - \frac{r^2\Omega^2\delta}{36} = -0.0225ra(r\Omega)^{7/4}\left(\frac{\nu}{\delta}\right)^{1/4}(1+a^2)^{3/8} \tag{4.161}$$

For Equation 4.160, the left-hand side is known from Equation 4.142, and the right-hand side from Equation 4.114, giving for $u_{\phi,o} = \Omega r$ and $u_{\phi,\infty} = 0$

$$\frac{49}{720}\frac{d}{dr}(r^4 a\delta\Omega^2) = 0.0225r^2 a(r\Omega)^{7/4}\left(\frac{\nu}{\delta}\right)^{1/4}(1+a^2)^{3/8} \tag{4.162}$$

The equations are satisfied if the relationship between the boundary layer thickness and the axial location is

$$\delta = \beta r^{3/5} \tag{4.163}$$

From Equation (4.161),

$$\frac{343}{1656}\frac{d}{dr}(r^3 r^{3/5}\alpha^2\beta\Omega^2)-\frac{r^2 r^{3/5}\beta\Omega^2}{36} = -0.0225r\alpha(r\Omega)^{7/4}\left(\frac{\nu}{\delta}\right)^{1/4}(1+\alpha^2)^{3/8}$$

(4.164)

$$\frac{343}{1656}\frac{d}{dr}(r^{18/5}\alpha^2\beta\Omega^2)-\frac{r^{13/5}\beta\Omega^2}{36} = -0.0225r\alpha(r\Omega)^{7/4}\left(\frac{\nu}{r^{3/5}\beta}\right)^{1/4}(1+\alpha^2)^{3/8}$$

(4.165)

$$\frac{343}{1656}\times\frac{18}{5}r^{13/5}\alpha^2\beta\Omega^2-\frac{r^{13/5}\beta\Omega^2}{36} = -0.0225r\alpha(r\Omega)^{7/4}\left(\frac{\nu}{r^{3/5}\beta}\right)^{1/4}(1+\alpha^2)^{3/8}$$

(4.166)

$$\frac{343}{460}r^{13/5}\alpha^2\beta\Omega^2-\frac{r^{13/5}\beta\Omega^2}{36} = -0.0225r\alpha(r\Omega)^{7/4}\left(\frac{\nu}{r^{3/5}\beta}\right)^{1/4}(1+\alpha^2)^{3/8}$$

(4.167)

or

$$0.7457r^{13/5}\alpha^2\beta\Omega^2-0.0278r^{13/5}\beta\Omega^2 = -0.0225r\alpha(r\Omega)^{7/4}\left(\frac{\nu}{r^{3/5}\beta}\right)^{1/4}(1+\alpha^2)^{3/8}$$

(4.168)

Dividing Equation 4.167 through by $\Omega^2 r^{13/5}$ gives

$$\frac{343}{460}\alpha^2\beta-\frac{\beta}{36} = -0.0225\frac{r\alpha(r\Omega)^{7/4}}{\Omega^2 r^{13/5}}\left(\frac{\nu}{r^{3/5}\beta}\right)^{1/4}(1+\alpha^2)^{3/8}$$

(4.169)

$$\frac{343}{460}\alpha^2\beta-\frac{\beta}{36} = -0.0225\alpha\left(\frac{\nu}{\beta\Omega}\right)^{1/4}(1+\alpha^2)^{3/8}$$

(4.170)

From Equation 4.162

$$\frac{49}{720}\frac{d}{dr}(r^4\alpha r^{3/5}\beta\Omega^2) = 0.0225r^2\alpha(r\Omega)^{7/4}\left(\frac{\nu}{r^{3/5}\beta}\right)^{1/4}(1+\alpha^2)^{3/8}$$

(4.171)

$$\frac{49}{720}\frac{d}{dr}(r^{23/5}\alpha\beta\Omega^2) = 0.0225r^2\alpha(r\Omega)^{7/4}\left(\frac{\nu}{r^{3/5}\beta}\right)^{1/4}(1+\alpha^2)^{3/8}$$

(4.172)

$$\frac{49}{720}\times\frac{23}{5}r^{18/5}\alpha\beta\Omega^2 = 0.0225r^2\alpha(r\Omega)^{7/4}\left(\frac{\nu}{r^{3/5}\beta}\right)^{1/4}(1+\alpha^2)^{3/8}$$

(4.173)

$$\frac{1127}{3600} r^{18/5} \alpha\beta\Omega^2 = 0.0225 r^2 \alpha (r\Omega)^{7/4} \left(\frac{\nu}{r^{3/5}\beta}\right)^{1/4} (1+\alpha^2)^{3/8} \qquad (4.174)$$

or

$$0.3131 r^{18/5} \alpha\beta\Omega^2 = 0.0225 r^2 \alpha (r\Omega)^{7/4} \left(\frac{\nu}{r^{3/5}\beta}\right)^{1/4} (1+\alpha^2)^{3/8} \qquad (4.175)$$

Dividing Equation 4.174 through by $r^{18/5}\Omega^2$ gives

$$\frac{1127}{3600} \alpha\beta = 0.0225 \left(\frac{\nu}{\beta\Omega}\right)^{1/4} (1+\alpha^2)^{3/8} \qquad (4.176)$$

We now have two equations for α and β, Equations 4.170 and 4.176. Dividing Equation 4.170 by Equation 4.176 gives

$$\frac{\dfrac{343}{460}\alpha^2\beta - \dfrac{\beta}{36}}{\dfrac{1127}{3600}\alpha\beta} = -\alpha \qquad (4.177)$$

$$\frac{343}{460}\alpha^2\beta - \frac{\beta}{36} = -\frac{1127}{3600}\alpha^2\beta \qquad (4.178)$$

$$\frac{343}{460}\alpha^2 - \frac{1}{36} = -\frac{1127}{3600}\alpha^2 \qquad (4.179)$$

$$\left(\frac{343}{460} + \frac{1127}{3600}\right)\alpha^2 = \frac{1}{36} \qquad (4.180)$$

$$\alpha^2 = \frac{1656000}{36\left((3600 \times 343) + (460 \times 1127)\right)} \qquad (4.181)$$

$$\alpha = \sqrt{\frac{2300}{87661}} = 0.1620 \qquad (4.182)$$

β can be found from Equation 4.176. Multiplying through by

$$\frac{3600\beta^{1/4}}{1127\alpha} \qquad (4.183)$$

gives

$$\beta^{5/4} = 0.0225 \frac{3600}{1127} \left(\frac{\nu}{\Omega}\right)^{1/4} \frac{(1+\alpha^2)^{3/8}}{\alpha} \qquad (4.184)$$

Substituting for α gives

$$\beta^{5/4} = 0.4480 \left(\frac{\nu}{\Omega}\right)^{1/4} \qquad (4.185)$$

Hence

$$\beta = 0.5261 \left(\frac{\nu}{\Omega}\right)^{1/5} \qquad (4.186)$$

From Equation 4.163, the boundary layer thickness is given by

$$\delta = \beta r^{3/5} = 0.5261 \left(\frac{\nu}{\Omega}\right)^{1/5} r^{3/5} = 0.5261 r \left(\frac{\nu}{r^2 \Omega}\right)^{1/5} \qquad (4.187)$$

or expressed as a ratio,

$$\frac{\delta}{r} = 0.5261 \left(\frac{\mu}{\rho \Omega r^2}\right)^{1/5} = 0.5261 (x^2 \mathrm{Re}_\phi)^{-0.2} \qquad (4.188)$$

Substitution of Equations 4.182 and 4.188 in Equation 4.145 gives

$$\dot{m}_o = \frac{49\pi}{60} \times 0.162 \times 0.5261 \rho \Omega r^3 \mathrm{Re}_\phi^{-1/5} = 0.2186 \rho \Omega r^3 \mathrm{Re}_\phi^{-1/5} \qquad (4.189)$$

Dividing through by μr to give the nondimensional mass flow rate gives

$$\frac{\dot{m}_o}{\mu r} = 0.2186 \left(\frac{\rho \Omega r^2}{\mu}\right) (x^2 \mathrm{Re}_\phi)^{0.8} \qquad (4.190)$$

or

$$\frac{\dot{m}_o}{\mu r} = 0.2186 (x^2 \mathrm{Re}_\phi)^{0.8} \qquad (4.191)$$

Equation 4.191 is the same as Equation 4.52 previously presented de facto in Section 4.4. Here, however, the derivation behind this equation has been developed.

The moment on one side of a rotating disc due to frictional drag is given by

$$T_q = 2\pi b^2 \rho \int_0^1 u_r u_\phi dz \qquad (4.192)$$

The integral term has already been determined as part of Equation 4.142

$$\int_0^1 u_r u_\phi dz = \frac{49}{720} a r^2 \Omega^2 \delta = 0.0681 a r^2 \Omega^2 \delta \qquad (4.193)$$

So

$$T_q = 2\pi b^2 \rho 0.0681 \times 0.162 r^2 \Omega^2 0.5261 r \mathrm{Re}_\phi^{-1/5} \qquad (4.194)$$

$$T_q = 0.03647 r^5 \Omega^2 \rho \mathrm{Re}_\phi^{-1/5} \qquad (4.195)$$

For a disc of radius b the torque for one side is given by

$$T_q = 0.03647 b^5 \Omega^2 \rho \mathrm{Re}_\phi^{-1/5} \qquad (4.196)$$

The moment coefficient, for one side of a disc, using the definition given in Equation 4.38 is given by

$$C_m = \frac{0.03647b^5\Omega^2\rho Re_\phi^{-1/5}}{\frac{1}{2}\rho\Omega^2 b^5} \qquad (4.197)$$

which gives

$$C_m = 0.07294 Re_\phi^{-1/5} \qquad (4.198)$$

The constant in Equation 4.198 is sometimes quoted as 0.07294, 0.073, or 0.07288, depending on the number of significant figures and approximation used in its determination.

The nondimensional shear stress can be determined by substituting for the boundary layer thickness and α using Equations 4.188 and 4.182 in Equation 4.114 to give

$$\frac{\tau_{\phi,o}}{\rho\Omega^2 b^2} = -0.02668x^{1.6}Re_\phi^{-0.2} \qquad (4.199)$$

The velocity can in principle be modeled by an appropriate profile. For a general $1/n$ power law, the coefficients can be found from Equations 4.200 to 4.209 (Owen and Rogers, 1989):

$$\alpha^2 = \frac{4(3n+2)(2n+1)(n+1)}{n^2(16n^3 + 85n^2 + 145n + 66)} \qquad (4.200)$$

$$\frac{\delta}{r} = \gamma\left(\frac{\rho\Omega r^2}{\mu}\right)^{-\frac{2}{n+3}} \qquad (4.201)$$

$$\frac{\dot{m}_o}{\mu r} = \varepsilon_m\left(\frac{\rho\Omega r^2}{\mu}\right)^{\frac{n+1}{n+3}} \qquad (4.202)$$

$$C_m = \varepsilon_M Re_\phi^{-\frac{2}{n+3}} \qquad (4.203)$$

$$K_w = \frac{3(5n+11)}{4(n+2)(n+3)} \qquad (4.204)$$

For

$$\frac{u}{u_\tau} = C\left(\frac{\rho u_\tau z}{\mu}\right)^{1/n} \qquad (4.205)$$

where the coefficient C is given by the data from Wieghardt (1946), and

$$K = C^{-\frac{2n}{n+1}} \qquad (4.206)$$

TABLE 4.3 Power law coefficients

	$n = 7$	$n = 8$	$n = 9$	$n = 10$
C	8.74	9.71	10.6	11.5
K	0.0225	0.0176	0.0143	0.0118
α	0.1620	0.1430	0.1280	0.1159
K_w	0.3833	0.3477	0.3182	0.2933
γ	0.5263	0.4973	0.4791	0.4633
ε_m	0.2187	0.1869	0.1643	0.1460
ε_M	0.0729	0.0561	0.0448	0.0365

and

$$\gamma^{\frac{n+3}{n+1}} = K \frac{2(n+1)(n+2)(n+3)(2n+1)(1+\alpha^2)^{(n-1)/2(n+1)}}{3n^2(5n+11)\alpha} \tag{4.207}$$

$$\varepsilon_m = \frac{2n^2\pi\alpha\gamma}{(n+1)(2n+1)} \tag{4.208}$$

$$\varepsilon_M = \frac{6n^2\pi\alpha\gamma}{(n+1)(n+2)(2n+1)} \tag{4.209}$$

Selected values for the coefficients C, K, α, K_w, γ, ε_m, and ε_M are given in Table 4.3 for $n = 7$, 8, 9, and 10.

4.4.2 Ekman Integral Equations

The Ekman integral equations involve a number of assumptions and therefore limitations in their application. They are useful, for example, in the consideration of turbulent flow in a rotor-stator wheelspace and are developed further in Chapter 5.

The equations of motion can be stated for a rotating frame of reference rotating with angular speed Ω' about the z-axis, which may or may not be the same as the angular speed of the principal rotating component Ω or flow structure ω, for two-dimensional axisymmetric flow. The relative velocity components are given by

$$u = u_r \tag{4.210}$$

$$v = u_\phi - \Omega'r \tag{4.211}$$

$$w = u_z \qquad (4.212)$$

Substituting for the relative velocities in Equations 4.6–4.9 gives

$$\frac{\partial u}{\partial r} + \frac{u}{r} + \frac{\partial w}{\partial z} = 0 \qquad (4.213)$$

$$u\frac{\partial u}{\partial r} + w\frac{\partial u}{\partial z} - \frac{v^2}{r} - 2\Omega'v - \Omega'^2 r = -\frac{1}{\rho}\frac{\partial p}{\partial r} + \nu\left(\nabla^2 u - \frac{u}{r^2}\right) \qquad (4.214)$$

$$u\frac{\partial v}{\partial r} + w\frac{\partial v}{\partial z} + \frac{uv}{r} + 2\Omega'u = \nu\left(\nabla^2 v - \frac{v}{r^2}\right) \qquad (4.215)$$

$$u\frac{\partial w}{\partial r} + w\frac{\partial w}{\partial z} = -\frac{1}{\rho}\frac{\partial p}{\partial z} + \nu\nabla^2 w \qquad (4.216)$$

where

$$\nabla^2 = \frac{\partial^2}{\partial r^2} + \frac{1}{r}\frac{\partial}{\partial r} + \frac{\partial^2}{\partial z^2} \qquad (4.217)$$

Neglecting the nonlinear terms such as $u\partial u/\partial r$, $w\partial u/\partial z$, v^2/r, $u\partial v/\partial r$, $w\partial v/\partial z$, uv/r, $u\partial w/\partial r$, $w\partial w/\partial z$ gives

$$-2\Omega'v = -\frac{1}{\rho}\frac{\partial}{\partial r}\left(p - \frac{1}{2}\Omega'^2 r^2\right) + \nu\left(\nabla^2 u - \frac{u}{r^2}\right) \qquad (4.218)$$

$$2\Omega'u = \nu\left(\nabla^2 v - \frac{v}{r^2}\right) \qquad (4.219)$$

$$0 = -\frac{1}{\rho}\frac{\partial p}{\partial z} + \nu\nabla^2 w \qquad (4.220)$$

If viscous forces are negligible, then

$$u = 0 \qquad (4.221)$$

$$\frac{\partial v}{\partial z} = 0 \qquad (4.222)$$

$$\frac{\partial w}{\partial z} = 0 \qquad (4.223)$$

This is called the geostrophic approximation and is useful in meteorological and oceanographical applications. It is a special case of the Taylor-Proudman theorem, introduced in Chapter 3, which states that for a system with solid boundaries perpendicular to the axis of rotation the flow is two-dimensional in planes perpendicular to the axis of rotation.

If Equations 4.218–4.220 are valid, then the boundary layer equations, Equations 4.60–4.63, simplify to:

$$-2\Omega'(v-v_\infty) = \frac{1}{\rho}\frac{\partial \tau_r}{\partial z} \tag{4.224}$$

$$2\Omega'u = \frac{1}{\rho}\frac{\partial \tau_\phi}{\partial z} \tag{4.225}$$

The mass flow in the boundary layer can be determined from Equation 4.73 assuming that there is no radial component of velocity outside the boundary layer.

Integration of Equation 4.225 gives

$$\frac{\tau_{\phi,o}}{\rho\Omega^2 b^2} = -\frac{1}{\pi}\frac{\Omega'}{\Omega}\frac{1}{\mathrm{Re}_\phi}\frac{\dot{m}_o}{\mu r} \tag{4.226}$$

The Ekman layer equations can be integrated to give

$$-2\Omega'\int_0^\delta (v-v_\infty)dz = -\frac{1}{\rho}\tau_{r,o} \tag{4.227}$$

$$2\Omega'\int_0^\delta u\,dz = -\frac{1}{\rho}\tau_{\phi,o} \tag{4.228}$$

where $v_\infty = v_{\phi,\infty}-\Omega'r$ is the relative tangential velocity outside the Ekman layer.

4.5. IMPINGING FLOW ON A ROTATING DISC

For the case of a uniform superimposed flow of air perpendicular to a rotating disc impinging on the surface (Figure 4.14), the moment coefficient has been found to be proportional to the ratio of the mean axial component of the flow outside the boundary layer and the tangential component at the outer radius of the disc.

Dorfman (1963) defined the ratio by

$$\sigma_{im} = \frac{2}{\pi}\frac{\bar{u}_{z,z=\infty}}{u_{\phi,r=b}} = \frac{2}{\pi}\frac{U}{\Omega b} \tag{4.229}$$

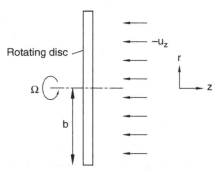

FIGURE 4.14 A uniform superimposed flow of air impinging on the surface of a rotating disc.

where

$u_{z,z=\infty} = U$ is the axial component of velocity outside of the boundary layer

$u_{\phi,r=b} = \Omega b$, is the tangential component of velocity evaluated at the outer radius.

Tifford and Chu (1954) obtained solutions to the laminar flow equations for the moment coefficient for impinging flow on a rotating disc for various values of σ, for $r \ll b$, which are presented in Table 4.4. Owen and Rogers (1989) note that these results need to be viewed with caution as their solution does not hold for the whole disc.

Truckenbrodt (1954) considered turbulent impinging flow on a rotating disc. As there is a radial pressure gradient, additional terms are required for the integral equation, and Equation 4.71 is replaced by

$$\frac{d}{dr}\left(r\int_0^\delta u_r(u_r-u_{r,\infty})dz\right) + r\frac{du_{r,\infty}}{dr}\int_0^\delta (u_r-u_{r,\infty})dz - \int_0^\delta u_\phi^2 dz = -\frac{r}{\rho}\tau_{r,o} \quad (4.230)$$

and as $u_{\phi,\infty} = 0$, Equation 4.72 becomes

$$\frac{d}{dr}\left(r^2\int_0^1 u_r u_\phi dz\right) = -\frac{r^2}{\rho}\tau_{\phi,o} \quad (4.231)$$

The profiles for the velocity components are assumed to have the form

$$u_r = \Omega r[\alpha(1-z_*) + \sigma z_*]z_*^{1/n} \quad (4.232)$$

TABLE 4.4 Moment coefficients for impinging flow on a rotating disc (Tifford and Chu, 1954)

σ_{im}	Moment coefficient
0	$C_m = 1.94\mathrm{Re}_\phi^{-0.5}$
0.1	$C_m = 2.03\mathrm{Re}_\phi^{-0.5}$
0.25	$C_m = 2.24\mathrm{Re}_\phi^{-0.5}$
0.5	$C_m = 2.65\mathrm{Re}_\phi^{-0.5}$
1.0	$C_m = 3.21\mathrm{Re}_\phi^{-0.5}$
1.5	$C_m = 4.20\mathrm{Re}_\phi^{-0.5}$
2.0	$C_m = 4.82\mathrm{Re}_\phi^{-0.5}$
4.0	$C_m = 6.67\mathrm{Re}_\phi^{-0.5}$
6.0	$C_m = 8.32\mathrm{Re}_\phi^{-0.5}$

TABLE 4.5 Variation of G_1, G_2, G_3, G_4, and G_5 with n (Owen and Rogers, 1989)

	$n = 7$	$n = 8$	$n = 9$	$n = 10$
G_1	0.068	0.063	0.058	0.054
G_2	0.029	0.026	0.024	0.022
G_3	0.029	0.012	−0.002	−0.013
G_4	0.026	0.020	0.016	0.013
G_5	1.056	1.032	1.014	1.000

$$u_\phi = \Omega r (1 - z_*^{1/n}) \tag{4.233}$$

Using a procedure similar to that described in Section 4.4.1, the following relationships can be developed for the moment coefficient.

$$C_m = \gamma (G_1 \alpha + G_2 \sigma_{im}) \mathrm{Re}_\phi^{-\frac{2}{n+3}} \tag{4.234}$$

where

$$\alpha^2 + G_3 \alpha \sigma_{im} = G_4 + G_5 \sigma_{im}^2 \tag{4.235}$$

$$\gamma = K^{\frac{n+1}{n+3}} \left[\frac{(n+3)(1+\alpha^2)^{\frac{n-1}{2(n+3)}}}{(5n+11)(G_1 \alpha + G_2 \sigma_{im})} \right]^{\frac{n+1}{n+3}} \tag{4.236}$$

where K is given by Equation 4.206 and values for the coefficients G_1, G_2, G_3, G_4 and G_5 are given in Table 4.5.

4.6. CONCLUSIONS

This chapter has presented an overview of the flow associated with isolated rotating discs and rotating fluids near a surface. Consideration has been given to the flow characteristics, whether laminar, transitional, or turbulent. Design orientated correlations for bulk parameters such as boundary layer thickness, mass flow, and moment coefficients have also been given. The insight resulting from the descriptions, along with the correlations, provide a basis for modeling disc flows and undertaking preliminary design calculations. Flow over an isolated rotating disc can be used as a starting point for analysis of flow in a rotor-stator disc system where a disc rotates in close proximity to a stationary disc. This topic is developed in Chapter 5.

REFERENCES

Bayley, F.J., and Owen, J.M. Flow between a rotating and a stationary disc. *Aeronautical Quarterly* 20 (1969), pp. 333–354.

Blasius, H. Das Ähnlichkeitsgesetz bei reibungsvorgängen in flüssikeiten. *Forsch. Arb. Ing. Wes.* No. 134, Berlin (1913).

Bödewadt, U.T. Die drehströmung über festem grunde. *Zeitschrift für angewandte Mathematik Mechanik.*, 20 (1940), pp. 241–253.

Childs, P.R.N., and Noronha, M.B. The impact of machining techniques on centrifugal compressor performance. ASME Paper 97-GT-456, 1997.

Cochran, W.G. The flow due to a rotating disc. *Proceedings Cambridge Phil. Society*, 30 (1934), pp. 365–375.

Dorfman, L.A. Resistance of a rotating rough disc. *Zh. Tekh. Fiz.*, 28 (1958), pp. 380–386. Translation. *Sov. Phys. Tech. Phys.*, 3 (1958), pp. 353–367.

Dorfman, L.A. *Hydrodynamic resistance and the heat loss of rotating solids.* Oliver & Boyd, 1963.

ESDU. 07004 Flow in rotating components—discs, cylinders and cavities. ESDU Fluid Mechanics, Internal Flow Series, Volume 4c (Flow in Rotating Machinery), 2007.

Gregory, N., Stuart, J.T., and Walker, W.S. On the stability of three dimensional boundary layers with application to the flow due to a rotating disc. *Philosophical Transactions of the Royal Society*, 248 (1955), pp. 155–198.

Kanaev, A.A. Frictional power loss of a disc rotating in a liquid. *Zh. Tekh. Fiz.*, 23 (1953), pp. 317–322.

Kármán, Th. von. Über laminare und turbulente reibung [On laminar and turbulent friction]. *Zeitschrift für angewandte Mathematik Mechanik*, 1 (1921), pp. 233–252.

Nydahl, J.E. Heat transfer for the Bödewadt problem. Colorado State University Dissertation, Fort Collins, Colorado, 1971.

Owen, J.M., and Rogers, R.H. *Flow and heat transfer in rotating-disc systems.* Volume 1: *Rotor-stator systems.* Research Studies Press, 1989.

Rogers, R.H., and Lance, G.N. The rotationally symmetric flow of a viscous fluid in the presence of an infinite rotating disc. *Journal of Fluid Mechanics*, 7 (1960), pp. 617–631.

Schlicting, H. *Boundary layer theory.* 7th Edition. McGraw-Hill, 1979.

Theodorsen, T. and Regier, A. Experiments on drag of revolving discs, cylinders and streamline rods at high speeds. *NACA Report* 793 (1944).

Tifford, A.N., and Chu, S.T. On the flow and temperature fields in forced flow against a rotating disc. *Proc. 2nd US Nat. Cong. Appl. Mech.* (published by ASME), (1954), pp. 793–800.

Truckenbrodt, E. Die turbulente strömung an einer angeblasenen rotierenden scheibe. *Zeitschrift für angewandte Mathematik Mechanik*, 34 (1954), pp. 150–162.

Wieghardt, K. Turbulente grenzschichten. *Göttinger Monographie*, Part B5 (1946).

Chapter 5

Rotor-Stator Disc Cavity Flow

A rotating disc adjacent to a stationary disc forms a cavity known as a rotor-stator cavity or wheelspace. The proximity of the discs, flow conditions around the periphery, and flow supplied to the cavity all have a significant influence on the conditions in the cavity.

5.1. INTRODUCTION

As introduced in Chapter 4, the flow associated with rotating disc systems is important to a wide range of engineering applications from computer memory disc drives, automotive disc brakes, cutting wheels and saws, gear wheels, and the discs used to support turbine and compressor blades. A common configuration in engineering is the rotor-stator wheelspace formed between coaxial rotating and stationary discs as illustrated schematically in Figure 5.1. Such a configuration is also referred to as a wheelspace, a rotor-stator disc cavity, a rotor-stator cavity, or a rotor-stator disc system. Rotor-stator disc cavity flow characteristics vary with the proximity of the stator and rotor, the presence of a stationary or rotating shroud, and the supply of any imposed flow.

In aircraft and electric power generation gas turbine engines alike, it is common practice for turbine disc cavities to be purged with coolant air taken from appropriate stages of the compression system. This air reduces the thermal load of the disc and prevents, or limits the effects of, the ingress of hot mainstream gas into the cavity between the rotating disc and the adjacent stationary casing. However, use of this air for turbine rim sealing is detrimental to engine cyclic performance, and turbine efficiency can be adversely affected by the seal air efflux into the main annulus. The lifetime of the rotor, and the cyclic and component performance of the engine, are therefore strongly dependent on turbine rim sealing efficiency. The combined requirements to be able to determine the power required to overcome frictional drag, local flow characteristics, and associated heat transfer have provided a sustained impetus to investigate a number of rotor-stator wheelspace configurations, some of which were illustrated in Figures 4.1(e) to 4.1(j). As an example of the complexity involved in attempting to determine the flow field, it has been found that the flow in the

Rotating Flow, DOI: 10.1016/B978-0-12-382098-3.00005-6

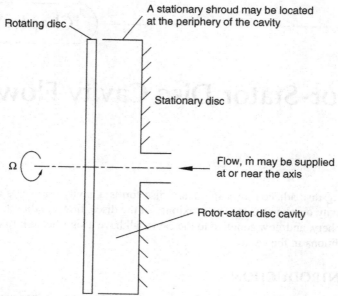

FIGURE 5.1 A rotor-stator disc cavity.

rotor-stator wheelspace of Figure 5.2 is three-dimensional and time dependent and that CFD modeling is sensitive to the numerical strategy and type of turbulence model adopted.

Prior to the advent of a large quantity of experimental data and numerical solutions, various theories were developed for the flow structure in a rotor-stator wheelspace. Batchelor (1951) proposed that there was a core of fluid between the rotating disc and the stator, which rotated at an angular velocity with a value between that of the stator and the rotor. This model implied the existence of boundary layers on both the rotor and the stator. In order to supply the rotor boundary layer, fluid is entrained from the core and pumped radially outward on the rotor. At the stator there is radial inflow of fluid and an efflux from the stator boundary layer to the core. In what has become known as the Batchelor-Stewartson controversy, Stewartson (1953) proposed that the flow was more like that of the free disc where the tangential velocity in the rotor boundary layer reduces from the rotor speed to zero away from the boundary layer with no core rotation. Characteristic velocity profiles for Batchelor and Stewartson models are illustrated in Figure 5.3.

Velocity measurements undertaken by Picha and Eckert (1958) showed that core rotation is minimal when the wheelspace is open to atmosphere, but when there is a stationary shroud at the outer radius a core of rotating fluid does exist. For a given rotor speed, the angular velocity of the core rotating fluid was found to reduce as the rotor-stator gap was increased. Subsequent numerical

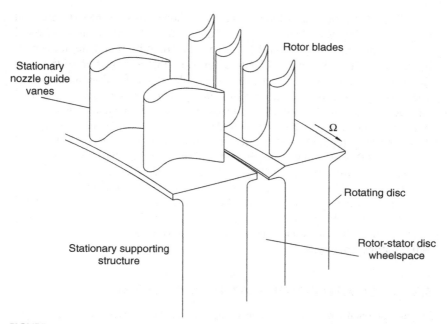

FIGURE 5.2   Schematic section of a turbine stage illustrating a rotor-stator disc wheelspace.

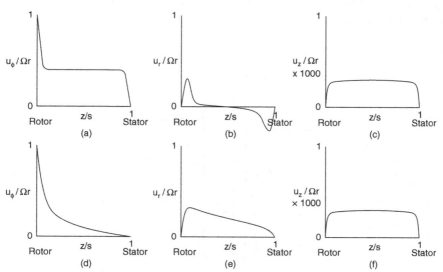

FIGURE 5.3   Characteristic velocity profiles in a rotor-stator wheelspace: (a) to (c) Batchelor flow; (d) to (f) Stewartson flow.

modeling—for example, Mellor, Chapple, and Stokes (1968), Chew and Vaughan (1988), and Poncet, Chauve, and Schiestel (2005)—has provided further supporting evidence for the existence of a rotating fluid core, for certain conditions, with a tangential velocity component that is a fraction of the rotor speed. It is now generally accepted that both the Batchelor flow model, characterized by three zones with two boundary layers separated by a central rotating core, and the Stewartson model characterized by the tangential velocity tending to zero outside the rotor boundary layer, are valid depending on the flow conditions (e.g., see Zandbergen and Dijkstra, 1987; Poncet, Chauve, and Schiestel, 2005). An important parameter in determining the flow structure is the supply of any superposed flow to the cavity. In general, for a rotor-stator cavity core, rotation will decrease with an increase in superposed radial outflow, as provided by a net flow of fluid supplied at the axis to the cavity; flow that can initially be characterized by the Batchelor model with a rotating core can tend toward that of the Stewartson model with no or very little core rotation as the radial outflow increases.

5.2. ENCLOSED ROTOR-STATOR DISC SYSTEMS

For an enclosed rotor-stator disc system, with no superposed flow supplied or extracted at the axis (Figure 5.4), the Batchelor flow model normally best describes the flow structure with fluid pumped radially outward in the rotor boundary layer, moving axially across to the stator in a boundary layer on the cylindrical outer shroud. In between the rotor and stator boundary layers there is an inviscid core of fluid that rotates at some fraction of the rotor angular velocity.

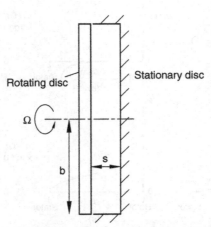

FIGURE 5.4 An enclosed rotor-stator wheelspace. A perfect seal is assumed at the periphery between the stationary casing and the rotor rim.

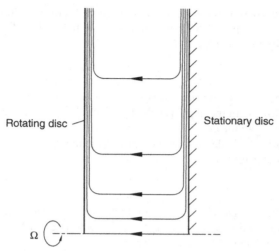

FIGURE 5.5   Batchelor's model for rotor-stator wheelspace flow.

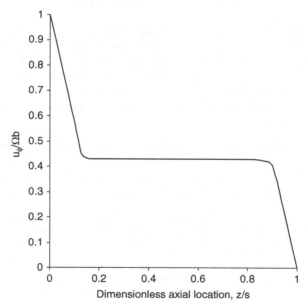

FIGURE 5.6   Typical data for the variation in tangential velocity with axial location across a rotor-stator wheelspace.

Figure 5.5 shows typical streamlines for a cross section of a rotor-stator wheelspace. Figure 5.6 shows typical data for the corresponding variation of the tangential velocity between the rotor and stator discs, with tangential shear between the disc speed in the rotor boundary to a constant value in the core and then reducing to zero in the boundary layer on the stator disc.

The speed of rotation of the core, ω, in the case of turbulent flow, is typically approximately 40% that of the rotor, that is, $\omega = 0.4\Omega$. This fraction can be characterized by the swirl ratio, β.

$$\beta = \frac{\omega}{\Omega} \tag{5.1}$$

In general, the core rotation decreases with a net radial outflow and increases with radial inflow, with the angular velocity of inflow entering the cavity being a very important parameter.

The gap ratio, G, is defined as the ratio of the axial clearance between the discs and the outer radius.

$$G = \frac{s}{b} \tag{5.2}$$

Daily, Ernst, and Asbedian (1964) found that the swirl ratio was 0.475, 0.45, and 0.42 for gap ratios, $G = s/b$, of 0.0276, 0.069, and 0.1241, respectively. This can be correlated approximately by

$$\beta = 0.49 - 0.57\frac{s}{b} = 0.49 - 0.57G \tag{5.3}$$

In addition to being a function of proximity of the geometry, the flow characteristics in a rotor-stator system also depend on the nature of the flow, that is, on whether it is laminar or turbulent. Experimental measurements by Daily and Nece (1960) for an enclosed rotor-stator wheelspace were used to identify four flow regimes categorized as a function of the gap ratio, G, of the rotor-stator wheelspace geometry and the rotational Reynolds number. These flow regimes are

- Regime 1: laminar flow, small clearance, merged boundary layers
- Regime 2: laminar flow, large clearance, separate boundary layers
- Regime 3: Turbulent flow, small clearance, merged boundary layers
- Regime 4: Turbulent flow, large clearance, separate boundary layers

Daily and Nece (1960) determined relationships for the moment coefficient for each of these regimes, Equations 5.6 to 5.9, for a range of rotational Reynolds numbers and gap ratios, 5.4 and 5.5.

$$1 \times 10^3 < Re_\phi < 1 \times 10^7 \tag{5.4}$$

$$0.0127 < G < 0.217 \tag{5.5}$$

For laminar flow in an enclosed rotor-stator wheelspace with no superposed flow with merged boundary layers, Regime 1, the moment coefficient, for one face of the disc, is given by

$$C_m = \frac{\pi}{GRe_\phi} \tag{5.6}$$

For laminar flow in an enclosed rotor-stator wheelspace with no superposed flow with separate boundary layers, Regime 2, the moment coefficient for one face of the disc is given by

$$C_m = 1.85G^{0.1}\text{Re}_\phi^{-0.5} \qquad (5.7)$$

For turbulent flow in an enclosed rotor-stator wheelspace with no superposed flow with merged boundary layers, Regime 3, the moment coefficient for one face of the disc is given by

$$C_m = 0.040G^{-0.167}\text{Re}_\phi^{-0.25} \qquad (5.8)$$

For turbulent flow in an enclosed rotor-stator wheelspace with no superposed flow with separate boundary layers, Regime 4, the moment coefficient for one face of the disc is given by

$$C_m = 0.0510G^{0.1}\text{Re}_\phi^{-0.2} \qquad (5.9)$$

Equations 5.6–5.9 have been used to define the demarcation between the various zones in Figure 5.7.

For regime 2, Dijkstra and von Heijst (1983) solved the Navier-Stokes equations and predicted a core to disc tangential velocity ratio, or swirl ratio, of $\beta = \omega/\Omega = 0.313$. Their study noted that the presence of shrouds on either the rotor or the stator can cause significant variation in the value of β. Owen (1988) proposed an improved approximation treating the rotor and stator separately, with the Navier-Stokes equations applied to the rotating disc and Ekman-layer equations to the stationary disc and found a swirl ratio, β of 0.426. Morse

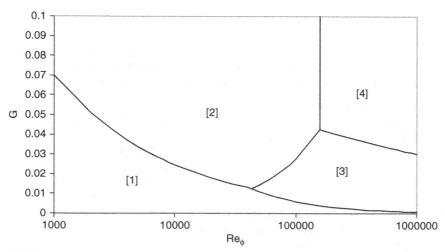

FIGURE 5.7   Regimes of flow for an enclosed rotor-stator wheelspace (Daily and Nece, 1960). [1] Laminar flow and small clearance; [2] Laminar flow and large clearance; [3] Turbulent flow and small clearance; [4] Turbulent flow and large clearance.

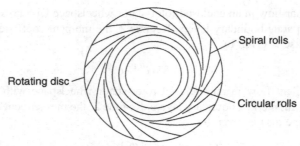

FIGURE 5.8   Schematic visualization of transitional disc flow for $G = 0.1145$, $Re_\phi = 2.09 \times 10^4$ showing the coexistence of circular and spiral rolls (after Schouveiler, Gal, and Chauve, 2001).

(1989) investigated the transition from laminar to turbulent flow for an enclosed rotor-stator system, solving the Reynolds averaged Navier-Stokes equations using a low turbulence Reynolds number k-ε model. For a rotational Reynolds number of 1×10^7 the swirl ratio was about 0.43 in accordance with the value predicted by Owen (1988). Interestingly, the first signs of transition predicted by Morse (1989) for the enclosed rotor-stator case were at a local rotational Reynolds number, $x^2 Re_\phi$, of 2.6×10^6.

Schouveiler, Gal, and Chauve (2001) used flow visualization and ultrasonic anemometry to observe the detailed instability for flow between a rotating and stationary disc in an enclosed cavity. Two scenarios for transition were observed.

1. For separate boundary layers, circular and spiral waves destabilize the stationary boundary layer, with transition occurring by the mixing of these waves.
2. For merged boundary layers, finite turbulence spots or spirals appear within the laminar domain, with their occurrence increasing with rotational Reynolds number.

A number of flow structures were identified, depending on the regime, including stationary and nonstationary circular rolls and spiral rolls and wave turbulence, with the coexistence of certain combinations of these for particular conditions.

- For larger gap ratios, $0.0714 \le G \le 0.1429$, the destabilizing flow regime sequence was: basic laminar flow; circular rolls; coexisting circular and spiral rolls (illustrated schematically in Figure 5.8); spiral rolls.
- For intermediate gap ratios, $0.0179 \le G \le 0.0714$, the destabilizing flow regime sequence was: basic laminar flow; spiral rolls; coexisting outward travelling spiral rolls, stationary spiral rolls and inward traveling circular rolls.
- For small gap ratios the $0.0071 \le G \le 0.0179$, the destabilizing flow regime sequence was: basic laminar flow; spiral vortices; coexisting spiral vortices and solitary waves; coexisting spiral vortices, solitary waves and turbulence spots; turbulence spots.

Digital memory disc drives are a limiting factor in battery life for devices such as personal music players and portable computers. The faster the speed of the magnetic medium discs, the faster the memory can be accessed. As the

FIGURE 5.9   Photograph of a typical computer memory hard disc drive.

diameter increases, the amount of memory available also increases. However, from Equation 4.38 the moment coefficient and hence the power consumed increases with the fifth power of the radius, and the power required to overcome the viscous drag increases with the cube of the rotational speed. Careful selection and optimization of the disc speed and diameter is therefore necessary in determining the configuration of digital memory disc drives. The photographs in Figures 5.9 and 5.10 show typical configurations for a 7200-rpm, 3.5-inch hard disc drive. The following example illustrates the typical calculations involved in determining the power consumed by a simple disc drive.

Example 5.1.

Determine the power required to overcome frictional drag for a hard disc drive of 3.5-inch diameter running at 7200 rpm. The gap between the disc and casing can be taken as 3 mm.

FIGURE 5.10 Side view of a typical computer memory hard disc drive.

Solution

The angular velocity of the disc is

$$\Omega = 7200 \times \frac{2\pi}{60} = 754.0 \ \ rad/s$$

The outer radius of the disc is $0.5 \times 3.5 \times 25.4 \times 10^{-3} = 0.04445$ m

Assuming that the disc is running in air and that the density and viscosity for air can be taken as 1.205 kg/m³ and 1.817×10^{-5} Pa s, respectively, the rotational Reynolds number is given by

$$Re_\phi = \frac{1.205 \times 754.0 \times 0.04445^2}{1.817 \times 10^{-5}} = 9.880 \times 10^4$$

With this value for the rotational Reynolds number, the flow can be taken as laminar.

The boundary layer thickness can be estimated using the relationship for a free disc, Equation 4.23,

$$\delta \approx 5.5 \sqrt{\frac{1.817 \times 10^{-5}/1.205}{754.0}} = 0.7778 \times 10^{-3} \ m$$

With a gap of 3 mm, this provides an indication that the boundary layers are not merged; therefore the appropriate equation for the moment coefficient, if there is no throughflow and the cavity is modeled as a closed rotor-stator system, is Equation 5.7, Regime 2.

$$G = \frac{0.003}{0.04445} = 0.06749$$

$$C_m = 1.85 G^{0.1} Re_\phi^{-0.5} = 1.85(0.06749)^{0.1}(9.880 \times 10^4)^{-0.5} = 4.495 \times 10^{-3}$$

The moment can be determined from Equation 4.38.

$$T_q = 0.5\rho\Omega^2 b^5 C_m = 0.5 \times 1.205 \times 754.0^2 \times 0.04445^5 \times 4.484 \times 10^{-3} = 2.672 \times 10^{-4} \, N\,m$$

The power required to spin this disc at 7200 rpm to overcome frictional drag on each disc face can be determined by

$$Power = T_q\Omega = 2.672 \times 10^{-4} \times 754.0 \approx 0.2 \, W$$

As a disc has two faces, the total power is approximately 0.4 W.

Comment

If, say, a 3 V-battery with a power rating of 10 Ampere hours is used, then 0.4 W of power can be sustained for a total of $10/(0.4/3) = 75$ hours. A typical current figure for personal portable computers is that the disc drive consumes about 8% of the total power (Intel Corporation, 2004). Taking this figure, the above battery life powering the whole of the portable computer might have a battery life of $10/((0.4 \times 12.5)/3) = 6$ hours. In this example, although the power consumption to overcome frictional drag may initially seem low, it does have an impact on battery life and therefore is important in determining the performance of a product. It should be noted that no account has been made in this example for any egress or ingress flows at the periphery of the cavity, the presence of multiple discs, or any external cooling flow.

In a gas turbine engine, the flow is highly turbulent, and rotational Reynolds numbers of the order of 2×10^7 to 3×10^7 can be reached on, for example, the high-pressure turbine disc. Of the two turbulent regimes considered by Daily and Nece (1960), regime 4 tends to be the more relevant, as the disc cavity spacing is usually sufficiently large to accommodate two separate boundary layers, with the core rotating at about 43% of the disc speed. It should be noted, however, that the introduction of any superimposed air supply into the cavity would modify this behavior.

Example 5.2.

Determine the moment coefficient and moment, and thereby power, required to overcome frictional drag for a disc with a diameter of 400 mm, rotating in air at 12,000 rpm in an enclosed cavity with a gap between the rotor and stator of 15 mm. The density and viscosity for air can be taken as 1.205 kg/m^3 and 1.817×10^{-5} Pa s, respectively.

Solution

The rotational Reynolds number for these conditions is given by

$$Re_\phi = \frac{1.205 \times (12000 \times 2\pi/60) \times 0.2^2}{1.817 \times 10^{-5}} = 3.334 \times 10^6$$

With this value for the rotational Reynolds number, as it is significantly above 200,000, the flow can be taken as turbulent.

The boundary layer thickness at the outer radius of the disc can be determined approximately using the relationship developed by von Kármán for the case of turbulent flow over a rotating free disc, Equation 4.51:

$$\delta = 0.526r\left[\left(\frac{r}{b}\right)Re_\phi\right]^{-0.2} = 0.526 \times 0.2(3.334 \times 10^6)^{-0.2} = 5.217 \times 10^{-3} \, m \approx 5.2 \, mm$$

This provides an indication that the boundary layers are not merged, and therefore the appropriate equation for the moment coefficient is Equation 5.9, Regime 4.

$$G = \frac{0.015}{0.2} = 0.075$$

$$C_m = 0.0510G^{0.1}Re_\phi^{-0.2} = 0.051(0.075)^{0.1}(3.334 \times 10^6)^{-0.2} = 1.952 \times 10^{-3}$$

The moment can be determined from Equation 4.38:

$$T_q = 0.5\rho\Omega^2 b^5 C_m = 0.5 \times 1.205 \times 1257^2 \times 0.2^5 \times 1.952 \times 10^{-3} = 0.5946 \, N\,m$$

The power required to spin this disc at 12,000 rpm to overcome frictional drag on each disc face can be determined by

$$Power = T_q\Omega = 747.4 \quad W$$

As there will be two disc faces, the total power required to spin the disc will be approximately 1.5 kW.

Comment

In this example, the power required to drive the disc is about 1.5 kW. In some machines, such as an electrical machine or a turbine, this quantity of power represents a significant mechanical efficiency loss. It should be noted that the moment, $T_q = 0.5 \times 0.051\rho\Omega^2 b^5 (s/b)^{0.1}(\rho\Omega b^2/\mu)^{-0.2} = 0.0255\rho^{0.8}\Omega^{1.8}s^{0.1}b^{4.5}\mu^{0.2}$, is a function of the 1.8 power of the angular velocity and the 4.5 power of the disc radius. The power is therefore a function of the 2.8 power of the angular velocity. Doubling the speed results in increasing the power required to overcome frictional drag by a factor of 6.964, and doubling the radius results in increasing the power required by a factor of 22.63.

5.3. INVISCID EQUATIONS OF MOTION

For the case of a single disc rotating in a fluid of a large extent, or for a rotor-stator wheelspace system where the clearance between the rotor and stator discs is sufficiently large, viscous and turbulence effects can be considered as negligible except in the thin boundary layers near the discs. This allows a considerable simplification of the equations of motion. The subscript ∞ is used here to refer to conditions far from the rotating disc in the case of a single rotating disc.

The subscript c is used to refer to conditions inside the rotating inviscid core for a rotor-stator wheelspace.

Considering steady-state axisymmetric conditions, the continuity equation, Equation 4.6, remains unchanged by these assumptions. Equations 4.7 to 4.9 become

$$u_{r,\infty}\frac{\partial u_{r,\infty}}{\partial r} + u_{z,\infty}\frac{\partial u_{r,\infty}}{\partial z} - \frac{u_{\phi,\infty}^2}{r} = -\frac{1}{\rho}\frac{\partial p_\infty}{\partial r} \tag{5.10}$$

$$u_{r,\infty}\frac{\partial u_{\phi,\infty}}{\partial r} + u_{z,\infty}\frac{\partial u_{\phi,\infty}}{\partial z} + \frac{u_{r,\infty}u_{\phi,\infty}}{r} = 0 \tag{5.11}$$

$$u_{r,\infty}\frac{\partial u_{z,\infty}}{\partial r} + u_{z,\infty}\frac{\partial u_{z,\infty}}{\partial z} = -\frac{1}{\rho}\frac{\partial p_\infty}{\partial z} \tag{5.12}$$

The radial velocity is zero or very small throughout the core region in the case of a rotor-stator system and is taken as zero in the case of a single disc, distant from the disc. From the continuity equation, Equation 4.6, it therefore follows that the axial velocity far from the disc is independent of the axial location, and from Equation 5.11 the tangential component of velocity is also independent of the axial location, that is, for $u_{r,\infty} = 0$,

$$\frac{\partial u_{\phi,\infty}}{\partial z} = 0 \tag{5.13}$$

and

$$\frac{\partial u_{z,\infty}}{\partial z} = 0 \tag{5.14}$$

The Taylor-Proudman theorem states that, for a system with solid boundaries perpendicular to the axis of rotation, the flow is two-dimensional in planes perpendicular to the axis of rotation; the theorem is therefore valid for the conditions in the core with $u_{r,\infty} = 0$. Equation 5.10 provides a relationship between the radial pressure gradient and the tangential velocity

$$\frac{1}{\rho}\frac{\partial p_\infty}{\partial r} = \frac{u_{\phi,\infty}^2}{r} \tag{5.15}$$

or for the case of a rotor-stator system with an inviscid rotating core with tangential velocity $u_{\phi,c}$,

$$\frac{1}{\rho}\frac{\partial p}{\partial r} = \frac{u_{\phi,c}^2}{r} \tag{5.16}$$

5.4. ROTOR-STATOR WITH RADIAL OUTFLOW

If a rotor-stator disc cavity is unshrouded and therefore open to the external environment, the pumped flow from the rotating disc will exit the cavity to the external surroundings ($r > b$), and ingress from the external environment can

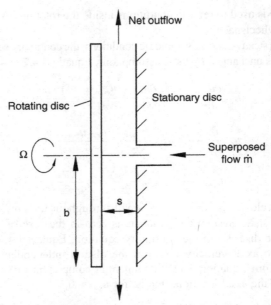

FIGURE 5.11   Rotor-stator wheelspace with a superposed supply of air.

occur to supply the entrainment demands of the rotating flow. At the outer radius of the cavity, there will therefore be both inflow and outflow of fluid. A supply of fluid to the cavity can be provided at or near the axis, and this flow will exit the cavity with a net radial velocity component. The supply of such a superposed flow and the associated radial outflow can alter the core rotation in the cavity and at high gap ratios can completely suppress it, giving a flow structure more comparable to that described by "Stewartson flow," where the tangential velocity in the rotor boundary layer reduces from the rotor speed to zero away from the boundary layer with no core rotation. As a result, the flow structure for an open rotor-stator system is more complex than that for an enclosed system.

In general, if a superposed supply of air is provided, it will alter the flow structure in the rotor-stator cavity, tending in the case of outflow to reduce the core rotation and increase the moment on the rotor. The moment coefficient can range from a value below that of the equivalent free disc when there is no superposed flow to a value well above. In order to reduce the amount of ingress, an alternative supply of fluid to meet the demands of entrainment can be supplied, typically near the axis of the cavity, as indicated in Figure 5.11.

The case of an unshrouded rotor-stator cavity with radial outflow has been studied for laminar merged boundary layers by Soo (1958); for laminar separated boundary layers by Vaughan (1987); for turbulent flow merged boundary layers by Dorfman (1961); and for turbulent flow separated boundary layers by Daily, Ernst, and Asbedian (1964) and Vaughan (1987).

Bayley and Owen (1969) studied the effects of the radial outflow in a rotor-stator cavity by comparing experimental measurements of pressure, velocity, and moment coefficients with values obtained from integral solutions of the boundary layer equations. Tests were performed at rotational Reynolds numbers up to 3.8×10^6 and gap ratios of between 0.008 and 0.03. Reducing the superposed flow rate or increasing the rotational Reynolds number increased the tangential velocity in the cavity. As the gap between the discs was increased, the tangential velocity of the "inviscid" core region decreased, and ultimately the core region disappeared as the effects of the stator lessened and the flow began to behave as if adjacent to a free disc. The moment coefficient increased with an increase in the superposed flow rate but was seen to drop, ultimately to free disc levels, with increasing rotational Reynolds number. Narrower gaps resulted in higher moment coefficients. The pressure distribution predicted by momentum integral techniques agreed well with experimental data in the larger gaps, but as the clearance was reduced the effects of the outflow became over predicted. Conversely, the moment coefficient was better predicted at small clearances.

For the range of conditions given in 5.17 to 5.20, Owen and Haynes (1976) obtained the correlation given in Equation 5.21 where C_{m1} and C_{m2} are defined in Equations 5.22 and 5.23.

$$\frac{a}{b} = 0.1333 \tag{5.17}$$

$$0.01 \leq G \leq 0.18 \tag{5.18}$$

$$14000 \leq Cw \leq 98000 \tag{5.19}$$

$$0 < \text{Re}_\phi \leq 4 \times 10^6 \tag{5.20}$$

$$C_m = (C_{m1}^6 + C_{m2}^6)^{1/6} \tag{5.21}$$

$$C_{m1} = 0.0553 \left(\frac{Cw}{G}\right)^{0.8} \frac{1}{\text{Re}_\phi} \tag{5.22}$$

$$C_{m2} = 0.0655 Re_\phi^{-0.186} \left(1 + 12.4 \frac{Cw}{Re_\phi}\right) \tag{5.23}$$

For a rotor-stator system with superposed flow, then for steady-state conditions assuming radial outflow at the rotor and inflow at the stator, the pumped radial outflow, $\dot{m}_o$, will be balanced by the superposed flow supplied at the axis, $\dot{m}_{superposed}$, and any ingress at the periphery, $\dot{m}_s$. This can be modeled by

$$\dot{m}_o + \dot{m}_s = \dot{m}_{superposed} \tag{5.24}$$

Appropriate attention to the sign convention in Equation 5.24 is necessary. For example, the pumped radial outflow might be 60 g/s, the inflow at the stator 20 g/s, and the superposed flow 40 g/s. Using the convention for Equation 5.24, $\dot{m}_o =$ 60 g/s, $\dot{m}_s = -20$ g/s and $\dot{m}_{superposed} = 40$ g/s, giving in this case $60 - 20 = 40$ g/s.

For the case of a rotating disc, with angular velocity Ω, in a rotating fluid rotating with an angular velocity of $\omega = \beta\Omega$, the integral momentum equations assuming 1/7th power law profiles, Equations 4.129, 4.135, and 4.142 are valid with $u_{\phi,o} = \Omega r$ and $u_{\phi,c} = \beta\Omega r$. A solution with the form given in Equation 5.25 can be assumed.

$$\frac{\delta}{r} = \gamma(x^2\mathrm{Re}_\phi)^{-0.2}(1-\beta)^{0.6} \tag{5.25}$$

Assuming solid body rotation, $\beta =$ a constant, Equations 4.129, 4.135, and 4.142 can be written as (Owen and Rogers, 1989)

$$\frac{\dot{m}_o}{\mu r} = \frac{49\pi}{60}(x^2\mathrm{Re}_\phi)^{0.8}(1-\beta)^{1.6}\alpha\gamma \tag{5.26}$$

$$\frac{343}{1656}\left(\frac{18}{5} + 2\frac{x\,d\alpha}{\alpha\,dx} + \frac{x\,d\gamma}{\gamma\,dx}\right) - \frac{1+8\beta}{36(1-\beta)\alpha^2} = -0.0225\frac{(1+\alpha^2)^{3/8}}{(1-\beta)\alpha\gamma^{5/4}} \tag{5.27}$$

$$\frac{49}{720}\left(\frac{23+37\beta}{5(1-\beta)} + \frac{x\,d\alpha}{\alpha\,dx} + \frac{x\,d\gamma}{\gamma\,dx}\right) = 0.0225\frac{(1+\alpha^2)^{3/8}}{(1-\beta)\alpha\gamma^{5/4}} \tag{5.28}$$

A solution to these equations exists, with α and γ independent of x. α and γ, however, depend on the swirl ratio β. α and γ are given by Equations 5.29 and 5.30, respectively.

$$\alpha = \left[\frac{2300(1+8\beta)}{49(1789-409\beta)}\right]^{0.5} \tag{5.29}$$

$$\gamma = \left[\frac{81(1+\alpha^2)^{3/8}}{49(23+37\beta)\alpha}\right]^{0.8} \tag{5.30}$$

Equation 5.29 breaks down with the numerator becoming zero or negative if $\beta < -1/8$, and the denominator tending to zero or negative if $\beta \geq 4.374$. Values for α and γ and several related parameters are given in Table 5.1 as a function of the swirl ratio β for $0 < \beta < 4$.

The nondimensional mass flow in the boundary layer is given by

$$\frac{\dot{m}_o}{\mu r} = \frac{49\pi}{60}\alpha\gamma(1-\beta)^{8/5}(x^2\mathrm{Re}_\phi)^{0.8} = \varepsilon_m(x^2\mathrm{Re}_\phi)^{0.8} \tag{5.31}$$

TABLE 5.1 Variation of α and γ and other parameters with β for a rotating disc in a rotating fluid

β	α	γ	ε_m	ε_m'	ε_M	K_w
0	0.1620	0.5261	0.2186	0.2280	0.0729	0.3833
0.1	0.2198	0.3680	0.1754	0.1707	0.0679	0.4450
0.2	0.2674	0.2855	0.1371	0.1332	0.0604	0.5067
0.3	0.3095	0.2333	0.1047	0.1031	0.0517	0.5683
0.4	0.3483	0.1968	0.0777	0.0781	0.0425	0.6300
0.5	0.3849	0.1698	0.0553	0.0571	0.0333	0.6917
0.6	0.4200	0.1490	0.0371	0.0395	0.0243	0.7533
0.7	0.4540	0.1324	0.0225	0.0249	0.0159	0.8150
0.8	0.4875	0.1189	0.0113	0.0129	0.0086	0.8767
0.9	0.5205	0.1077	0.0036	0.0033	0.0029	0.9383
1.0	0.5533	0.0983	0.0000	-0.0060	0.0000	1.0000
2.0	0.9065	0.0498	-0.1159	-0.1109	-0.1629	1.6167
3.0	1.4450	0.0310	-0.3489	-0.3594	-0.6775	2.2333
4.0	3.1818	0.0200	-0.9446	-0.7234	-2.3409	2.8500

where

$$\varepsilon_m = \frac{49\pi}{60}\alpha\gamma(1-\beta)^{8/5} \tag{5.32}$$

The moment coefficient is given by

$$C_m = \frac{49\pi}{4140}(1-\beta)^{8/5}(23+37\beta)\alpha\gamma\mathrm{Re}_\phi^{-0.2} = \varepsilon_M \mathrm{Re}_\phi^{-0.2} \tag{5.33}$$

where

$$\varepsilon_M = \frac{49\pi}{4140}(1-\beta)^{8/5}(23+37\beta)\alpha\gamma \tag{5.34}$$

The shear stress is related to the rotational Reynolds number and mass flow rate by

$$\frac{\tau_{\phi,o}}{\rho\Omega^2 b^2} = -\frac{1}{\pi}\frac{K_w}{\mathrm{Re}_\phi}\frac{\dot{m}_o}{\mu r} \tag{5.35}$$

where K_w is the stress flow rate parameter given by

$$K_w = \frac{23 + 37\beta}{60} \tag{5.36}$$

Values of ε_m, ε_M, and K_w are given in Table 5.1 as a function of β.

An approximation for ε_m is useful in the consideration of rotor-stator disc wheelspaces. This is given by $\varepsilon_m \approx \varepsilon_m'$, where

$$\varepsilon_m' = 0.154\beta^{1.6} - 0.386\beta^{0.8} + 0.228 \text{ for } \beta < 1 \tag{5.37}$$

$$\varepsilon_m' = -0.164\beta^{1.6} + 0.308\beta^{0.8} - 0.15 \text{ for } \beta \geq 1 \tag{5.38}$$

Values for ε_m' and a comparison with ε_m are given in Table 5.1. Examination of the values for ε_m' and ε_m shows that the approximation is within 11% for $\beta \leq 3$.

An alternative approach for determining the rotor boundary layer mass flow is solution of the Ekman boundary layer equations developed in Section 4.4.2. The axes are taken to be rotating with angular velocity Ω, giving $v = 0$ at $z = 0$ and an inviscid rotating core velocity

$$v_c = -(1-\beta)\Omega r = \omega r - \Omega r \tag{5.39}$$

The relative velocity profiles are

$$u = -(1-\beta)\alpha\Omega r \left(\frac{z}{\delta}\right)^{1/7}\left[1 - \left(\frac{z}{\delta}\right)\right] \tag{5.40}$$

$$v = -(1-\beta)\Omega r \left(\frac{z}{\delta}\right)^{1/7} \tag{5.41}$$

The radial and tangential shear stress components on the rotating disc are given by Equations 4.115 and 4.116, respectively, with

$$v_o - v_c = -(1-\beta)\Omega r \tag{5.42}$$

As a result

$$\alpha = 0.553 \tag{5.43}$$

$$\frac{\delta}{r} = 0.0983(1-\beta)^{0.6}(x^2 \text{Re}_\phi)^{-0.2} \tag{5.44}$$

$$\frac{\dot{m}_o}{\mu r} = 0.1395(1-\beta)^{1.6}(x^2 \text{Re}_\phi)^{0.8} \tag{5.45}$$

This approximate solution is valid when $(1 - \beta)$ is small and the values for α and γ are identical with those listed in Table 5.1 for a swirl ratio of $\beta = 1$.

For the case of a fluid rotating with an angular velocity of ω over a stationary disc, Equations 4.129, 4.135, and 4.142 are valid with $u_{\phi,o} = 0$ and $u_{\phi,\infty} = \omega r$. Taking a solution with the form

$$\frac{\delta}{r} = \gamma(x^2 \text{Re}'_\phi)^{-0.2} \tag{5.46}$$

where Re'_ϕ is a modified rotational Reynolds number defined by

$$\text{Re}'_\phi = \frac{\rho \omega b^2}{\mu} \tag{5.47}$$

Assuming solid body rotation, $\beta = $ a constant, Equations 4.129, 4.135, and 4.142 can be written as (Owen and Rogers, 1989)

$$\frac{\dot{m}_s}{\mu r} = -\frac{49\pi}{60}(x^2 \text{Re}'_\phi)^{0.8} \alpha \gamma \tag{5.48}$$

$$\frac{343}{1656}\left(\frac{18}{5} + 2\frac{x\,d\alpha}{\alpha\,dx} + \frac{x\,d\gamma}{\gamma\,dx}\right) + \frac{2}{9\alpha^2} = 0.0225\frac{(1+\alpha^2)^{3/8}}{\alpha\gamma^{5/4}} \tag{5.49}$$

$$\frac{49}{720}\left(-\frac{37}{5} + \frac{x\,d\alpha}{\alpha\,dx} + \frac{x\,d\gamma}{\gamma\,dx}\right) = -0.0225\frac{(1+\alpha^2)^{3/8}}{\alpha\gamma^{5/4}} \tag{5.50}$$

These equations have no solution with α constant. An alternative approach is solution of the Ekman boundary layer equations presented in Section 4.4.2. The axes are taken to be rotating with angular velocity ω such that the rotor speed, referred to the rotating frame of reference, is given by $v_o = -\omega r$ and $v_\infty = 0$. This gives

$$\alpha = 0.553 \tag{5.51}$$

$$\frac{\delta}{r} = 0.0983(x^2 \text{Re}'_\phi)^{-0.2} \tag{5.52}$$

$$\frac{\dot{m}_s}{\mu r} = -0.1395(x^2 \text{Re}'_\phi)^{0.8} \tag{5.53}$$

Assuming an inviscid core and using 1/7th power law profiles to model the rotor and stator boundary layer flows, the mass flow in the rotor and stator boundary layers is given by Equations 5.31 and 5.48, respectively. These are restated in Equations 5.54 and 5.55 with the subscripts o and s referring, respectively, to the rotor and stator.

$$\frac{\dot{m}_o}{\mu r} = \frac{49\pi}{60}\alpha_o\gamma_o(1-\beta)^{1.6}(x^2 \text{Re}_\phi)^{0.8} \tag{5.54}$$

$$\frac{\dot{m}_s}{\mu r} = -\frac{49\pi}{60}\alpha_s\gamma_s\beta^{0.8}(x^2 \text{Re}_\phi)^{0.8} \tag{5.55}$$

Dividing through Equation 5.24 by μb gives

$$\frac{\dot{m}_o}{\mu b} + \frac{\dot{m}_s}{\mu b} = \frac{\dot{m}_{superposed}}{\mu b} \qquad (5.56)$$

or

$$\frac{\dot{m}_o}{\mu b} + \frac{\dot{m}_s}{\mu b} = Cw \qquad (5.57)$$

From Equations 5.54 and 5.55, multiplying through by $x = r/b$,

$$\frac{\dot{m}_o}{\mu b} = x\frac{49\pi}{60}\alpha_o\gamma_o(1-\beta)^{1.6}(x^2\mathrm{Re}_\phi)^{0.8} \qquad (5.58)$$

$$\frac{\dot{m}_s}{\mu b} = -x\frac{49\pi}{60}\alpha_s\gamma_s\beta^{0.8}(x^2\mathrm{Re}_\phi)^{0.8} \qquad (5.59)$$

Substitution in Equation 5.57 gives

$$x\frac{49\pi}{60}\alpha_o\gamma_o(1-\beta)^{1.6}(x^2\mathrm{Re}_\phi)^{0.8} - x\frac{49\pi}{60}\alpha_s\gamma_s\beta^{0.8}(x^2\mathrm{Re}_\phi)^{0.8} = Cw \qquad (5.60)$$

On rearrangement

$$\frac{49\pi}{60}[\alpha_o\gamma_o(1-\beta)^{1.6} - \alpha_s\gamma_s\beta^{0.8}] = \frac{Cw}{\mathrm{Re}_\phi^{0.8}x^{2.6}} \qquad (5.61)$$

or

$$\alpha_o\gamma_o(1-\beta)^{1.6} - \alpha_s\gamma_s\beta^{0.8} = \frac{60}{49\pi}Cw\mathrm{Re}_\phi^{-0.8}\left(\frac{r}{b}\right)^{-2.6} \qquad (5.62)$$

For the Ekman layer assumption for the stator flow, the coefficients α_s and γ_s are both approximated by $\alpha_s = \alpha(1)$ and $\gamma_s = \gamma(1)$. Hence from Table 5.1, $\alpha_s = 0.5533$ and $\gamma_s = 0.0983$. For $0 < \beta < 0.5$, $\alpha_o\gamma_o$ can be approximated by (see Owen, 1988)

$$\alpha_o\gamma_o = 0.0852(1-0.51)\beta \qquad (5.63)$$

Substitution in Equation 5.62 gives a relationship between the swirl ratio and the nondimensional superimposed flow rate.

$$(1-\beta)^{1.6}(1-0.51\beta) - 0.638\beta^{0.8} = 4.57Cw\mathrm{Re}_\phi^{-0.8}\left(\frac{r}{b}\right)^{-2.6} \qquad (5.64)$$

The term $Cw\mathrm{Re}_\phi^{-0.8}$ is known as the turbulent flow parameter, λ_t.

$$\lambda_t = Cw\mathrm{Re}_\phi^{-0.8} \qquad (5.65)$$

Equation 5.64 can be stated in the form

$$(1-\beta)^{1.6}(1-0.51\beta) - 0.638\beta^{0.8} = 4.57\lambda_t\left(\frac{r}{b}\right)^{-2.6} \qquad (5.66)$$

Examination of Equation 5.64 shows that the rotating core flow is suppressed, $\beta = 0$, when

$$Cw = 0.219Re_\phi^{0.8}(r/b)^{2.6} \tag{5.67}$$

This value for the nondimensional flow rate corresponds to the entrainment flow for a free disc as predicted using the von Kármán (1921) modeling approach, Equation 4.52. With no superimposed flow, $Cw = 0$ and hence the turbulent flow parameter $\lambda_t = 0$. Solution of Equation 5.66 with $\lambda_t = 0$ gives a value for the swirl ratio of $\beta = 0.4258$. The ratio β^* is defined as the value for the swirl ratio when the superposed flow, Cw, is zero. With no superimposed flow supplied to the cavity the core swirl ratio, β^*, is therefore 0.4258. β and β^* represent useful quantities for quantifying the swirl fraction with and without superposed flow.

If the Ekman layer assumption is made for both the rotor and the stator flow, the coefficients are both approximated by $\alpha_s = \alpha_o = \alpha(1) = 0.5533$ and $\gamma_s = \gamma_o = \gamma(1) = 0.0983$, from Table 5.1. Equation 5.61 with these coefficients becomes

$$\frac{49\pi}{60}\alpha\gamma\left[(1-\beta)^{1.6}-\beta^{0.8}\right] = 0.1395\left[(1-\beta)^{1.6}-\beta^{0.8}\right] = \frac{Cw}{Re_\phi^{0.8}x^{2.6}} \tag{5.68}$$

If the dimensionless mass flow is zero, from Equation 5.68 with $Cw = 0$, $\beta = \beta^* = 0.382$.

Consideration of Equations 5.60–5.62, 5.64, or 5.66 reveals that as the turbulent flow parameter decreases, which would occur if less cooling air is introduced into the cavity or alternatively if the rotational speed is increased, more fluid is entrained near the stator in order to satisfy disc pumping. This causes an increase in the tangential velocity of the inviscid core and an increase in the radial pressure gradient providing the centripetal forces on the core, resulting in the cavity pressure being reduced further below that of the external environment. The variation of β/β^* with $\lambda_t/x^{13/5}$ is illustrated in Figure 5.12 for both Equations 5.61 and 5.68, along with a correlation from Daily, Ernst, and Asbedian (1964), Equation 5.69, and a correlation given by Dadkhah (1989).

$$\frac{\beta}{\beta^*} = \left(1 + 12.74\frac{\lambda_t}{x^{13/5}}\right)^{-1} \tag{5.69}$$

The correlation from Daily, Ernst, and Asbedian (1964), given in Equation 5.69, was based on velocity measurements carried out at three radial locations, $x = 0.469$, 0.648, and 0.828, for a rotational Reynolds number of 6.9×10^5 and two gap ratios, $G = 0.052$ and 0.069. An additional correlation for the core tangential velocity is given by Dadkhah (1989) in Equation 5.70, based on experimental data obtained using a pitot tube at $r/b = 0.661$ for a cavity with a gap ratio of $G = 0.1$. Comparison of the analytical and empirical correlation

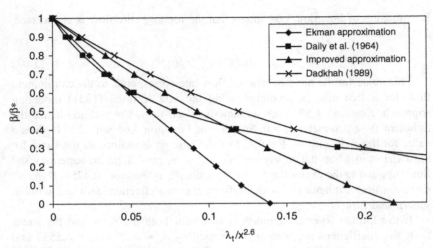

FIGURE 5.12   Variation of β/β^* with $\lambda_t/x^{13/5}$.

results shows that results from the integral momentum technique, Equation 5.61, diverge significantly from the experimental data for values of $\lambda_t/x^{13/5}$ above about 0.06, while the improved approximation, Equation 5.68, is satisfactory only for $\lambda_t/x^{13/5}$ less than approximately 0.15.

$$\frac{\beta}{\beta^*} = 0.087e^{[5.2(0.486-\lambda_t x^{-13/5})-1]} \tag{5.70}$$

Coren et al. (2009) present experimental data and a corresponding correlation for a shrouded rotor-stator system with superposed flow for a cavity using a tapered disc with a typical gap ratio of 0.1 at rotational Reynolds numbers in the range 3×10^6 to 2.5×10^7, nondimensional mass flow rate in the range 3×10^4 < Cw < 1×10^5 and values of the turbulent flow parameter, λ_t, between 0.05 and 0.5. The correlation developed for the moment coefficient for one side of a disc is given by

$$C_m = 0.26Cw^{0.37}\text{Re}_\phi^{-0.57} + 0.0014 \tag{5.71}$$

5.5. ROTOR-STATOR WITH RADIAL INFLOW

If fluid is drawn inward to the space between a rotating disc and a stationary disc at a sufficiently large rate, for example, by imposing a large pressure drop between the periphery and axis of the system, the outflow otherwise generated by disc pumping should be largely suppressed in the absence of an asymmetric unsteady boundary condition at the outer radius. Bayley and Conway (1964) investigated this effect for the case of an open rotor-stator system with a net radial inflow of superposed flow as illustrated in Figure 5.13. Moment

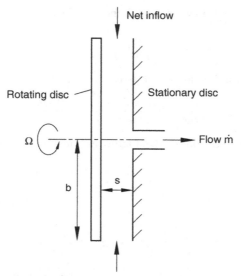

FIGURE 5.13   A rotor-stator wheelspace with a net radial inflow of superposed flow.

coefficients for the stator disc, the static pressure distribution, and velocity were measured for a range of gaps, rotational speeds, and superposed flow rates.

Increasing the flow rate or the rotational Reynolds number was found to increase the moment coefficient, whereas increasing the gap had the reverse effect. It was noted that for low-speed or high-flow rates the effect of changing the rotor-stator axial gap diminished. The pressure distribution on the stator disc was also found to be dependent on the same variables as the stator moment coefficient. An integral solution of the boundary layer equations to determine minimum superposed radial inflow to suppress the outflow due to disc pumping was undertaken.

The boundary layer equation for radial equilibrium is

$$u_r \frac{\partial u_r}{\partial r} + u_z \frac{\partial u_r}{\partial z} - \frac{u_\phi^2}{r} = -\frac{1}{\rho}\frac{\partial p}{\partial r} + \frac{\mu}{\rho}\frac{\partial^2 u_r}{\partial z^2} \qquad (5.72)$$

At the rotating disc $u_r = 0$ and $u_\phi = \Omega r$. If u_r is assumed to be constant across the region concerned, then $\partial u_r/\partial z = 0$ giving

$$-\Omega^2 r = -\frac{1}{\rho}\frac{\partial p}{\partial r} + \frac{\mu}{\rho}\frac{\partial^2 u_r}{\partial z^2} \qquad (5.73)$$

For inviscid flow between the boundary layers

$$u_r \frac{\partial u_r}{\partial r} - \frac{u_\phi^2}{r} = -\frac{1}{\rho}\frac{\partial p}{\partial r} \qquad (5.74)$$

If the quantity of fluid flowing through the boundary layers is neglected,

$$u_r = \frac{\dot{m}}{\rho A} = \frac{\dot{m}}{\rho \times 2\pi rs} \tag{5.75}$$

Substituting for u_r in Equation 5.74 gives

$$\frac{\dot{m}}{2\pi\rho rs} \times -\frac{\dot{m}}{2\pi\rho r^2 s} - \frac{u_\phi^2}{r} = -\frac{1}{\rho}\frac{\partial p}{\partial r} \tag{5.76}$$

Substituting for the pressure gradient terms in Equation 5.72 gives

$$-\Omega^2 r = -\frac{1}{r}\left(\frac{\dot{m}}{2\pi\rho rs}\right)^2 - \frac{u_\phi^2}{r} + \frac{\mu}{\rho}\frac{\partial^2 u_r}{\partial z^2} \tag{5.77}$$

Rearranging gives

$$\frac{\dot{m}}{2\pi\rho rs} = \sqrt{\Omega^2 r^2 - u_\phi^2 + \frac{\mu r}{\rho}\frac{\partial^2 u_r}{\partial z^2}} \tag{5.78}$$

$$\frac{\dot{m}}{b^2\rho\Omega s} = 2\pi\left(\frac{r}{b}\right)^2\sqrt{1 - \frac{u_\phi^2}{\Omega^2 r^2} + \frac{\mu r}{\rho\Omega^2 r^2}\frac{\partial^2 u_r}{\partial z^2}} \tag{5.79}$$

$$\frac{\dot{m}/\mu b}{(\rho\Omega b^2/\mu)(s/b)} = 2\pi\left(\frac{r}{b}\right)^2\sqrt{1 - \left(\frac{u_\phi}{\Omega r}\right)^2 + \frac{b^2}{\Omega r(\rho\Omega b^2/\mu)}\frac{\partial^2 u_r}{\partial z^2}} \tag{5.80}$$

or

$$\frac{Cw}{Re_\phi G} = 2\pi x^2\sqrt{1 - \beta^2 + \frac{b}{x\Omega Re_\phi}\frac{\partial^2 u_r}{\partial z^2}} \tag{5.81}$$

If the term $\partial^2 u_r/\partial z^2$ is assumed to always have a negative value on the rotating disc, then the maximum possible value for the terms within the square root symbol in Equation 5.81 is unity. That is, when

$$\frac{Cw}{Re_\phi G} = 2\pi \tag{5.82}$$

When $Cw = 2\pi GRe_\phi$, it is thus assumed that the radial velocity axial gradient at the periphery is either zero or has a negative value. Experiments, however, suggested a value much less, perhaps unity for the ratio Cw/GRe_ϕ, was sufficient; that is,

$$Cw = GRe_\phi \tag{5.83}$$

Cw, in the case of Equation 5.83, is the dimensionless rate of radially inward flow. The experiments were conducted for the following ranges of gap ratio, dimensionless flow, and rotational Reynolds number.

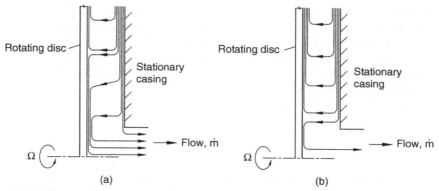

FIGURE 5.14 (a) Rotor-stator cavity with large inflow rate. (b) Rotor-stator cavity with small inflow rate.

$$0.004 \leq G \leq 0.06 \tag{5.84}$$

$$0 \leq |Cw| \leq 1 \times 10^4 \tag{5.85}$$

$$0 \leq Re_\phi \leq 4 \times 10^6 \tag{5.86}$$

Flow visualization and detailed velocity measurements using Laser Doppler Anemometry (LDA) were made by Pincombe (1989) for laminar and turbulent inflow for a shrouded rotor-stator cavity with superposed radial inflow for a cavity with $G = 0.1$ and $G_c = 0.02$, $175 < Cw < 2366$ and $Re_\phi \leq 2.5 \times 10^6$. The flow visualization data indicated a region of core flow that the inflow had difficulty penetrating. Analysis of the measurements identified the presence of a point of radial stagnation on the rotor disc that was seen to move radially with the rate of inflow, shown schematically in Figure 5.14.

Below the point of radial stagnation, the flow was radially inward, and the core tangential velocity was greater than the local disc speed. The flow field in the cavity was effectively described by three flow regimes:

- Inflow dominated, where the point of radial stagnation was at the tip of the rotor
- Rotationally dominated, where the point of radial stagnation was at the hub of the rotor
- An intermediate regime, where the flow combines both rotational and inflow effects

Inlet swirl was found to have significant effects toward the tip of the rotor, but these effects diminished toward the hub. The opposite was true for the effects of the inflow rate.

Torque measurements on rotating and stationary discs with a net radial inflow of air were investigated by Graber, Daniels, and Johnson (1987). The torque coefficient was found to tend toward a constant value on the stator disc as the flow rate was increased. The torque on the rotor disc, however, was seen to continue to reduce as more flow was supplied. The torque measured on the stator was always lower than that on the rotor. The velocity and pressure fields were seen to behave in a similar fashion to that reported by Bayley and Conway (1964) and Pincombe (1989).

Debuchy et al. (1998) and Laroche et al. (1999) present results from a combined experimental and numerical study for a rotor-stator cavity with inflow. The studies showed the critical importance of inlet conditions, in particular the tangential velocity inside the cavity in determining conditions. Poncet, Chauve, and Schiestel (2005) present results from a two-component LDA experimental and numerical investigation of a rotor-stator cavity with superposed radial inflow and outflow. The results show an increase in the core rotation with an increase in superposed inflow. Good agreement between the experimental and numerical data using a low Reynolds number second-order full-stress transport closure Reynolds stress modeling (RSM) was found. It is possible for the core flow to rotate faster than the disc if the inflow is high enough; the results from Poncet, Chauve, and Schiestel (2005) demonstate this for certain conditions.

The moment coefficient for a rotor-stator wheelspace with superposed flow for $G = 0.0276$, 0.069, and 0.124 was correlated by Daily, Ernst, and Asbedian (1964) by

$$C_m = 0.051 G^{0.1} \mathrm{Re}_\phi^{-0.2}(1 + 13.9 \beta^* \lambda_t G^{-1/8}) \tag{5.87}$$

for

$$2 \times 10^6 < \mathrm{Re}_\phi < 1 \times 10^7 \tag{5.88}$$

$$0 < \lambda_t < 0.06 \tag{5.89}$$

5.6. ROTOR-STATOR WITH STATIONARY SHROUD AND RADIAL OUTFLOW

The inclusion of a shroud at the outer radius of a rotor-stator cavity will normally act to restrict the interchange of fluid between the cavity and external environment.

Bayley and Owen (1970) investigated the impact of a shroud attached at the periphery of the stator disc, as illustrated in Figure 5.15. For rotational Reynolds numbers up to approximately 4.2×10^6, comparable to those experienced gas turbine engines, the shroud reduced the moment on the rotor for the zero superposed flow cases, when compared with an unshrouded stator, but increased the moment when a superposed outflow was supplied.

The reduced moment at no flow and increased moment when superposed flow was supplied at the axis was attributed to the relative strengths of two

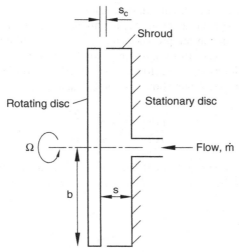

FIGURE 5.15   A shrouded rotor-stator system.

mechanisms. The first of these mechanisms was increased angular momentum conservation due to the better sealing of the cavity, and the second was increased viscous shear as the flow exits the cavity due to the narrow gap. At zero and very-low-flow superposed flow rates, the conservation of angular momentum in the cavity was dominant, but as the flow was increased the shear in the rim region at the periphery became more significant. Changes of the shroud to rotor gap, s_c, showed that for larger clearances the moment on the rotor dropped. Changing the overall gap between the rotor and stator discs, s, had little impact on disc torque. As with the unshrouded case (Bayley and Owen, 1969), the moment coefficient dropped with increasing rotational Reynolds number.

The presence of the shroud increased the pressure in the cavity above the atmospheric conditions, and the pressure inside the cavity increased as the shroud clearance ratio and rotational Reynolds numbers decreased. However, at a radius ratio below about 0.8 the radial static pressure gradient was not affected significantly by variations of the shroud clearance ratio. This suggested that the pressure could be modeled by superposition of the pressure drop in the shroud region and the pressure drop in the region with a radius ratio equal to or less than 0.8, which is unaffected by the shroud. The solution procedure adopted was addition of the nondimensional pressure drop across the shroud, ΔC_p, to the distribution occurring for $x \leq 0.8$, where the pressure drop has been defined in terms of the pressure coefficient given in Equation 5.90.

$$C_p = \frac{(p - p_a)\rho b^2}{\mu^2} \tag{5.90}$$

where p_a is the external ambient static pressure.

For small values of shroud clearance to gap ratio, that is, $G_c/G \ll 1$, where $G_c = s_c/b$, the changes in the pressure and momentum of the fluid as it leaves the wheelspace are likely to be large in comparison to viscous effects. Axial components of velocity can therefore be neglected, and the pressure gradient, assuming axisymmetric steady flow, can be calculated from

$$-\frac{1}{\rho}\frac{dp}{dr} = u_r\frac{\partial u_r}{\partial r} - \frac{u_\phi^2}{r} \qquad (5.91)$$

The radial component of velocity in the cavity for a superposed mass flow rate $\dot{m}$, supplied at the axis to the cavity, was assumed to be represented as an average value by

$$u_r = \frac{k_1\dot{m}}{2\pi\rho rs} \qquad (5.92)$$

and at the outlet of the shroud by

$$u_r = \frac{k_1\dot{m}}{2\pi\rho bs_c} \qquad (5.93)$$

The tangential component of velocity was taken as

$$u_\phi = k_2\Omega r \qquad (5.94)$$

where k_1 and k_2 are constants.

Equation 5.91 can be integrated to give a relationship between the pressure drop, superposed mass flow, and rotational Reynolds number.

From Equation 5.92,

$$\frac{\partial u_r}{\partial r} = -\frac{k_1\dot{m}}{2\pi\rho r^2 s} \qquad (5.95)$$

Integration of Equation 5.91 between r and b gives

$$-\frac{1}{\rho}\int_r^b \frac{dp}{dr}dr = \int_r^b \left(\frac{k_1\dot{m}}{2\pi\rho rs} \times -\frac{k_1\dot{m}}{2\pi\rho r^2 s}\right)dr - \int_r^b \frac{(k_2\Omega r)^2}{r}dr \qquad (5.96)$$

At a radius b, the pressure is equal to p_a:

$$-\frac{1}{\rho}(p_a-p) = \int_r^b \left(\left(\frac{k_1\dot{m}}{2\pi\rho s}\right)^2 r^{-3}\right)dr - \left[\frac{(k_2\Omega)^2 r^2}{2}\right]_r^b \qquad (5.97)$$

Recognizing that $s = s_c$ at $r = b$,

$$\frac{1}{\rho}(p-p_a) = \frac{1}{2}\left(\frac{k_1\dot{m}}{2\pi\rho}\right)^2\left(\frac{1}{s_c^2 b^2} - \frac{1}{s^2 r^2}\right) - \frac{(k_2\Omega)^2}{2}(b^2 - r^2) \tag{5.98}$$

Multiplying Equation 5.98 through by $\rho^2 b^2/\mu^2$, aiming toward presenting the equation in terms of recognizable dimensionless groups, gives

$$\frac{(p-p_a)\rho b^2}{\mu^2} = \frac{1}{2}\left(\frac{k_1\dot{m}}{2\pi\mu}\right)^2\frac{1}{b^2}\left(\frac{b^2}{s_c^2} - \frac{b^2}{r^2 s^2}\right) - \frac{1}{2}\left(\frac{k_2\rho\Omega}{\mu}\right)^2 b^2(b^2 - r^2) \tag{5.99}$$

$$\frac{(p-p_a)\rho b^2}{\mu^2} = \frac{1}{2}\left(\frac{k_1\dot{m}}{2\pi\mu}\right)^2\frac{1}{b^2}\left(\frac{b^2}{s_c^2} - \frac{b^2}{r^2 s^2}\right) - \frac{1}{2}\left(\frac{k_2\rho\Omega}{\mu}\right)^2 b^4\left(1 - \frac{r^2}{b^2}\right) \tag{5.100}$$

$$\frac{(p-p_a)\rho b^2}{\mu^2} = \frac{1}{2}\left(\frac{k_1\dot{m}}{2\pi\mu b}\right)^2\left(\frac{b^2}{s_c^2} - \frac{b^2}{r^2 s^2}\right) - \frac{1}{2}\left(\frac{k_2\rho\Omega b^2}{\mu}\right)^2\left(1 - \frac{r^2}{b^2}\right) \tag{5.101}$$

Substituting for the nondimensional mass flow rate, Cw, the gap ratios G_c and G, and the rotational Reynolds number, Re_ϕ, gives:

$$\Delta C_p = \frac{1}{2}\left(\frac{k_1 Cw}{2\pi}\right)^2\left[\frac{1}{G_c^2} - \left(\frac{b}{r}\right)^2\frac{1}{G^2}\right] - \frac{1}{2}k_2^2 Re_\phi^2\left[1 - \left(\frac{r}{b}\right)^2\right] \tag{5.102}$$

or

$$\Delta C_p = ACw^2 - BRe_\phi^2 \tag{5.103}$$

where

$$A = \frac{k_1^2}{8\pi^2}\left[\frac{1}{G_c^2} - \left(\frac{b}{r}\right)^2\frac{1}{G^2}\right] \tag{5.104}$$

and

$$B = \frac{1}{2}k_2^2\left[1 - \left(\frac{r}{b}\right)^2\right] \tag{5.105}$$

Values for the constants k_1 and k_2 were found for the range of geometry tested from the experimental data to be in range of $0.58 < k_1 < 0.66$ and $0.17 < k_2 < 0.21$ for $G_c = 0.0067$ and $r/b = 0.864$.

Examination of Equation 5.102 shows that if the rotational Reynolds number becomes high enough, then the pressure in the cavity will become negative relative to the external environment. This does not, however, indicate that this condition represents the onset of inflow through the shroud clearance, given the assumptions inherent to the model and the selection of an arbitrary radius ratio of 0.864 to define the nondimensional pressure difference. Setting $\Delta C_p = 0$ gives

$$Cw = \sqrt{\frac{B}{A}G_cRe_\phi}$$ (5.106)

Using the mean values for k_1 and k_2 gives the minimum nondimensional superposed flow rate required, Cw_{min}, to be supplied at the axis to prevent ingestion into the wheelspace as

$$Cw_{min} = 0.86G_cRe_\phi$$ (5.107)

An experimental study was undertaken by Bayley and Owen (1970) to determine the minimum coolant flow requirements to fully seal the cavity by measuring the pressure difference across the shroud. When the pressure difference between a location just under the shroud and the external environment was zero, they assumed ingress was suppressed. The minimum coolant flow necessary to seal against ingestion, Cw_{min}, was found to be strongly dependent on the shroud clearance, independent of gap ratio and well predicted by the empirical relationship given in Equation 5.108, using the same form as Equations 5.106 and 5.107.

$$Cw_{min} = 0.61G_cRe_\phi$$ (5.108)

for $G_c \ll G$, gap clearance ratios of $G_c = 0.0033$ and 0.0067 and gap ratios, $G = 0.06$, 0.12, and 0.18. The range of rotational Reynolds number was approximately between 4×10^5 and 4.2×10^6.

The difference in the constants between Equations 5.107 and 5.108 is a result of the arbitrary definition for ΔC_p. Although based on a relatively simple model, Equation 5.108 provides a reasonable estimate for the necessary superposed flow rate to prevent ingress for an axial clearance seal rotor-stator cavity in the absence of significant external asymmetries. It should be noted that this model does not take account of the rotational speed of the inviscid core reducing as Cw increases. As the core speed decreases, the radial pressure gradient in the cavity will also decrease, and thus less flow would be required to seal the system than is indicated by Equation 5.108. Owen and Rogers (1989) note that core rotation for a free disc is suppressed when the superposed flow rate is greater than the flow that would be entrained by the equivalent free disc, that is,

$$Cw > Cw_{free\ disc}$$ (5.109)

The flow entrained by a free disc, for turbulent flow, is given in Equation 4.52. Using this value in Equation 5.108 provides an indicative limitation for the validity of Equation 5.108, that is,

$$G_cRe_\phi^{-0.2} < 0.359$$ (5.110)

Example 5.3

For a shrouded rotor-stator cavity with radius 0.25 m, rotor speed 10,000 rpm, with a gap between the rotor and stator of 0.012 m, and a gap between the shroud and rotor of 1 mm, determine the flow rate necessary to be supplied to the cavity to prevent ingress. The density and viscosity for the air in the cavity can be taken as 1.55 kg/m^3 and 3.87×10^{-5} Pa s, respectively.

Solution

$$G_c = s_c/b = 0.001/0.25 = 0.004$$

$$Re_\phi = \frac{\rho \Omega b^2}{\mu} = \frac{1.55 \times 1047 \times 0.25^2}{3.87 \times 10^{-5}} = 2.621 \times 10^6$$

Checking the range of validity using (5.110),

$$G_c Re_\phi^{-0.2} = 0.004(2.621 \times 10^6)^{-0.2} = 2.081 \times 10^{-4}$$

This is much less than 0.359, so from Equation 5.108

$$Cw_{min} = 0.61 G_c Re_\phi = 0.61 \times 0.004 \times 2.621 \times 10^6 = 6395$$

The mass flow required to prevent ingress is therefore given by

$$\dot{m} = \mu \rho Cw_{min} = 3.87 \times 10^{-5} \times 1.55 \times 6395 = 0.3836 \; kg/s$$

Owen and Phadke (1980) extended the range of shroud clearances considered by Bayley and Owen (1970) using both flow visualization and pressure measurements to determine the minimum superposed flow rate to suppress ingress for five clearance ratios in the range $0.0025 \leq G_c \leq 0.04$ for a gap clearance ratio of $G = 1$. For rotational Reynolds numbers less than $1 \times 10^{6,}$ the correlation given in Equation 5.111 was obtained.

$$Cw_{min} = 0.14 G_c^{0.66} Re_\phi \qquad (5.111)$$

Phadke and Owen (1982), investigating the effects of the shroud configuration on the minimum coolant flow requirements to prevent ingress, found that radial clearance seals performed better than axial clearance seals and that the radial seal performed better for larger overlaps. The radial clearance seal investigated was found to be less sensitive to the shroud clearance. It was also observed that for radial seals the pressure in the cavity increased with rotor speed rather than decreased as with the plain axial seal, provided Cw was considerably greater than Cw_{min}.

Rim seals with no external flow have also been investigated by Phadke and Owen (1988, part 1), Bhavani et al. (1992a, 1992b), Chew, Dadkhah, and

Turner (1992), and Khilnani et al. (1994). Chew (1991) proposed the model given in Equation 5.112 to predict the onset of ingestion for a wheelspace in the absence of external flow.

$$Cw_{\min} = 2\pi K_L \left(\frac{u_{dm}}{\Omega b}\right) G_c Re_{\phi} \qquad (5.112)$$

where K_L is a loss coefficient and u_{dm} is the mixed out radial velocity for near wall flow at the cavity outer radius. It is assumed that the fluid flowing out from the seal will have first been entrained into the boundary layer on the rotating disc and will have attained a radial velocity. The mixed out velocity u_{dm} is used to quantify this and is calculated from the integral solution of the boundary layer at $r = b$ such that radial momentum is conserved. It is assumed that the seal flow is supplied from near the wall region of the boundary layer and averaging is carried out only for the portion of the boundary layer supplying flow to the seal.

5.7. ROTOR-STATOR WITH STATIONARY SHROUD AND EXTERNAL FLOW

The cases detailed for a rotor-stator wheelspace with a shroud have typically been for the idealized condition where the circumferential boundary condition at the shroud is axisymmetric. In a gas turbine engine with axial blading, for example, there is a complicated mainstream flow above the seal gap with flow guided from stationary nozzle guide vanes on to rotating blades mounted on a rotating disc. The combined effects of boundary layer formation, wake interactions, rotation, and pressure gradients result in secondary flows, with passage cross-flows and endwall effects as illustrated in Figure 5.16 (see Langston, 1980; Gostelow, 1984).

Some clearance between rotating and stationary components is necessary in order to enable relative rotation and avoid excessive wear of components. The cavity formed between the stationary disc supporting the nozzle guide vanes and the rotor disc may be provided with a supply of cooling air in order to limit the ingestion into the cavity of hot mainstream annulus gas, which otherwise could cause overheating of the turbine discs. Typically, some form of peripheral rim seal is provided to limit the flow ingress or egress and limit mainstream annulus flow spoiling due to interactions with flow in the rim seal region. The basic configuration for a rotor-stator wheelspace with external flow is illustrated in Figure 5.17.

As noted, the flow in an axial bladed turbomachinery stage is highly complex, with secondary flows superposed on the mainstream flow path and unsteady effects. The flow conditions are not axisymmetric with, for instance, pressure and velocity varying with circumferential location and unsteady inter-actions between one blade row and another. Such asymmetries can significantly affect the conditions in the wheelspace and the associated ingestion. A large range of configurations are possible for rim seals, depending on a combination of flow and mechanical requirements as listed below and illustrated in Figure 5.18.

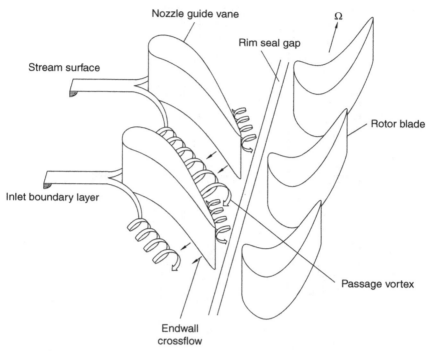

FIGURE 5.16   Schematic representation of the mainstream annulus cross-flow in a turbine passage.

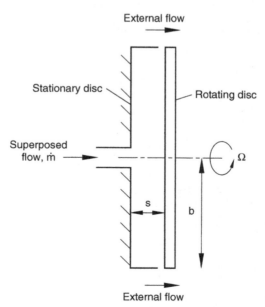

FIGURE 5.17   A rotor-stator wheelspace with external flow.

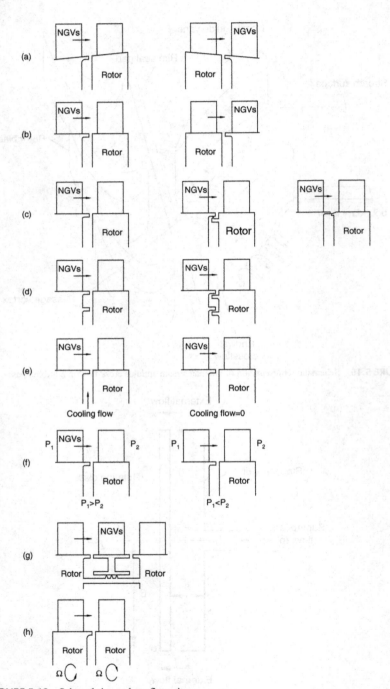

FIGURE 5.18 Selected rim seal configurations.

a. Hade angle: positive hade; negative hade; constant annulus line
b. Location of the rim seal relative to the rotor: upstream; downstream
c. Seal type: for example, axial; radial; shingling
d. Additional inner seal
e. Supply of cooling air; $Cw > 0$, $Cw = 0$
f. Pressure gradient: turbine, $p_1 > p_2$; compressor $p_1 < p_2$
g. Inner drive arm: e.g., turbine or compressor stator well
h. Contra-rotation

   The possible combinations identified in this list alone result in 576 options, indicating the challenge associated with rim seal design and quantifying ingress.
   Phadke and Owen (1988, Parts 2 and 3) investigated the influence of pressure asymmetries in an external flow on sealing performance for a range of axial and radial clearance seals and a mitred seal. In terms of ingestion into the rotor-stator wheelspace, a rotationally dominated flow regime was identified as well as an external flow-dominated regime. In the rotationally dominated flow regime, where the pressure of the rotating fluid in the rotor-stator cavity can drop below that of the external flow, the minimum flow required to prevent ingress was found to be related to the gap clearance ratio and rotational Reynolds number by

$$Cw_{min} \propto G_c^m Re_\phi^n \qquad (5.113)$$

   For the external flow-dominated regime, a simple model based on use of a discharge coefficient was developed.

$$Cw = 2\sqrt{2}\pi C_d G_c \sqrt{\frac{\rho \Delta p_s b^2}{\mu^2}} \qquad (5.114)$$

where Δp_s is the circumferentially averaged pressure drop across the shroud.
   The minimum flow required to prevent ingress

$$Cw_{min} = 2\pi K G_c \sqrt{p_{max}} \qquad (5.115)$$

where

$$K = \sqrt{2C_{rs}C_d} \qquad (5.116)$$

$$p_{max} = \frac{\rho \Delta p_{max} b^2}{\mu^2} = \frac{1}{2} C_{p,max} Re_z^2 \qquad (5.117)$$

and C_{rs} is an empirical constant, $Re_z = \rho \bar{u}_z b / \mu$ is an external flow Reynolds number, and $\bar{u}_z$ is the average axial component of velocity for the external flow. For the axial clearance seals and one of the radial clearance seals tested, the results correlated reasonably well with a value of $K = 0.6$ for $G_c = 0.005$, 0.01, and 0.02, $Re_\phi = 0$ and $Re_z \leq 1.1 \times 10^6$.

Campbell (1978) found that the flow requirement for rim sealing in a gas turbine engine was determined by the mainstream flow and circumferential pressure variations rather than by rotational effects. Abe, Kikuchi, and Takeuchi (1979), Bohn et al. (1995), Dadkhah, Turner, and Chew (1992), Hamabe and Ishida (1992), and Ishida and Hamabe (1997) demonstrated the influence of external flow on ingestion by means of nozzle guide vanes. Studies extending the geometry to a full stage were undertaken by Green and Turner (1992), Hills, Chew, and Turner (2001), Roy et al. (1999, 2000, 2001, 2005), Feiereisen, Paolillo, and Wagner (2000), Gallier, Lawless, and Fleeter (2000), and Bohn et al. (1999, 2000, 2003). Chew, Green, and Turner (1994) showed that the pressure asymmetry decayed axially downstream of the nozzle guide vanes. A potential flow approximation of the annulus pressure variation used by Chew, Green, and Turner (1994) and Hills et al. (1997) gives the first harmonic of the pressure asymmetry decay as

$$e^{-\xi z} \qquad (5.118)$$

where

$$\xi = \frac{N}{r} \frac{\sqrt{1-M^2}}{1-M_z^2} \qquad (5.119)$$

where ξ is a nondimensional decay rate, N is the number of blades, M is the Mach number, and M_z is the axial Mach number.

Daniels et al. (1992) performed experiments with an axisymmetric external flow and found that the external flow swirl level did not effect ingestion and that the amount of superposed cooling flow required to prevent ingestion was independent of rotational speed. This result contrasted with that of previous experience for rim seals in the absence of external flow where dependence, for the minimum flow required to prevent ingress, on the rotational Reynolds number had been identified. Kobayashi, Matsumato, and Shizuya (1984) also demonstrated a negligible influence of rotational speed on ingestion for a rim seal with external flow. The physical mechanisms involved in turbine rim seal ingestion are described in the review paper by Johnson et al. (1994). Okita et al. (2005) show the use of insight to rim sealing flow characteristics in the demonstration of a combination seal arrangement for minimizing ingestion.

The insights developed in the investigations reported reveal that the flow associated with a rim seal is three-dimensional and time dependent. The numerical methods and computational power needed to undertake such calculations for a complete turbine stage are now available. Jakoby et al. (2004) reported the findings from an investigation with unsteady cavity pressure measurements and the requirements to produce sufficiently accurate numerical predictions for both a forward- and backward-facing axial-bladed turbine cavity. A large-scale low-frequency pulsating rotating structure was identified

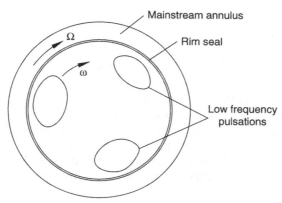

FIGURE 5.19   Rotor-stator cavity pressure distribution (Jakoby et al., 2004)

inside the forward rotor-stator cavity for certain conditions, comprising three low-pressure bubbles rotating at about 80% of the rotor speed with a pattern comparable to a "Mercedes star" (Figure 5.19).

The periodic flow pattern significantly influences the pressure distribution inside the cavity and the sealing effectiveness with ingestion following the low-pressure rotating structure. It was concluded that the low-frequency pulsations were not caused by the interaction of the vanes and rotating blades as the frequency of the pulsations (406 Hz for $Cw = 7000$) was an order of magnitude less than the blade-passing frequency (≈ 4.8 kHz). The pulsations disappeared if the superposed cooling exceeded a certain limit. In order to capture the unsteady phenomena involved, a full 360° CFD model was used. Similar flow structures are observed in rotating cavities, considered in Chapter 7, where simultaneous radial inflow and outflow in a rotating fluid requires Coriolis forces of opposite sign, a condition generated by alternate regions of cyclonic and anticyclonic circulation.

Although numerical modeling of the full three-dimensional time-dependent flow is possible, it is, at the time of writing, computationally expensive and the model set up requires significant expertise and insight. An alternative model approach is use of simpler techniques to capture the most significant aspects of the flow physics. A number of these techniques have been developed, modeling the egress and ingress flows around the circumference as a simple orifice flow with the flow moderated by discharge coefficients (e.g., Chew, Green, and Turner, 1994; Bayley and Childs, 1997; Reichert and Lieser, 1999; Hills, Chew, and Turner, 2001; Scanlon et al., 2004; Owen, 2009a, 2009b; Owen et al., 2010a, 2010b).

The model developed by Scanlon et al. (2004) provides a means to estimate the circumferential pressure variation for the mainstream turbine annulus; using this model, the flow ingested into a rim seal cavity can be determined. Assuming the flow to be one-dimensional, incompressible, and isentropic and the cavity pressure uniform, the mass flow through the annular gap modeled as a simple orifice is given by

$$\int d\dot{m} = C_d\sqrt{\frac{2p_{cav}}{RT}}\int\sqrt{\Delta p}\,dA_{gap} \tag{5.120}$$

where p_{cav} is the rotor-stator cavity pressure.

In order to determine the pressure drop at every point around the circumference of the rim seal, an assumption needs to be made about the form of the pressure asymmetry. Typical data for a representative rim seal showed that a parabolic profile for the annulus pressure asymmetry provided a reasonable approximation to available data. The pressure drop driving ingestion can therefore be represented by

$$\Delta p = (\phi_*^2 - C_p)(p_{max} - p_{min}) \tag{5.121}$$

where ϕ_* is the nondimensional angular position and C_p is a pressure coefficient.

$$\phi_* = \frac{N\phi}{\pi} - 1 \tag{5.122}$$

for $0 \leq \phi_* \leq 2\pi/N$ and

$$C_p = \frac{p_{cav} - p_{min}}{p_{max} - p_{min}} \tag{5.123}$$

p_{max} and p_{min} are the maximum and minimum annulus static pressure, respectively.

The mass flow into and out of the rotor-stator wheelspace can be estimated from Equations 5.124 and 5.125, respectively.

$$\int d\dot{m}_{out} = C_d A_{gap}\sqrt{\frac{2p_{cav}(p_{max}-p_{min})}{RT}}\int_0^{\sqrt{C_p}}\sqrt{(C_p - \phi_*^2)}\,d\phi_* \tag{5.124}$$

$$\int d\dot{m}_{in} = C_d A_{gap}\sqrt{\frac{2p_{cav}(p_{max}-p_{min})}{RT}}\int_{\sqrt{C_p}}^1\sqrt{(\phi_*^2 - C_p)}\,d\phi_* \tag{5.125}$$

The egress and ingress mass flows are given by Equations 5.126 and 5.127, respectively.

$$\dot{m}_{out} = C_d A_{gap}\sqrt{\frac{2p_{cav}(p_{max}-p_{min})}{RT}}\left(\frac{C_p\pi}{4}\right) \tag{5.126}$$

$$\dot{m}_{in} = C_d A_{gap}\sqrt{\frac{2p_{cav}(p_{max}-p_{min})}{RT}}\left[\frac{1}{2}\left(\sqrt{1-C_p} - C_p\cosh^{-1}\frac{1}{\sqrt{C_p}}\right)\right] \tag{5.127}$$

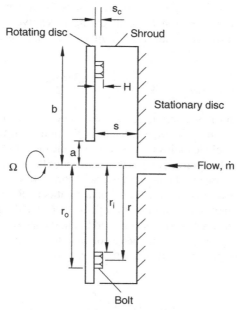

FIGURE 5.20   Rotor-stator wheelspace with protruding bolts.

The model works well for the majority of test data available but less well where unsteady pressure fluctuations were measured in the cavity, suggesting that these fluctuations when present are significant in determining the level of ingestion.

5.8.   STATIC AND ROTATING PROTRUSIONS

Many practical rotating disc applications involve protrusions and features, such as bolts and hooks, attached to either the stationary disc or the rotating disc in a rotor-stator cavity. The protrusions can significantly influence the local flow and heat transfer in the vicinity of the features, and as a result the power required to overcome viscous drag. A typical rotating disc configuration with fixing bolts is illustrated in Figure 5.20.

Zimmermann et al. (1986) subdivided the additional losses associated with protrusions into three categories:

- form drag
- boundary layer losses
- pumping losses

The tangential velocity of the bolts on the disc is larger than the speed of the fluid in the core in the rotor-stator gap, giving a relative speed differential between the bolts and the fluid. This results in a drag loss, which can be determined by

$$\Delta C_{m,form} = N C_D (1-\beta^2)\left(\frac{r}{b}\right)^3 \frac{H d_{bolt}}{b^2} K_f \qquad (5.128)$$

where N is the number of bolts, C_D is the drag coefficient of the bolts in the undisturbed flow, β is the core rotation factor in the middle of the cavity (i.e., at $z = s/2$), r is the pitch circle radius for the bolts, b is the outer radius of the disc, H is the height of the bolts or cover, d_{bolt} is the diameter of the bolts, and K_f is a form drag correction factor due to the interference with the wakes from adjacent bolts (Taniguchi, Sakamoto, and Arie, 1982). For a rotational Reynolds number of 4×10^6 and a nondimensional throughflow of $Cw = 2.6 \times 10^4$, the swirl ratio was assumed to be 0.42 based on test evidence and 0.6 in the absence of throughflow.

The boundary layer losses can be accounted for using the von Kármán relationship for turbulent flow over a rotating ring of inner radius r_i and outer radius r_o:

$$C_m = 0.073 \mathrm{Re}_\phi^{-0.2}\left(\frac{r_o}{b}\right)^{4.6}\left[1-\left(\frac{r_i}{r_o}\right)^{5.7}\right]^{0.8} \qquad (5.129)$$

If the bolts are assumed to act as the blades for a radial compressor, the pumping losses can be modeled using the Euler turbomachine equation,

$$Power = \Omega \dot{m}\left(u_{\phi,2} r_o - u_{\phi,1} r_i\right) \qquad (5.130)$$

where $\dot{m}$ is the mass flow pumped by the bolt heads, $u_{\phi,1}$ is the tangential velocity of the bolt-driven fluid at the inner bolt radius, and $u_{\phi,2}$ is the tangential velocity of the bolt-driven fluid at the outer bolt radius.

The total power required to overcome frictional drag can be determined by addition of the power loss for each of the three components listed to that of the power required to overcome frictional drag for a plain disc of the same diameter concerned.

Millward and Robinson (1989 performed a series of experiments to quantify the effect of both static and rotating protrusions and recesses. For a range of conditions with $Re_\phi \leq 1 \times 10^7$ and $Cw \leq 1 \times 10^4$ and a range of pitch to diameter ratios between 3 and 20, they developed the following correlation to account for the additional frictional drag over and above that of the plain disc.

$$C_m = 2.3\left(\frac{\dot{m}/\mu r}{\rho \Omega r^2/\mu}\right)^{(1.4-r/3a)}\left(\frac{l_{pitch}}{d_{bolt}}\right)^{0.44}\frac{r^3 NA}{b^5} \qquad (5.131)$$

where r is the radius of the bolt pitch circle, l_{pitch} is the pitch of the protrusions, d_{bolt} is the diameter of the bolts, N is the number of the bolts, and A is the projected cross-sectional area of the protrusion, a is the inner radius of the rotating disc, and b is the outer radius of the disc.

This equation can be restated in terms of the local nondimensional flow rate and local rotational Reynolds number as

$$C_m = 2.3 \left(\frac{Cw_{local}}{Re_{\phi,local}} \right)^{(1.4 - r/3a)} \left(\frac{l_{pitch}}{d_{bolt}} \right)^{0.44} \frac{r^3 NA}{b^5} \tag{5.132}$$

The recommended limits for the correlation (Equation 5.131 or 5.132) are:

$$1.5 < \frac{r}{a} < 2.25 \tag{5.133}$$

$$0.001 < \frac{Cw_{local}}{Re_{\phi,local}} < 0.1 \tag{5.134}$$

$$3 < \frac{l_{pitch}}{d_{bolt}} < 20 \tag{5.135}$$

Example 5.4.

Determine the windage for a disc of outer radius 0.3 m rotating at 10,500 rpm in air with a density and viscosity of 4 kg/m^3 and 3×10^{-5} Pa s, respectively. Twenty bolts with a width of 17 mm across the flats are located on a pitch circle radius of 0.2 m, and height of the bolt heads is 7.18 mm. The disc inner radius is 0.1 m. The disc is in a rotor-stator wheelspace, with a gap of 0.14 m between the rotor and stator discs, and the cavity is supplied with 200 g/s of air.

Solution

The circular pitch of the bolts is given by

$$l_{pitch} = \frac{2\pi r}{N} = \frac{2\pi \times 0.2}{20} = 0.06283 \ m$$

The projected cross-sectional area of the bolt depends on the orientation of the bolts. If the bolts are aligned to present the minimum target area, then

$$A = d_{bolt} \times H = 0.017 \times 0.00718 = 1.221 \times 10^{-4} \ m$$

The local nondimensional mass flow and rotational Reynolds numbers at r are:

$$Cw_{local} = \frac{\dot{m}}{\mu r} = \frac{0.2}{3 \times 10^{-5} \times 0.2} = 3.333 \times 10^4$$

$$Re_{local} = \frac{\rho \Omega r^2}{\mu} = \frac{4 \times 1100 \times 0.2^2}{3 \times 10^{-5}} = 5.867 \times 10^6$$

From Equation 5.132,

$$C_m = 2.3\left(\frac{3.33\times10^4}{5.867\times10^6}\right)^{(1.4-(0.2/3\times0.1))}\left(\frac{0.06283}{0.017}\right)^{0.44}\frac{0.2^3\times20\times1.221\times10^{-4}}{0.3^5}$$
$$= 7.412\times10^{-4}$$

The power required to overcome frictional drag due to the bolts is given by

$$Power = T_q\Omega = 0.5\rho\Omega^3 b^5 C_m = 0.5\times4\times1100^3\times0.3^5\times7.412\times10^{-4} = 4794\ W$$

For the rotating disc face,

$$Re_\phi = \frac{\rho\Omega b^2}{\mu} = \frac{4\times1100\times0.3^2}{3\times10^{-5}} = 1.320\times10^7$$

$$Cw = \frac{\dot{m}}{\mu b} = \frac{0.2}{3\times10^{-5}\times0.3} = 2.222\times10^4$$

$$\lambda_t = \frac{Cw}{Re_\phi^{0.8}} = \frac{2.222\times10^4}{(1.32\times10^7)^{0.8}} = 4.470\times10^{-2}$$

If, in the absence of superposed flow, the inviscid core is assumed to be rotating with a swirl fraction of $\beta^* = 0.42$, then the moment for the rotating disc, using Equation 5.87, is given by

$$C_m = 0.051G^{0.1}Re_\phi^{-0.2}(1+13.9\beta^*\lambda_t G^{-1/8}) = 0.002291$$

The power required to overcome frictional drag due to the disc face is given by

$$Power = T_q\Omega = 0.5\rho\Omega^3 b^5 C_m = 0.5\times4\times1100^3\times0.3^5\times0.002291 = 14820\ W$$

The total power required to overcome the frictional drag for both the disc face and the bolts is therefore 4.794 kW + 14.82 kW= 19.61 kW, with the bolts in this case responsible for about 24% of the windage.

5.9. THRUST ON A DISC

A frequent requirement in rotating machinery applications is the evaluation of the net thrust on a rotating component such as an impeller in order to specify an appropriate bearing solution. The schematic in Figure 5.21 illustrates the cross section of a centrifugal compressor. The cavity formed between the casing and the rear face of the impeller can be approximated by an enclosed rotor-stator cavity.

As described in Section 5.2, there will be thin boundary layers on both the rotor and stator disc faces. Disc pumping radial outflow will occur for the rotor and radial inflow for the stator. The region between the boundary layers is referred to as the inviscid core and can be considered to rotate with solid body rotation at a percentage of the impeller speed, Ω. From Equation 4.3 for axisymmetric flow, steady flow

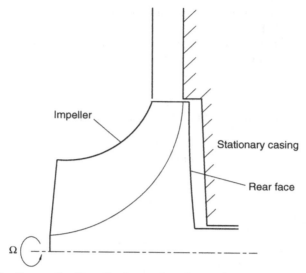

FIGURE 5.21   Cross section illustrating the rear face of a centrifugal compressor

$$\frac{\partial p}{\partial r} = \rho \frac{u_\phi^2}{r} - \rho u_r \frac{\partial u_r}{\partial r} - \rho u_z \frac{\partial u_r}{\partial z} + \mu \left(\nabla^2 u_r - \frac{u_r}{r^2} \right) \qquad (5.136)$$

The radial velocity is negligible if the core rotation controls the radial pressure gradient, the axial velocity is approximately zero in a rotor-stator system, and the viscous terms are negligible outside the boundary layers. Therefore, Equation 5.136 reduces to

$$\frac{\partial p}{\partial r} = \rho \frac{u_\phi^2}{r} \qquad (5.137)$$

or

$$\frac{\partial p}{\partial r} = \rho \omega^2 r \qquad (5.138)$$

where ω refers to the angular velocity of the core region. Usually, as referred to in Section 5.2, the speed of rotation of the core is approximately 40% that of the rotor, that is, $\omega = 0.4\Omega$. In general, the core rotation decreases with a net radial outflow and increases with radial inflow, with the angular velocity of inflow entering the cavity being a very important parameter.

The pressure field can be determined from Equation 5.138. Depending on conditions, the flow can be modeled as isothermal, isentropic, or incompressible. These different assumptions are developed below and followed by an example for the evaluation of the thrust on the backface of a centrifugal compressor impeller.

If isothermal flow is assumed, T = constant, and the density can be found from the ideal gas law

$$\rho = \frac{p}{RT} \tag{5.139}$$

Substitution of Equation 5.139 in Equation 5.138 gives

$$\frac{\partial p}{\partial r} = \frac{p}{RT}\omega^2 r \tag{5.140}$$

$$\frac{\partial p}{p} = \frac{\omega^2}{RT}r\partial r \tag{5.141}$$

Integration gives

$$\frac{p_2}{p_1} = e^{\left[\frac{\omega^2}{2RT}(r_2^2 - r_1^2)\right]} \tag{5.142}$$

Alternatively, if isentropic flow is assumed,

$$p \propto \rho^\gamma \tag{5.143}$$

where γ is the isentropic index, $\gamma = c_p/c_v$

$$p = k_3\rho^\gamma \tag{5.144}$$

$$\rho = \left(\frac{p}{k_3}\right)^{1/\gamma} \tag{5.145}$$

k_3 can be determined from reference conditions
From Equation 5.138,

$$\partial p = \rho\omega^2 r\partial r \tag{5.146}$$

$$\partial p = \frac{p^{\frac{1}{\gamma}}}{k_3^{\frac{1}{\gamma}}}\omega^2 r\partial r \tag{5.147}$$

So

$$p^{-\frac{1}{\gamma}}\partial p = k_4 r\partial r \tag{5.148}$$

where

$$k_4 = \frac{\omega^2}{k_3^{1/\gamma}} \tag{5.149}$$

Integration of Equation 5.148 gives

$$\left[\frac{p^{1-\frac{1}{\gamma}}}{1-\frac{1}{\gamma}}\right]_1^2 = \frac{k_4}{2}[r^2]_1^2 \tag{5.150}$$

$$(p_2-p_1)^{(\gamma-1)/\gamma} = \left(\frac{\gamma-1}{\gamma}\right)\frac{k_4}{2}(r_2^2-r_1^2) \tag{5.151}$$

$$p_2-p_1 = \left[\left(\frac{\gamma-1}{\gamma}\right)\frac{k_4}{2}(r_2^2-r_1^2)\right]^{\gamma/(\gamma-1)} \tag{5.152}$$

If the flow is assumed to be incompressible, ρ = constant, from Equation 5.138,

$$\partial p = \rho\omega^2 r\partial r \tag{5.153}$$

$$k_5 = \rho\omega^2 = \rho\beta^2\Omega^2 \tag{5.154}$$

$$[p]_1^2 = \frac{k_5}{2}(r_2^2-r_1^2) \tag{5.155}$$

$$p_2 = p_1 + \frac{k_5}{2}(r_2^2-r_1^2) \tag{5.156}$$

The above relationship, Equation 5.156, can be used to characterize the local pressure, p at radius r, in terms of a known pressure p_1, or a known pressure p_2, that is,

$$p = p_1 + \frac{k_5}{2}(r^2-r_1^2) \tag{5.157}$$

or

$$p = p_2 - \frac{k_5}{2}(r_2^2-r^2) \tag{5.158}$$

The force on an elemental area δA, is $\delta F = p\delta A = p \times 2\pi r\delta r$. If the outer pressure p_2 is known, then for a disc of inner radius a and outer radius b, assuming incompressible flow,

$$F = \int_a^b\left(p_2-\frac{\rho\beta^2\Omega^2}{2}(b^2-r^2)\right)2\pi rdr \tag{5.159}$$

$$F = \int_a^b(2\pi rp_2-\pi\rho\beta^2\Omega^2 rb^2 + \pi\rho\beta^2\Omega^2 r^3)dr \tag{5.160}$$

$$F = \left[\pi p_2 r^2-\pi\rho\beta^2\Omega^2 b^2\frac{r^2}{2} + \pi\rho\beta^2\Omega^2\frac{r^4}{4}\right]_a^b \tag{5.161}$$

$$F = \pi p_2(b^2-a^2)-\frac{\pi\rho\beta^2\Omega^2 b^2}{2}(b^2-a^2) + \frac{\pi\rho\beta^2\Omega^2}{4}(b^4-a^4) \tag{5.162}$$

In practice, for many practical applications it does not make much of a difference which of the above relationships is used to determine the pressure as a function of radius. From knowledge of the local pressure, the thrust on an elemental annular surface can be determined and this can be summed for the overall radius concerned to give the total thrust.

Example 5.5.

Determine the thrust on the backface of a centrifugal impeller if the pressure at the exducer tip is 2.5 bar. The impeller spins at 90,000 rpm and has a diameter of 140 mm. The shaft diameter is 15 mm. The density can be taken as 2 kg/m³.

Solution

From Equation 5.162, assuming incompressible flow and a swirl ratio for the cavity of 0.4, the thrust is given by

$$F = \pi 2.5 \times 10^5 (0.07^2 - 0.0075^2) - \frac{\pi 2 \times 0.4^2 9424^2 0.07^2}{2}(0.07^2 - 0.0075^2)$$

$$+ \frac{\pi 2 \times 0.4^2 9424^2}{4}(0.07^2 - 0.0075^4)$$

$$F = 3281 \text{ N}$$

5.10. CONCLUSIONS

An overview of the flow associated with a single disc rotating with proximity to a stationary casing has been given. The flow structures are typically nontrivial with complex interactions and different behavior, depending on whether the disc boundary layers are separate, the presence of a shroud at the outer periphery, and the supply or extraction of fluid from the cavity. Design-oriented correlations for bulk parameters such as boundary layer thickness, mass flow, and moment coefficients have been presented for most of the geometrical conditions considered. The insight resulting from the descriptions, along with the correlations, provide a basis for modeling disc flows, undertaking preliminary design calculations, and validating CFD models.

REFERENCES

Abe, T., Kikuchi, J., and Takeuchi, H. An investigation of turbine disc cooling (Experimental investigation and observation of hot gas flow into a wheelspace). 13th International Congress on Combustion Engines, Vienna, Paper GT30, 1979.

Batchelor, G.K. Note on the class of solutions of the Navier-Stokes equations representing steady rotationally symmetric flow. *Quarterly Journal of Applied Mathematics*, 4 (1951), pp. 29–41.

Bayley, F.J., and Childs, P.R.N. Prediction of ingress rates to turbine and compressor wheelspaces. *International Journal of Heat and Fluid Flow*, 18 (1997), pp. 218–228.

Bayley, F.J., and Conway, L. Fluid friction and leakage between a stationary and rotating disc. *Journal of Mechanical Engineering Science*, 6 (1964), pp. 164–172.

Bayley, F.J., and Owen, J.M. Flow between a rotating and a stationary disc. *Aeronautical Quarterly*, 20 (1969), pp. 333–354.

Bayley, F.J., and Owen, J.M. The fluid dynamics of a shrouded disc system with a radial outflow of coolant. *ASME Journal of Engineering for Power*, 92 (1970), pp. 335–341.

Bhavani, S.H., Khilnani, V.I., Tsai, L.C., Khodadi, J.M., Goodling, J.S., and Waggott, J. Effective sealing of a disc cavity using a double toothed rim seal. ASME Paper 92-GT-379, 1992a.

Bhavani, S.H., Khodadadi, J.M., Goodling, J.S., and Waggott, J. An experimental study of fluid flow in disc cavities. *Trans. ASME, Journal of Turbomachinery*, 114 (1992b), pp. 454–461.

Bohn, D., Decker, A., Ma, H., and Wolff, M. Influence of sealing air mass flow on the velocity distribution in and inside the rim seal of the upstream cavity of a 1.5-stage turbine. ASME-Paper GT2003-38459, 2003.

Bohn, D., Johann, E., and Kruger, U. Experimental and numerical investigations of aerodynamic aspects of hot gas ingestion in rotor-stator systems with superposed cooling mass flow. ASME Paper 95-GT-143, 1995.

Bohn, D., Rudzinski, B., and Sürken, N. Influence of rim seal geometry on hot gas ingestion into the upstream cavity of an axial turbine stage. ASME Paper 99-GT-248, 1999.

Bohn, D., Rudzinski, B., Sürken, N., and Gärtner, W. Experimental and numerical investigation of the influence of rotor blades on hot gas ingestion into the upstream cavity of an axial turbine stage. ASME Paper 2000-GT-284, 2000.

Campbell, D. Gas turbine disc sealing design. *Proceedings of the AGARD Conference on Seal Technology in Gas Turbine Engines, AGARD-CP-237* (1978).

Chew, J.W. A theoretical study of ingress for shrouded rotating disc systems with radial outflow. *Trans. ASME, Journal of Turbomachinery*, 113 (1991), pp. 91–97.

Chew, J.W., and Vaughan, C.M. Numerical predictions for the flow induced by an enclosed rotating disc. ASME Paper 88-GT-127, 1988.

Chew, J.W., Dadkhah, S., and Turner, A.B. Rim sealing of rotor-stator wheelspaces in the absence of external flow. *Trans. ASME, Journal of Turbomachinery*, 114 (1992), pp. 433–438.

Chew, J.W., Green, T., and Turner, A.B. Rim sealing of rotor-stator wheelspaces in the presence of external flow. ASME Paper 94-GT-126, 1994.

Coren, D., Childs, P.R.N., and Long, C.A. Windage sources in smooth-walled rotating disc systems. *Proceedings of the Institution of Mechanical Engineers, Part C*, 23 (2009), pp. 873–888.

Dadkhah, S. Ingestion and sealing performance of rim seals in rotor-stator wheelspaces. DPhil Thesis, University of Sussex, UK, 1989.

Dadkhah, S., Turner, A.B., and Chew, J.W. Performance of radial clearance rim seals in upstream and downstream rotor-stator wheelspaces. Trans. *ASME, Journal of Turbomachinery*, 114 (1992), pp. 439–445.

Daily, J.W., and Nece, R.E. Chamber dimension effects on induced flow and frictional resistance of enclosed rotating discs. *Journal of Basic Engineering*, 82 (1960), pp. 217–232.

Daily, J.W., Ernst, W.D., and Asbedian, V.V. Enclosed rotating discs with superposed throughflow: Mean steady and periodic unsteady characteristics of induced flow. MIT Department of Civil Engineering, Hydrodynamics Laboratory Report No. 64, 1964.

Daniels, W.A., Johnson, B.V., Graber, D.J., and Martin, R.J. Rim seal experiments and analysis for turbine applications. *Trans. ASME, Journal of Turbomachinery*, 114 (1992), pp. 426–432.

Debuchy, R., Dyment, A., Muhe, H., and Micheau, P. Radial inflow between a rotating and a stationary disc. *European Journal of Mechanics—B/Fluids*, 17 (1998), pp. 791–810.

Dijkstra, D., and van Heijst, G.J.F. The flow between two finite rotating discs enclosed by a cylinder. *Journal of Fluid Mechanics*, 128 (1983), pp. 123–154.

Dorfman, L.A. Effect of radial flow between the rotating disc and housing on their resistance and heat transfer. *Izv. Akad. Nauk.-SSSR, OTN, Mekh.i Mash.*, No. 4 (1961), pp. 26–32.

Feiereisen, J.M., Paolillo, R.E., and Wagner, J. UTRC turbine rim seal ingestion and platform cooling experiments. AIAA Paper 2000-3371, 2000.

Gallier, K.D., Lawless, P.B., and Fleeter, S. Investigation of seal purge flow effects on the hub flow field in a turbine stage using particle image velocimetry. AIAA Paper 2000-3370, 2000.

Gostelow, J.P. *Cascade aerodynamics*. Pergamon Press, 1984.

Graber, D.J., Daniels, W.A., and Johnson, B.V. Disc pumping test. Air force Wright aeronautical Lab., Report No. AFWAL-TR-87-2050, 1987.

Green, T., and Turner, A.B. Ingestion into the upstream wheelspace of an axial turbine stage. ASME Paper 92-GT-303, 1992.

Hamabe, K. and Ishida, K. Rim seal experiments and analysis of a rotor-stator system with non-axisymmetric main flow. ASME Paper 92-GT-160, 1992.

Hills, N.J., Chew, J.W. and Turner, A.B. Computational and mathematical modelling of turbine rim seal ingestion. ASME Paper 2001-GT-0204, 2001.

Hills, N.J., Green, T., Turner, A.B., and Chew, J.W. Aerodynamics of turbine rim seal ingestion. ASME Paper 97-GT-268, 1997.

Intel Corporation. The battery life challenge—balancing performance and power. White Paper, Intel Centrino mobile technology, Enabling battery life, Intel Corporation, 2004.

Ishida, K., and Hamabe, K. Effect of main flow pressure asymmetry on seal characteristics of a gas turbine rotor-stator system. ASME Paper 97-GT-205, 1997.

Jakoby, R., Lindblad, K., deVito, L., Bohn, D.E., Funcke, J., and Decker, A. Numerical simulation of the unsteady flow field in an axial gas turbine rim seal configuration. ASME Paper GT2004-53829, 2004.

Johnson, B., Mack, G., Paolillo, R., and Daniels, W. Turbine rim seal gas path flow ingestion mechanisms. AIAA Paper 94-2703, 1994.

Kármán, Th. von. Über laminare und turbulente reibung [On laminar and turbulent friction]. *Zeitschrift für angewandte Mathematik Mechanik*, 1 (1921), pp. 233–252.

Khilnani, V.I., Tsai, L.C., Bhavnani, S.H., Khodadi, J.M., Goodling, J.S., and Waggott, J. Main-stream ingress suppression in gas turbine disc cavities. *Trans. ASME, Journal of Turbomachinery*, 116 (1994), pp. 339–346.

Kobayashi, N., Matsumato, M., and Shizuya, M. An experimental investigation of a gas turbine disc cooling system. *Trans. ASME, Journal of Engineering for Gas Turbines and Power*, 106 (1984), pp. 136–141.

Langston, L.S. Crossflows in a turbine cascade passage. *Trans. ASME, Journal of Engineering for Power*, 102 (1980), pp. 866–874.

Laroche, E., Desportes, S., Djaoui, M., Debuchy, R., and Paté, L. A combined experimental and numerical investigation of the flow in a heated rotor-stator cavity with a radial inflow. ASME Paper 99-GT-170, 1999.

Mellor, G.L., Chapple, P.J., and Stokes, V.K. On the flow between a rotating and a stationary disc. *Journal of Fluid Mechanics* (1968), pp. 95–112.

Millward, J.A., and Robinson, P.H. Experimental investigation into the effects of rotating and static bolts on both windage heating and local heat transfer coefficients in a rotor-stator cavity. ASME Paper 89-GT-196, 1989.

Morse, A.P. Assessment of laminar-turbulent transition in closed disc geometries. ASME Paper 89-GT-179, 1989.

Okita, Y., Nishiura, M., Yamawaki, S., and Hironaka, Y. A novel cooling method for turbine rotor-stator rim cavities affected by mainstream ingress. *Trans. ASME, J. Engineering for Gas Turbines and Power*, 127 (2005), pp. 798–806.

Owen, J.M. An approximate solution for the flow between a rotating and a stationary disc. ASME Paper 88-GT-293, 1988.

Owen, J.M. Prediction of ingestion through turbine rim seals. Part 1: Rotationally-induced ingress. ASME Paper GT 2009-59121, 2009a.

Owen, J.M. Prediction of ingestion through turbine rim seals. Part 2: Externally-induced and combined ingress. ASME Paper GT 2009-59122, 2009b.

Owen, J.M., and Haynes, C.M. Design formulae for the heat loss and frictional resistance of air cooled rotating discs. *Improvements in Fluid Mechanics and Systems for Energy Conversion*, IV, pp. 127–160 Hoepli, Milan 1976.

Owen, J.M., and Phadke, U.P. An investigation of ingress for a simple shrouded rotating disc system with a radial outflow of coolant. ASME Paper 80-GT-49, 1980.

Owen, J.M., and Rogers, R.H. *Flow and heat transfer in rotating-disc systems*. Volume 1: *Rotor-stator systems*. Research Studies Press, 1989.

Owen, J.M., Zhou, K., Wilson, M., Pountney, O., and Lock, G. Prediction of ingress through turbine rim seals. Part 1: Externally-induced ingress. ASME Paper GT2010-23346, 2010a.

Owen, J.M., Pountney, O., and Lock, G. Prediction of ingress through turbine rim seals. Part 2: Combined ingress. ASME Paper GT2010-23349, 2010b.

Phadke, U.P. and Owen, J.M. Aerodynamic aspects of the sealing of gas turbine rotor-stator systems. Part 1: The behaviour of simple shrouded rotating disc systems in a quiescent environment. *Int. J. Heat and Fluid Flow*, 9 (1988), pp. 98–105.

Phadke, U.P., and Owen, J.M. Aerodynamic aspects of the sealing of gas turbine rotor-stator systems. Part 2: The performance of simple seals in a quasi-axisymmetric external flow. *Int. J. Heat and Fluid Flow*, 9 (1988), pp. 106–112.

Phadke, U.P., and Owen, J.M. Aerodynamic aspects of the sealing of gas turbine rotor-stator systems. Part 3: The effect of non-axisymmetric external flow on seal performance. *Int. J. Heat and Fluid Flow*, 9 (1988), pp. 113–117.

Phadke, U.P., and Owen, J.M. An investigation of ingress for an air cooled shrouded rotating disc system with radial clearance seals. ASME Paper 82-GT-145, 1982.

Picha, K.G., and Eckert, E.R.G. Study of the air flow between coaxial discs rotating with arbitrary velocities in an open or closed space. *Proceedings of the 3rd U.S. National Congress of Applied Mechanics*,(1958), pp. 791–798.

Pincombe, J.R. Flow visualisation and velocity measurements in a rotor-stator system with a forced radial inflow. Thermo-Fluid Mechanics Research Centre Technical Note 88/TFMRC/TN61, University of Sussex, UK, 1989.

Poncet, S., Chauve, M-P., and Schiestel, R. Batchelor versus Stewartson flow structures in a rotor-stator cavity with throughflow. *Physics of Fluids*, 17 (2005).

Reichert, A.W., and Lieser, D. Efficiency of air-purged rotor-stator seals in combustion turbine engines. ASME Paper 99-GT-250, 1999.

Roy, R.P., Devasenathipathy, S., Xu, G., and Zhao, Y. A study of the flow field in a model rotor-stator disc cavity. ASME Paper 99-GT-246, 1999.

Roy, R.P., Feng, J., Narzary, D., and Paolillo, R.E. Experiment on gas ingestion through axial flow turbine rim seals. *Trans. ASME, Journal of Engineering for Gas Turbines and Power*, 127 (2005), pp. 573–582.

Roy, R.P., Xu, G., and Feng, J. Study of main stream gas ingestion in a rotor-stator disc cavity. AIAA Paper 2000-3372, 2000.

Roy, R.P., Xu, G., Feng, J., and Kang, S. Pressure field and main stream gas ingestion in rotor-stator disc cavity. ASME Paper 2001-GT-564, 2001.

Scanlon, T., Wilkes, J., Bohn, D., and Gentilhomme, O. A simple method for estimating ingestion of annulus gas into a turbine rotor-stator cavity in the presence of external pressure variations. ASME Paper GT2004-53097, 2004.

Schouveiler, L., Gal, P. Le, and Chauve, M.P. Instabilities of the flow between a rotating and a stationary disc. *Journal of Fluid Mechanics*, 443 (2001), pp. 329–350.

Soo, S.L. Laminar flow over an enclosed rotating disc. *Trans. ASME* (1958), pp. 287–296.

Stewartson, K. On the flow between two rotating coaxial discs. *Proceedings of the Cambridge Philosophical Society*, 49 (1953), pp. 333–341.

Taniguchi, S., Sakamoto, H., and Arie, M. Interference between two circular cylinders of finite height vertically immersed in a turbulent boundary layer. *Journal of Fluids Engineering*, 104 (1982), p. 529.

Theodorsen, T., and Regier, A. Experiments on drag of revolving discs, cylinders and streamline rods at high speeds. NACA Report 793, 1944.

Vaughan, C.M. A numerical investigation into the effect of an external flow field on the sealing of a rotor-stator cavity. D.Phil. Thesis, University of Sussex, 1987.

Wieghardt, K. Turbulente grenzschichten. *Göttinger Monographie*, Part B5, 1946.

Zandbergen, P.J., and Dijkstra, D. Von Kármán swirling flows. *Annual Review of Fluid Mechanics*, 19 (1987), pp. 465–491.

Zimmermann, H., Firsching, A., Dibelius, G.H., and Ziemann, M. Friction losses and flow distribution for rotating discs with shielded and protruding bolts. ASME Paper 86-GT-158, 1986.

Chapter 6

Rotating Cylinders, Annuli, and Spheres

Flow over rotating cylinders is important in a wide number of applications from shafts and axles to spinning projectiles. Also considered here is the flow in an annulus formed between two concentric cylinders where one or both of the cylindrical surfaces is or are rotating. As with the rotating flow associated with discs, the proximity of a surface can significantly alter the flow structure.

6.1. INTRODUCTION

An indication of the typical flow configurations associated with rotating cylinders and for an annulus with surface rotation is given in Figure 6.1. For the examples given in 6.1a, 6.1c, 6.1e, 6.1g, and 6.1i, a superposed axial flow is possible, resulting typically in a skewed, helical flow structure.

This chapter describes the principal flow phenomena, and develops and presents methods to determine parameters such as the drag associated with a rotating cylinder and local flow velocities. Flow for a rotating cylinder is considered in Section 6.2. Laminar flow between two rotating concentric cylinders, known as rotating Couette flow, is considered in Section 6.3. Rotation of annular surfaces can lead to instabilities in the flow and the formation of complex toroidal vortices, known, for certain flow conditions, as Taylor vortices. Taylor vortex flow represents a significant modeling challenge and has been subject to a vast number of scientific studies. The relevance of these to practical applications stems from the requirement to avoid Taylor vortex flow, the requirement to determine or alter fluid residence time in chemical processing applications, and the desire to understand fluid flow physics and develop and validate fluid flow models. Flow instabilities in an annulus with surface rotation and Taylor vortex flow are considered in Section 6.4. One of the most important practical applications of rotating flow is the journal bearing. Rotation can lead to the development of high pressures in a lubricant and the separation of shaft and bearing surfaces, thereby reducing wear. The governing fluid flow equation is called the Reynolds equation, and this equation, along with a

Rotating Flow, DOI: 10.1016/B978-0-12-382098-3.00006-8

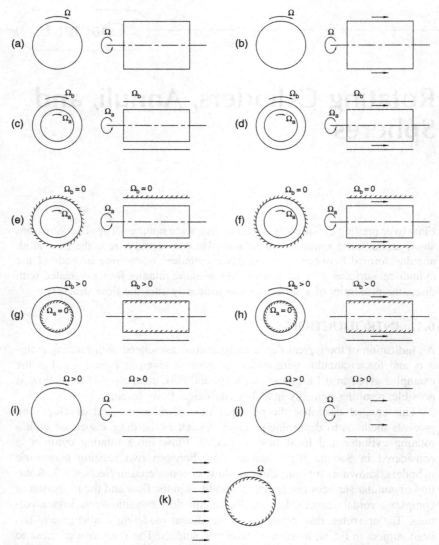

FIGURE 6.1 Selected cylinder and annulus flow configurations: (a) rotating cylinder, (b) rotating cylinder with superposed axial flow, (c) annulus with inner and outer cylinder rotation, (d) annulus with inner and outer cylinder rotation and throughflow, (e) annulus with inner cylinder rotation, (f) annulus with inner cylinder rotation and axial throughflow, (g) annulus with outer cylinder rotation, (h) annulus with outer cylinder rotation and axial throughflow, (i) internal flow within a rotating cylinder, (j) internal flow within a rotating cylinder with axial throughflow, (k) rotating cylinder with superposed cross-flow.

procedure for hydrodynamic bearing design, is developed in Section 6.5. A rotating cylinder with cross flow, along with the related case of a spinning sphere in a cross flow, is considered in Section 6.6.

6.2. ROTATING CYLINDER FLOW

A boundary layer will form on a rotating body of revolution due to the no-slip condition at the body surface as illustrated in Figure 6.2. At low values of the rotational Reynolds number, the flow will be laminar. As the rotational Reynolds number rises, the flow regime will become transitional and then turbulent. The approximate limit for laminar flow is for a rotational Reynolds number somewhere between 40 and 60.

The flow about a body of revolution rotating about its axis and simultaneously subjected to a flow in the direction of the axis of rotation is relevant to a number of applications, including certain rotating machinery and the ballistics of projectiles with spin. Various parameters such as the drag, moment coefficient, and the critical Reynolds number are dependent on the ratio of the circumferential to free-stream velocity. For example, the drag tends to increase with this ratio. The physical reason for this dependency is due to processes in the boundary layer where the fluid due to the no-slip condition co-rotates in the immediate vicinity of the wall and is therefore subject to the influence of strong "centrifugal forces." As a result, the processes of separation and transition from laminar to turbulent flow are affected by these forces and therefore drag too.

The boundary layer on a rotating body of revolution in an axial flow consists of the axial component of velocity and the circumferential component due to the

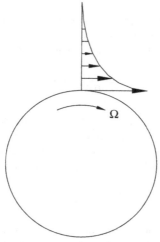

FIGURE 6.2   Boundary layer flow over a rotating cylinder.

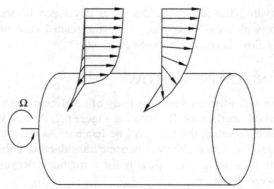

FIGURE 6.3 Boundary layer flow over a rotating cylinder with superposed axial flow.

no-slip condition at the body surface. The result is a skewed boundary layer as illustrated in Figure 6.3 for the case of a cylindrical body.

The thickness of the boundary layer increases as a function of the rotation parameter, λ_m, which is given by

$$\lambda_m = \frac{\Omega a}{u_{z,\infty}} \qquad (6.1)$$

where Ω is the angular velocity of the cylinder, a is the radius of the cylinder, and $u_{z,\infty}$ is the free-stream axial velocity.

Steinheuer (1965) found the following relationship between the axial and rotational velocity for the boundary on a rotating cylinder with axial flow

$$\frac{u_\phi}{\Omega a} = 1 - \frac{u_z}{u_{z,\infty}} \qquad (6.2)$$

A common requirement in practical applications is the need to quantify the power required to overcome the frictional drag of a rotating shaft. The moment coefficient for a rotating cylinder can be expressed by

$$C_{mc} = \frac{T_q}{0.5\pi\rho\Omega^2 a^4 L} \qquad (6.3)$$

where a, L, and Ω are the radius, length, and angular velocity of the cylinder, respectively, as illustrated in Figure 6.4.

For laminar flow, Lamb (1932) gives the moment coefficient for a rotating cylinder, based on the definition given in Equation 6.3, as

$$C_{mc} = \frac{8}{\text{Re}_\phi} \qquad (6.4)$$

Extensive experiments on rotating cylinders for the case of both laminar and turbulent flow were undertaken by Theodorsen and Regier (1944) for a range of

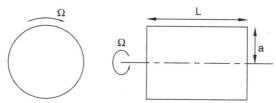

FIGURE 6.4   Rotating cylinder principal dimensions.

cylinders with diameters ranging from 12.7 mm to 150 mm and lengths between 150 mm and 1.2 m in oil, kerosene, water and air at rotational Reynolds numbers up to approximately 5.6×10^6. The data follow the laminar trend within approximately 15% up to a rotational Reynolds number of approximately 60. An empirical correlation was produced for turbulent flow for smooth cylinders, with the contribution due to the end discs removed, given in Equation 6.5.

$$\frac{1}{\sqrt{C_{mc}}} = -0.8572 + 1.250\ln(\mathrm{Re}_\phi \sqrt{C_{mc}}) \qquad (6.5)$$

It is not possible to separate out the moment coefficient algebraically; this equation must be solved using an iterative technique for a given rotational Reynolds number. Equation 6.5 is restated in Equation 6.6, and a starting value of, say, $C_{mc} = 0.02$ generally results in a converged solution to within three significant figures within about four or five iterations.

$$C_{mc} = \left(\frac{1}{-0.8572 + 1.25\ln(\mathrm{Re}_\phi \sqrt{C_{mc}})}\right)^2 \qquad (6.6)$$

Values for the drag coefficient as a function of the rotational Reynolds number for both Equations 6.4 and 6.6 are presented in Figure 6.5.

The power required to overcome frictional drag can be determined by

$$Power = T_q \times \Omega \qquad (6.7)$$

Substituting for the torque in Equations 6.7 using Equation 6.3 gives

$$Power = 0.5\pi\rho\Omega^3 a^4 L C_{mc} \qquad (6.8)$$

The power required to overcome frictional drag for a rotating cylinder as a function of a rotational Reynolds number for a range of radii between 40 mm and 400 mm, assuming a cylinder length of 0.4 m and a density of 1.2047 kg/m^3, are presented in Figure 6.6. For the case of a rotating drum with end discs, the contribution to the overall power requirement to overcome frictional drag due to the rotating discs would also need to be accounted for, and the techniques presented in Chapters 4 and 5 can be adopted as appropriate.

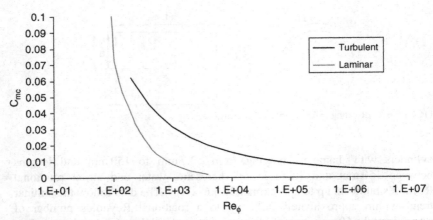

FIGURE 6.5 Moment coefficients for a rotating cylinder as a function of the rotational Reynolds number.

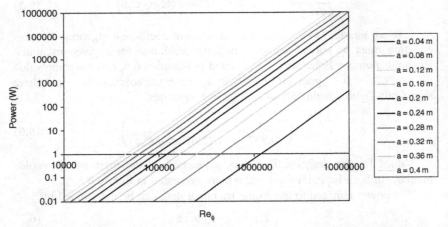

FIGURE 6.6 Power requirement for a smooth rotating cylinder, of length 400 mm in air with density 1.2047 kg/m^3.

Example 6.1.

Determine the power required to overcome frictional drag for a 250-mm-long shaft with a diameter of 100 mm rotating at 10,500 rpm in air with a density and viscosity of 4 kg/m^3 and 3×10^{-5} Pa s, respectively. Compare the figure for the frictional drag to that of the power required to overcome frictional drag for both the discs and the bolts for the disc also spinning at 10,500 rpm in Example 5.4, Chapter 5.

Solution

The rotational Reynolds number for the shaft, from Equation (4.1), is

$$\mathrm{Re}_\phi = \frac{\rho \Omega b^2}{\mu} = \frac{4 \times 1100 \times 0.05^2}{3 \times 10^{-5}} = 3.667 \times 10^5$$

This value is significantly higher than 60, so the flow can be taken to be turbulent.

Assuming an initial guess for the moment coefficient in Equation 6.6 of $C_{mc} = 0.02$,

$$C_{mc} = \left(\frac{1}{-0.8572 + 1.25 \ln(3.667 \times 10^5 \sqrt{0.02})} \right)^2 = 0.006187$$

Using the updated value for the moment coefficient in Equation 6.6

$$C_{mc} = \left(\frac{1}{-0.8572 + 1.25 \ln(3.667 \times 10^5 \sqrt{0.006187})} \right)^2 = 0.006968$$

Using the updated value for the moment coefficient in Equation 6.6

$$C_{mc} = \left(\frac{1}{-0.8572 + 1.25 \ln(3.667 \times 10^5 \sqrt{0.006968})} \right)^2 = 0.006882$$

Using the updated value for the moment coefficient in Equation 6.6

$$C_{mc} = \left(\frac{1}{-0.8572 + 1.25 \ln(3.667 \times 10^5 \sqrt{0.006882})} \right)^2 = 0.006891$$

Using the updated value for the moment coefficient in Equation 6.6

$$C_{mc} = \left(\frac{1}{-0.8572 + 1.25 \ln(3.667 \times 10^5 \sqrt{0.006891})} \right)^2 = 0.006890$$

This value is within one significant figure of the previous value and can be assumed to be converged.

The power required to overcome frictional drag, using Equation 6.8 is given by

$$\text{Power} = 0.5 \times \pi \times 4 \times 1100^3 \times 0.05^4 \times 0.25 \times 0.006890 \approx 90 \text{ W}$$

The power required to over frictional drag for the rotating shaft is comparatively small, less than 1% of that required to overcome the frictional drag for both the disc face and the bolts for the disc configuration of Example 5.4.

Example 6.2.

Determine the power required to overcome frictional drag for a 300-mm-long shaft with a diameter of 120 mm rotating at 15,000 rpm in air with a density and viscosity of 5.2 kg/m^3 and 3×10^{-5} Pa s, respectively.

Solution

The rotational Reynolds number for the shaft, from Equation 4.1, is

$$\text{Re}_\phi = \frac{\rho \Omega b^2}{\mu} = \frac{5.2 \times 1571 \times 0.06^2}{3 \times 10^{-5}} = 9.80 \times 10^5$$

This value is significantly higher than 60, so the flow can be taken to be turbulent. Assuming an initial guess for the moment coefficient in Equation 6.6 of $C_{mc} = 0.02$,

$$C_{mc} = \left(\frac{1}{-0.8572 + 1.25\ln(9.80 \times 10^5 \sqrt{0.02})} \right)^2 = 0.005144$$

Using the updated value for the moment coefficient in Equation 6.6

$$C_{mc} = \left(\frac{1}{-0.8572 + 1.25\ln(9.80 \times 10^5 \sqrt{0.005144})} \right)^2 = 0.005833$$

Using the updated value for the moment coefficient in Equation 6.6

$$C_{mc} = \left(\frac{1}{-0.8572 + 1.25\ln(9.80 \times 10^5 \sqrt{0.005833})} \right)^2 = 0.005764$$

Using the updated value for the moment coefficient in Equation 6.6

$$C_{mc} = \left(\frac{1}{-0.8572 + 1.25\ln(9.80 \times 10^5 \sqrt{0.005764})} \right)^2 = 0.005770$$

Using the updated value for the moment coefficient in Equation 6.6

$$C_{mc} = \left(\frac{1}{-0.8572 + 1.25\ln(9.80 \times 10^5 \sqrt{0.005770})} \right)^2 = 0.005769$$

This value is within one significant figure of the previous value and can be assumed to be converged.

The power required to overcome frictional drag, using Equation 6.8, is given by

$$\text{Power} = 0.5 \times \pi \times 5.2 \times 1571^3 \times 0.06^4 \times 0.3 \times 0.005769 = 710.4 \text{ W}$$

In this case, the power required to overcome frictional drag for the rotating shaft is quite significant, and as in most rotating machines, putting heat into the fluid surrounding a shaft is an unwanted side effect that generally needs to be minimized by minimizing the radius of the shaft and its length.

For rough cylinders, Theodorsen and Regier (1944) found that the moment coefficient was dependent on the relative grain size and grain spacing applied to the cylinder surface, and that beyond a critical Reynolds number the drag coefficient remains constant and independent of the rotational Reynolds number. For

$$e > 3.3 \frac{\nu}{\sqrt{\tau_o/\rho}} \tag{6.9}$$

where e is the roughness grain size and τ_o is the shear stress at the surface, the moment coefficient is given by

$$\frac{1}{\sqrt{C_{mc}}} = 1.501 + 1.250\ln\frac{a}{e} \tag{6.10}$$

The center line average roughness, R_a, is related to the sand-grain roughness by $R_a \approx k_s/2$ (see Childs and Noronha, 1997), allowing the following substitution.

$$\frac{1}{\sqrt{C_{mc}}} = 0.8079 + 1.250\ln\frac{a}{R_a} \tag{6.11}$$

6.3.   ROTATING COUETTE FLOW

The term *Couette flow* describes flow between two surfaces that are in close proximity, such that flow is dominated by viscous effects and inertial effects are negligible. In cylindrical coordinates this involves flow in an annulus, as illustrated in Figure 6.7, and the Navier-Stokes equations can be solved exactly by analytical techniques, subject to a number of significant assumptions, which severely limit the application of the resulting solution. Nevertheless, the approach is instructive and also forms the basis for a technique to determine the viscosity of fluid.

The analysis presented here for rotating Couette flow assumes laminar flow and is valid provided the Taylor number given in a form based on the mean annulus radius in Equation 6.12, is less than the critical Taylor number, Ta_{cr}. If the critical Taylor number is exceeded, toroidal vortices can be formed in the annulus. The formation of such vortices is considered in Section 6.4. The critical Taylor number is dependent on a number of factors, including the rotation ratio and annulus dimensions. For the case of a narrow gap annulus, with a stationary outer cylinder the critical Taylor number is 41.19.

$$Ta_m = \frac{\Omega r_m^{0.5}(b-a)^{1.5}}{\nu} \tag{6.12}$$

where $r_m = (a + b)/2$.

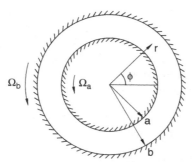

FIGURE 6.7   Rotating Couette flow.

If the flow is assumed to be contained between two infinite concentric cylinders, as illustrated in Figure 6.7, with either cylinder rotating at speed Ω_a and Ω_b, respectively, with no axial or radial flow under steady conditions, the continuity equation reduces to Equation 6.13 and the radial and tangential and components of the Navier-Stokes equations reduce to Equations 6.14 and 6.15.

$$\frac{\partial u_\phi}{\partial \phi} = 0 \tag{6.13}$$

$$-\frac{\rho u_\phi^2}{r} = -\frac{dp}{dr} \tag{6.14}$$

$$0 = \mu \left(\frac{d^2 u_\phi}{dr^2} + \frac{1}{r} \frac{du_\phi}{dr} - \frac{u_\phi}{r^2} \right) \tag{6.15}$$

or

$$\frac{d^2 u_\phi}{dr^2} + \frac{d}{dr} \left(\frac{u_\phi}{r} \right) = 0 \tag{6.16}$$

These equations can be solved with appropriate boundary conditions to give the velocity distribution as a function of radius and the torque on the inner and outer cylinders:

The boundary conditions are as follows:

at $r = a$, $u_\phi = \Omega a$

at $r = b$, $u_\phi = \Omega b$, and $p = p_b$

Equation 6.16 is in the familiar form of an ordinary differential equation, which has a standard solution of the form

$$u_\phi = C_1 r + \frac{C_2}{r} \tag{6.17}$$

The constants C_1 and C_2 are given by

$$C_1 = \frac{\Omega_b b^2 - \Omega_a a^2}{b^2 - a^2} \tag{6.18}$$

$$C_2 = \frac{(\Omega_a - \Omega_b) a^2 b^2}{b^2 - a^2} \tag{6.19}$$

The velocity given in Equation 6.17 can be substituted in Equation 6.14 to give the pressure distribution

$$\frac{dp}{dr} = \rho \left(C_1^2 r + \frac{2 C_1 C_2}{r} + \frac{C_2^2}{r^3} \right) \tag{6.20}$$

This equation can be integrated giving

$$p = \rho \left(C_1^2 \frac{r^2}{2} + 2C_1 C_2 \ln r - \frac{C_2^2}{2r^2} \right) + C \tag{6.21}$$

where the constant of integration C can be evaluated by specifying the value of pressure $p = p_a$ at $r = a$ or $p = p_b$ at $r = b$ for any particular problem.

The viscous shear stress at $r = a$ is given by

$$\tau = \mu \frac{r \partial (u_\phi / r)}{\partial r} \tag{6.22}$$

Hence for $r = a$, from Equations 6.17 and 6.19

$$\tau = -2\mu \frac{C_2}{r^2} = -2\mu \frac{C_2}{a^2} = -2\mu \frac{(\Omega_a - \Omega_b) b^2}{b^2 - a^2} \tag{6.23}$$

The torque per unit length L is equal to $\tau A a / L$. The surface area of the cylinder is $2\pi a L$, so the torque per unit length is

$$T_q = \frac{\tau (2\pi a L) a}{L} = 2\pi a^2 \tau \tag{6.24}$$

Substituting for the shear stress from Equation 6.23 gives the torque per unit length as

$$T_q = 4\pi \mu \frac{a^2 b^2}{b^2 - a^2} (\Omega_b - \Omega_a) \tag{6.25}$$

Similarly, the torque per unit length for the outer cylinder is given by

$$T_q = -4\pi \mu \frac{a^2 b^2}{b^2 - a^2} (\Omega_b - \Omega_a) \tag{6.26}$$

It should be noted that these equations for the torque per unit length are valid only if the flow remains entirely circumferential.

It is possible to make use of Equation 6.26 in the measurement of viscosity, known as viscometry, using a device made up of two concentric cylinders, arranged vertically with height L, with the test substance held between the cylinders (Mallock, 1888, 1896; Couette, 1890). The inner cylinder is locked in a stationary position and the outer cylinder is rotated as indicated in Figure 6.8.

From measurements of the angular velocity of the outer cylinder and the torque on the inner cylinder, the viscosity can be determined, as in Equation 6.27.

$$\mu = \left| T_q \frac{b^2 - a^2}{4\pi \Omega_b a^2 b^2 L} \right| \tag{6.27}$$

Bilgen and Boulos (1973) give the following equation for the moment coefficient for an annulus with inner cylinder rotation with no axial pressure gradient for laminar flow.

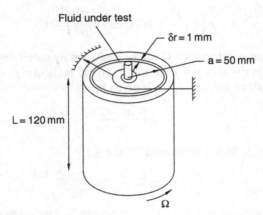

FIGURE 6.8 Typical viscometer.

$$C_{mc} = 10\left(\frac{b-a}{a}\right)^{0.3} \text{Re}_{\phi m}^{-1} \qquad (6.28)$$

where the rotational Reynolds number, $Re_{\phi m}$, is based on the annulus gap,

$$\text{Re}_{\phi m} = \frac{\rho \Omega a (b-a)}{\mu} \qquad (6.29)$$

for

$$\text{Re}_{\phi m} < 64 \qquad (6.30)$$

The equivalent relationship for the moment coefficient from the linear theory, Equations 6.3 and 6.25, is

$$C_{mc} = \frac{8b^2}{b^2-a^2}\text{Re}_{\phi}^{-1} = \frac{8b^2}{a(b+a)}\text{Re}_{\phi m}^{-1} \qquad (6.31)$$

Equation 6.28 compares favorably with Equation 6.31. Bilgen and Boulos (1973) report that the maximum mean deviation of Equation 6.28 from their experimental data was ±5.8%.

Example 6.3.

Determine the power required to overcome frictional drag for a cylinder with a radius of 0.25 m rotating at 10 revolutions per minute within an annulus, with the outer cylinder rotating if the length of the annulus is 0.4 m and the gap between the inner and outer cylinder is 5 mm. The annulus is filled with oil of viscosity 0.02 Pa s and density 900 kg/m³.

Solution

The rotational Reynolds number from Equation 6.29 is

$$Re_{\phi m} = \frac{\rho \Omega a (b-a)}{\mu} = \frac{900 \times 10 \times (2\pi/60) \times 0.25 \times 0.005}{0.02} = 58.9$$

As this is less than 64, the empirical correlation given in Equation 6.28 can be used.

$$C_{mc} = 10 \left(\frac{b-a}{a}\right)^{0.3} Re_{\phi m}^{-1} = 10 \left(\frac{0.005}{0.25}\right)^{0.3} (58.9)^{-1} = 0.0525$$

The power required to overcome frictional drag, from Equation 6.8 is

$$Power = 0.5\pi\rho\Omega^3 a^4 L C_{mc} = 0.5\pi \times 900 \times 1.047^3 \times 0.25^4 \times 0.4 \times 0.0525 = 0.1331 \, W$$

Example 6.4.

The torque indicated by a viscometer is 0.01 N m. If the diameter of the inner cylinder is 100 mm, the annular gap is 1 mm, the length is 120 mm, and the outer cylinder rotates at 40 rpm, determine the viscosity of the fluid contained in the viscometer.

Solution

The angular velocity of the outer cylinder is

$$\Omega_b = 40 \times \frac{2\pi}{60} = 4.188 \, rad/s$$

From Equation 6.27,

$$\mu = T_q \frac{b^2 - a^2}{4\pi\Omega_b a^2 b^2 L} = 0.01 \frac{0.051^2 - 0.05^2}{4\pi \times 4.188 \times 0.05^2 \times 0.051^2 \times 0.12} = 0.02459 \, Pa \, s$$

Examination of the tangential velocity in rotating Couette flow, Equation 6.17, yields a number of cases of special interest. For the case where the inner radius a and the angular velocity of the inner cylinder tend to zero, Equation 6.17 gives

$$u_\phi = \Omega_a r = constant \tag{6.32}$$

Equation 6.32 implies that rotation of a circular pipe filled with fluid will induce solid body motion of the fluid within it.   If the outer cylinder becomes very large and does not rotate, then in the limit as $r \to \infty$ with $\Omega_b = 0$

$$u_\phi = \frac{a^2 \Omega_a}{r} \tag{6.33}$$

Equation 6.33 represents a potential or free vortex, driven in a viscous fluid by the no-slip condition. Equation 6.17, and hence Equation 6.33, were derived assuming an infinite annulus length. The fluid contained within an infinite annulus will therefore have infinite momentum. It therefore takes, in theory, an infinite length of time for potential vortex flow to be generated by rotation of the inner cylinder alone.   Velocity distributions for the case of inner cylinder only rotation and outer cylinder only rotation are given in Figures 6.9 and 6.10 for a range of radius ratios a/b.

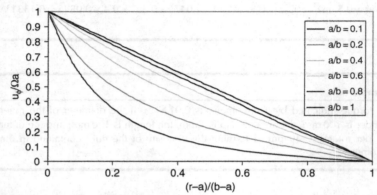

FIGURE 6.9 Velocity distribution for rotating Couette flow in a concentric annulus with inner cylinder only rotation for various values of radius ratio.

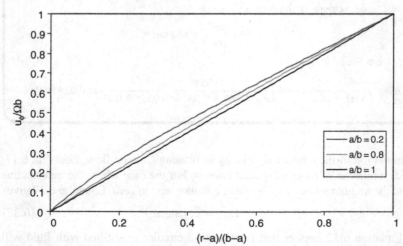

FIGURE 6.10 Velocity distribution for rotating Couette flow in a concentric annulus with outer cylinder only rotation.

Example 6.5.

Calculate the tangential velocity at a radius of 0.045 m for an annulus with an inner cylinder radius of 0.04 m and an annular gap of 0.01 m filled with oil with a density of 900 kg/m^3 and a viscosity of 0.022 Pa s for the following conditions.

  i.   The inner cylinder only is rotating with an angular velocity of 3 rad/s
 ii.   The outer cylinder only is rotating with an angular velocity of 3 rad/s
iii.   If the inner and outer cylinders are rotating with angular velocities of 3 rad/s and 6 rad/s respectively

Solution

The rotational Reynolds number from Equation 6.29 is

$$\text{Re}_{\phi m} = \frac{\rho \Omega a (b{-}a)}{\mu} = \frac{900 \times 3 \times 0.04 \times 0.01}{0.022} = 49.09$$

As this value is less than 64, the flow is likely to be laminar, and Equation 6.16 can be used to model the tangential velocity in the annulus.
   (i) From Equation 6.18,

$$C_1 = \frac{\Omega_b b^2 - \Omega_a a^2}{b^2 - a^2} = \frac{\Omega_a a^2}{b^2 - a^2} = -\frac{3 \times 0.04^2}{0.05^2 - 0.04^2} = -5.333$$

From Equation 6.19,

$$C_2 = \frac{(\Omega_a - \Omega_b) a^2 b^2}{b^2 - a^2} = \frac{\Omega_a a^2 b^2}{b^2 - a^2} = \frac{3 \times 0.04^2 \times 0.05^2}{0.05^2 - 0.04^2} = 0.01333$$

Hence from Equation 6.17,

$$u_\phi = -5.333r + \frac{0.01333}{r}$$

So at $r = 0.045$ m,

$$u_\phi = -5.333 \times 0.045 + \frac{0.01333}{0.045} = 0.05624 \text{ m/s}$$

   (ii) From Equation 6.18,

$$C_1 = \frac{\Omega_b b^2 - \Omega_a a^2}{b^2 - a^2} = \frac{\Omega_b a^2}{b^2 - a^2} = \frac{3 \times 0.05^2}{0.05^2 - 0.04^2} = 8.333$$

From Equation 6.19,

$$C_2 = \frac{(\Omega_a - \Omega_b) a^2 b^2}{b^2 - a^2} = -\frac{\Omega_b a^2 b^2}{b^2 - a^2} = -\frac{3 \times 0.04^2 \times 0.05^2}{0.05^2 - 0.04^2} = -0.01333$$

Hence from Equation 6.17,

$$u_\phi = 8.333r - \frac{0.01333}{r}$$

So at $r = 0.045$ m,

$$u_\phi = 8.333 \times 0.045 - \frac{0.01333}{0.045} = 0.0787 \text{ m/s}$$

(iii) From Equation 6.18,

$$C_1 = \frac{\Omega_b b^2 - \Omega_a a^2}{b^2 - a^2} = \frac{6 \times 0.05^2 - 3 \times 0.04^2}{0.05^2 - 0.04^2} = 11.33$$

From Equation 6.19,

$$C_2 = \frac{(\Omega_a - \Omega_b) a^2 b^2}{b^2 - a^2} = \frac{(3-6) \times 0.04^2 \times 0.05^2}{0.05^2 - 0.04^2} = -0.01333$$

Hence from Equation 6.17,

$$u_\phi = 11.33 r - \frac{0.01333}{r}$$

So at $r = 0.045$ m,

$$u_\phi = 11.33 \times 0.045 - \frac{0.01333}{0.045} = 0.2137 \text{ m/s}$$

Example 6.6.

Determine the pressure at a radius of 0.045 m and at a radius of 0.041 m for the conditions of part (i) identified in the previous example if the pressure at the outer radius of the annulus is 1.12 bar absolute.

Solution

From Equation 6.21,

$$C = p_b - \rho \left(C_1^2 \frac{b^2}{2} + 2 C_1 C_2 \ln b - \frac{C_2^2}{2 b^2} \right)$$

$$= 1.12 \times 10^5 - 900 \left((-5.333)^2 \frac{0.05^2}{2} + 2 \times -5.333 \times 0.01333 \ln 0.05 - \frac{0.01333^2}{2 \times 0.05^2} \right)$$

$$= 111616.546 \text{ Pa}$$

(The result to three decimal places has been deliberately stated here for the constant of integration, C, in order to enable the variation of pressure with radius to be distinguished.)

From Equation 6.21 at a radius of $r = 0.045$ m,

$$p = 111999.9 \text{ Pa}$$

From Equation 6.21 at a radius of $r = 0.041$ m,

$$p = 111999.328 \text{ Pa}$$

(The result is stated to three decimal places in order to illustrate the subtle variation of static pressure with radius.)

6.4. FLOW INSTABILITIES AND TAYLOR VORTEX FLOW

The equations presented for Couette flow in an annulus with rotation characterize a system in which dynamic equilibrium exists between the radial forces and
the radial pressure gradient. However, when it is not possible for the radial
pressure gradient and the viscous forces to dampen out and restore changes in
centrifugal forces caused by small disturbances in the flow, the fluid motion is
unstable and results in a secondary flow. A simple criterion for determining the
onset of inertial instability was developed by Rayleigh (1916).

The Rayleigh instability criterion examines the balance between radial forces
and the radial pressure gradient in a circular pathline flow field in order to identify
whether a displaced fluid has a tendency to return to its original location. In essence
the criterion determines whether the force due to inward radial pressure is adequate
to maintain inward centripetal acceleration for an arbitrary element of fluid.

In this analysis, an element of fluid is assumed to have been perturbed, or
moved, to a new radius, quickly enough so that its original fluid properties,
density, temperature, and so on, are maintained. In a Rayleigh stability analysis
for a rotating flow, the property that is conserved is momentum.

Consider a toroidal element of rotating fluid contained initially between
$r + dr$ and $z + dz$ which is displaced from an initial radius r_1 to a final radius r_2
such that $r_2 > r_1$. Local disturbances in a fluid can occur for many reasons from,
for example, small-scale fluctuations in fluid properties and conditions, dirt in
the flow, and vibration. At its new radial position, it will now be rotating faster
than the new environment, and the radial pressure gradient associated with the
flow will be insufficient to balance the radial acceleration associated with
the displaced fluid element. As a result, the fluid will tend to move further
outward, providing the mechanism for the unstable flow. In a similar manner, a
fluid element at a smaller radius will tend to move radially inward.

This qualitative argument is repeated here in a quantitative form for an
elemental toroid of fluid initially at radius r_1 with angular velocity Ω_1, which is
displaced to a radius r_2 without interacting with the remainder of the fluid and
ignoring the stabilizing tendency of viscosity.

Angular momentum is the product of linear moment and the perpendicular
distance of the particle from the axis. Angular momentum must be conserved
for an axisymmetric disturbance, so the velocities are related by

$$mu_{\phi 1}r_1 = mu_{\phi f}r_2 \tag{6.34}$$

or

$$m\Omega_1 r_1^2 = m\Omega_2 r_2^2 \tag{6.35}$$

The element's tangential velocity component after displacement will be

$$u_{\phi f} = \frac{u_{\phi 1}r_1}{r_2} \tag{6.36}$$

The centripetal acceleration of the displaced element of fluid is

$$-\frac{u_{\phi f}^2}{r_2} = -\left(\frac{u_{\phi 1} r_1}{r_2}\right)^2 \frac{1}{r_2} = -\frac{(u_{\phi 1} r_1)^2}{r_2^3} \tag{6.37}$$

The radial acceleration of the fluid element will be countered by the pressure gradient

$$\frac{\partial p}{\partial r} = -\rho \frac{u_{\phi 2}^2}{r_2} \tag{6.38}$$

So if the radial force of the fluid ring exceeds the pressure gradient, then the ring will have a tendency to continue moving outward and the associated motion will be unstable.
That is, if

$$\rho \frac{u_{\phi 1}^2 r_1^2}{r_2^3} > \rho \frac{u_{\phi 2}^2}{r_2} \tag{6.39}$$

or if

$$u_{\phi 1}^2 r_1^2 > u_{\phi 2}^2 r_2^2 \tag{6.40}$$

the flow will have a tendency to be unstable.

Conversely, the conditions for stability are if

$$\frac{u_{\phi 2}^2}{r_2} > \frac{(r_1 u_{\phi 1})^2}{r_2^3} \tag{6.41}$$

or

$$(r_2 u_{\phi 2})^2 > (r_1 u_{\phi 1})^2 \tag{6.42}$$

then the flow field is stable to the perturbation and a displaced element of fluid will return to its initial radius.

The condition for stability, that is,

$$\frac{d}{dr}(r u_\phi)^2 > 0 \tag{6.43}$$

or

$$\frac{d}{dr}|\Omega r^2| > 0 \tag{6.44}$$

is usually referred to as the Rayleigh criterion.

According to the Rayleigh criterion, an inviscid flow is thus unstable if the mean velocity distribution is such that the product $u_\phi r$ decreases radially outward with conservation of angular momentum about the axis of rotation.

Applying this to Couette flow, then for stability $\Omega_a > 0$. For two cylinders rotating in the same direction, the flow is either stable everywhere or unstable

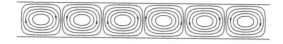

FIGURE 6.11   Taylor vortices.

everywhere. From the solution for the velocity distribution for Couette flow, $u_\phi = Ar + B/r$, then from Equation 6.44, gives

$$\frac{\Omega_b}{\Omega_a} > \left(\frac{a}{b}\right)^2 \tag{6.45}$$

or

$$\Omega_b b^2 > \Omega_a a^2 \tag{6.46}$$

For two cylinders rotating in opposite directions, the region close to the inner cylinder is unstable and the region close to the outer cylinder is stable.

The Rayleigh criterion based on the simple displaced particle concept provides an indication of the onset of instabilities. In real flows, the onset of instability is modified by the action of viscosity.

Experiments, for example, Taylor (1923), revealed that the simple axisymmetric laminar flow in an annulus with a combination of inner and outer cylinder rotation was replaced by a more complicated eddying flow structure. If the angular velocity exceeded a critical value, a steady axisymmetric secondary flow in the form of regularly spaced vortices in the axial direction, as illustrated in Figures 6.11 and 6.12, was generated. Alternate vortices rotate in opposite directions. The physical mechanism driving the vortices can be determined using insight from the Rayleigh stability criterion.

Considering the case of an annulus with the inner cylinder rotating at a higher speed than the outer cylinder, a fluid element at a low radius, if disturbed and forced radially outward, will have a radial outward force at its new radial location that will exceed the local radial pressure gradient. Thus the fluid element will have a tendency to continue to move with a net radial outward motion. Once it reaches the outer radius, it must go somewhere and will therefore flow along the outer cylinder. Continuity will cause particles to fill the space vacated the fluid element. The net result is the formation of a

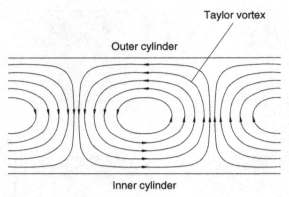

FIGURE 6.12   Taylor vortices cells.

circulating motion as indicated in Figure 6.12. The reasoning behind the multiple vortex cells is that if two fluid elements are both moving radially outward, then when they reach the outer radius, they cannot both go in the same direction along the outer cylinder; therefore the flow divides in opposite directions, causing two contra-rotating vortex cells to be formed.

Taylor (1923) formulated the stability problem taking into account the effects of viscosity by assuming an axisymmetric infinitesimal disturbance and solving the dynamic conditions under which instability occurs. Whereas the Rayleigh criterion, for the case where the outer cylinder is stationary, predicts that the flow for an inviscid fluid is unstable at infinitesimally small angular velocities of the inner cylinder, Taylor's linear stability analysis showed that viscosity delayed the onset of secondary flow.

Assuming the annular gap, $b - a$, is small compared to the mean radius, $r_m = (a + b)/2$, known as the small gap approximation, Taylor's solution for the angular velocity at which laminar flow breaks down and the flow instabilities grow leading to the formation of secondary flow vortices, called the critical speed, is given by

$$\Omega_{cr} = \pi^2 \nu \sqrt{\frac{(a+b)}{2S(b-a)^3 a^2 [1-(\Omega_b/\Omega_a)b^2/a^2](1-\Omega_b/\Omega_a)}} \qquad (6.47)$$

where

$$S = 0.0571\left[\frac{1+\Omega_b/\Omega_a}{1-\Omega_b/\Omega_a}+0.652\left(1-\frac{1}{x}\right)\right]+0.00056\left[\frac{1+\Omega_b/\Omega_a}{1-\Omega_b/\Omega_a}+0.652\left(1-\frac{1}{x}\right)\right]^{-1}$$

$$(6.48)$$

where $x = a/b$.

For the case of a stationary outer cylinder, $\Omega_b/\Omega_a = 0$, Equations 6.47 and 6.48 reduce to

$$\Omega_{cr} = \pi^2 \nu \sqrt{\frac{a+b}{2S(b-a)^3 a^2}} \qquad (6.49)$$

$$S = 0.0571[1-0.652(b-a)/a] + 0.00056[1-0.652(b-a)/a]^{-1} \qquad (6.50)$$

For the case of a narrow gap and stationary outer cylinder, $x \to 1$, $\Omega_b/\Omega_a = 0$ Equations 6.47 and 6.48 give

$$Ta_{m,cr} = \sqrt{1697} = 41.19 \qquad (6.51)$$

where $Ta_{m,cr}$ is the critical Taylor number based on the mean annulus radius and the corresponding Taylor number based on the mean annulus radius is defined by

$$Ta_m = \frac{\Omega r_m^{0.5}(b-a)^{1.5}}{\nu} \qquad (6.52)$$

The critical angular velocity for this case is given by

$$\Omega_{cr} = \frac{41.19\nu}{r_m^{0.5}(b-a)^{1.5}} \qquad (6.53)$$

For an annulus with a finite gap,

$$\Omega_{cr} = \frac{41.19\nu F_g}{r_m^{0.5}(b-a)^{1.5}} \qquad (6.54)$$

where F_g is a geometrical factor defined by

$$F_g = \frac{\pi^2}{41.19\sqrt{S}}\left(1-\frac{b-a}{2r_m}\right)^{-1} \qquad (6.55)$$

and S is given in an alternative form by

$$S = 0.0571\left(1-0.652\frac{(b-a)/r_m}{1-(b-a)/2r_m}\right) + 0.00056\left(1-0.652\frac{(b-a)/r_m}{1-(b-a)/2r_m}\right)^{-1} \qquad (6.56)$$

The wavelength for the instability is approximately

$$\lambda = 2(b-a) \qquad (6.57)$$

The instability criterion developed by Taylor agrees very well with experimental data such as Schultz-Grunow (1963), Donnelly (1958), Donnelly and Fultz (1960), Donnelly and Simon (1960) and Donnelly and Schwarz (1965). The critical Taylor number and stability criterion for a wide range of radius ratios and angular velocity ratios have been widely studied and are reviewed in Andereck and Hayot (1992). Data for the resulting critical Taylor number for some of these studies are summarized in Table 6.1.

TABLE 6.1 Critical Taylor number as a function of radius ratio and angular velocity ratio. * Experimental. (Data summarized from Moalem and Cohen, 1991; Di Prima and Swinney, 1981)

a/b	Ω_b/Ω_a	$2\pi(b-a)/\lambda$	$Ta_{m,cr}$	Reference
1	–	3.12	41.18	Walowit et al. (1964)
0.975	0	3.13	41.79	Roberts (1965)
0.9625	0	3.13	42.09	Roberts (1965)
0.95	0	3.13	42.44	Roberts (1965)
0.95	−3 to 0.85	3.14	42.44	Sparrow et al. (1964)
0.95	−0.25 to 0.9025	3.12	42.45	Walowit et al. (1964)
0.925	0	3.13	43.13	Roberts (1965)
0.9	0	3.13	43.87	Roberts (1965)
0.9	−0.25 to 0.81	3.13	43.88	Walowit et al. (1964)
0.8975	0	3.13	44.66	Roberts (1965)
0.85	0	3.13	45.50	Roberts (1965)
0.80	−0.25 to 0.64	3.13	47.37	Walowit et al. (1964)
0.75	0	3.14	49.52	Roberts (1965)
0.75	−2 to 0.53	3.14	49.53	Sparrow et al. (1964)
0.70	−0.5 to 0.49	3.14	52.04	Walowit et al. (1964)
0.65	0	3.14	55.01	Roberts (1965)
0.60	−0.25 to 0.36	3.15	58.56	Walowit et al. (1964)
0.50	0	3.16	68.19	Roberts (1965)
0.5	−1.5 to 0.235	3.16	68.19	Sparrow et al. (1964)
0.5	−0.5 to 0.25	3.16	68.18	Walowit et al. (1964)
0.40	−0.25 to 0.16	3.17	83.64	Walowit et al. (1964)
0.36	0	3.19	92.72	Roberts (1965)
0.35	−0.85 to 0.116	3.21	95.38	Sparrow et al. (1964)
0.30	−0.125 to 0.09	3.20	118.89	Walowit et al. (1964)
0.28	0	3.22	120.43	Roberts (1965)
0.25	0 to 0.06	3.24	136.40	Sparrow et al. (1964)

0.25		3.33	122.91	Sparrow et al. (1964)
0.20	0	3.25	176.26	Roberts (1965)
0.20	−0.03 to 0.04	3.23	176.33	Walowit et al. (1964)
0.15	−0.15 to 0.021	3.31	250.10	Sparrow et al. (1964)
0.10	0 to 0.01	3.36	423.48	Sparrow et al. (1964)
0.10	−	3.30	422.79	Walowit et al. (1964)
0.942	−2.89 to 0.864			Taylor (1923)*
0.880	−3.25 to 0.765			Taylor (1923)*
0.873	−3 to 0.76			Coles (1965)*
0.854	−0.41 to 0.41			Nissan et al. (1963)*
0.848	−0.68 to 0			Donnelly and Fultz (1960)*
0.763	−0.705 to 0.498			Nissan et al. (1963)*
0.743	−2.73 to 0.552			Taylor (1923)*
0.698	−0.975 to 0.435			Nissan et al. (1963)*
0.584	−0.555 to 0.294			Nissan et al. (1963)*
0.9625	−1.5 to 0.25			Donnelly and Schwarz (1965)*
0.95	−1.5 to 0.25			Donnelly and Schwarz (1965)*
0.925	0			Lewis (1927)*
0.9	0			Lewis (1927)*
0.875	0			Lewis (1927)*
0.85	0			Lewis (1927)*
0.5	0			Lewis (1927)*

Example 6.7.

An annulus has inner and outer radii of 0.15 m and 0.16 m, respectively. If the annulus is filled completely with a liquid with a viscosity of 0.02 Pa s and density of 890 kg/m^3, determine the value of angular velocity at which the onset of Taylor vortices is likely to occur.

Solution

The mean radius is

$$r_m = \frac{0.15 + 0.16}{2} = 0.155 \, m$$

The annular gap, $a - b$, is 0.01 m. As this is small in comparison to the mean radius, the small gap approximation can be applied.

$$\nu = \frac{\mu}{\rho} = 2.25 \times 10^{-5}$$

From Equation 6.53,

$$\Omega_{cr} = \frac{41.19\nu}{r_m^{0.5}(b-a)^{1.5}} = \frac{41.19 \times 2.25 \times 10^{-5}}{\sqrt{0.155}(0.01)^{1.5}} = 2.35 \, rad/s$$

For an annulus with inner cylinder rotation, further flow regimes can occur after the first transition to the regular Taylor vortex flow illustrated in Figures 6.11 and 6.12. Extensive experimental investigations by Coles (1965), Schwarz, Springnett, and Donnelly (1964), and Snyder (1969, 1970) have revealed that many regular flow configurations can exist beyond the first transition to Taylor vortex flow; these flow configurations are generally called wavy vortex flow. These wavy vortex flow configurations are characterized by a pattern of traveling waves, which are periodic in the circumferential direction and appear just above the critical Taylor number. The transition to wavy vortex flow is not unique and is subject to strong hysteresis. As the rotational Reynolds number is further increased, experimental studies by Gollub and Swinney (1975) and Fenstermacher, Swinney, and Gollub (1979) have shown that at the onset of wavy vortex flow the velocity spectra contains a single fundamental frequency that corresponds to the traveling circumferential waves. As the rotational Reynolds number is increased, a second fundamental frequency appears at $Re_{\phi m}/Re_{\phi m,critical} = 19.3$. This second frequency has been identified as a modulation of the waves. In addition to the spectral components, a broad band characterizing weakly turbulent flow is found at $Re_{\phi m}/Re_{\phi m,critical} \approx 12$. The transition from Taylor vortex flow to wavy spiral flow has been explored numerically by Hoffmann et al. (2009).

Taylor vortex flow theory can be applied to a number of applications. Hendriks (2010) reports on the interaction between disc boundary layer flow and Taylor vortex flows in disc drives and the resultant flow-induced vibrations.

Bilgen and Boulos (1973) give the following equation for the moment coefficient for an annulus with inner cylinder rotation with no axial pressure gradient for transitional flow.

$$C_{mc} = 2 \left(\frac{b-a}{a} \right)^{0.3} Re_{\phi m}^{-0.6} \tag{6.58}$$

for

$$\frac{b-a}{a} > 0.07 \tag{6.59}$$

and

$$64 < Re_{\phi m} < 500 \tag{6.60}$$

For turbulent flow, Bilgen and Boulos (1973) give the following two equations for the moment coefficient for an annulus with inner cylinder rotation with no axial pressure gradient.

$$C_{mc} = 1.03 \left(\frac{b-a}{a}\right)^{0.3} Re_{\phi m}^{-0.5} \tag{6.61}$$

for

$$Re_{\phi m} \leq 1 \times 10^4 \tag{6.62}$$

$$C_{mc} = 0.065 \left(\frac{b-a}{a}\right)^{0.3} Re_{\phi m}^{-0.2} \tag{6.63}$$

for

$$Re_{\phi m} > 1 \times 10^4 \tag{6.64}$$

Example 6.8.

Determine the power required to overcome frictional drag for a 200-mm-long shaft with a diameter of 100 mm rotating in an annulus with an outer diameter of 120 mm. The rotational speed of the inner shaft is 10,000 rpm. The annulus is filled with air with a density and viscosity of 4 kg/m^3 and 3×10^{-5} Pa s, respectively. Compare the figure for the frictional drag to that of the power required to overcome equivalent cylindrical shaft also spinning at 10,000 rpm in Example 6.1.

Solution

From Equation 6.29,

$$Re_{\phi m} = \frac{\rho \Omega a (b-a)}{\mu} = \frac{4 \times 1047 \times 0.05 \times 0.01}{3 \times 10^{-5}} = 69800$$

As the rotational Reynolds number is greater than 10,000, then from Equation 6.63,

$$C_{mc} = 0.065 \left(\frac{b-a}{a}\right)^{0.3} Re_{\phi m}^{-0.2} = 0.065 \left(\frac{0.01}{0.05}\right)^{0.3} (69800)^{-0.2} = 4.310 \times 10^{-3}$$

The power required to overcome frictional drag, from Equation 6.8, is

$$Power = 0.5 \pi \rho \Omega^3 a^4 L C_{mc} = 0.5\pi \times 4 \times 1047^3 \times 0.05^4 \times 0.2 \times 4.31 \times 10^{-3} = 38.85\,W$$

The power required to overcome drag for the rotating shaft located in the annulus is about 40% of that of the rotating cylinder of Example 6.1.

Experiments using hot wire anemometry and flow visualization, for the case of a throughflow of superposed axial air with inner cylinder by Kaye and Elgar (1958) revealed the existence of four modes of flow:

- purely laminar flow
- laminar flow with Taylor vortices
- turbulent flow with vortices
- turbulent flow

Kaye and Elgar's (1958) tests were performed using an annulus with inner and outer radii of 34.9 mm and 47.5 mm, respectively, at axial Reynolds numbers up to 2000 and rotational Reynolds numbers up to 3000. A schematic of the resulting flow regimes as a function of the axial and rotational Reynolds numbers is given in Figure 6.13.

The Rayleigh inviscid criterion for rotational instability remains valid for the case of an annulus with axial throughflow and a fully developed tangential velocity as shown by Chandrasekhar (1960a). Using linear stability theory to predict the onset of instability for the case an annulus with a narrow gap, $a/b{\to}1$, Chandrasekhar (1960b) and Di-Prima (1960) found that the onset of instability increased monotonically with the axial velocity. The critical Taylor number for the case of inner cylinder rotation with axial throughflow is given by

$$Ta_{cr}^2 = Ta_{cr,Re_{z}=0}^2 + 26.5Re_z^2 \qquad (6.65)$$

where the axial Reynolds number Re_z is given by

$$Re_z = \frac{\rho u_z(b-a)}{\mu} \qquad (6.66)$$

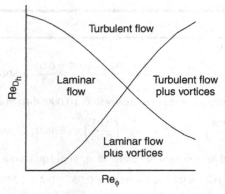

FIGURE 6.13 Schematic representation of the four modes of flow in an annulus with axial throughflow and inner cylinder rotation (after Kaye and Elgar, 1958).

Schwarz et al. (1964) gave the following relationship for the critical Taylor number for an annulus with inner cylinder rotation and axial throughflow based on their experimental measurements, for a narrow gap annulus with $Re_z < 25$.

$$Ta_{cr}^2 = \frac{2(a/b)^2(b-a)^4}{1-(a/b)^2}\frac{\Omega^2}{\nu^2} \qquad (6.67)$$

6.5. JOURNAL BEARINGS

The term *bearing* typically refers to contacting surfaces through which a load is transmitted. Bearings may roll or slide or do both simultaneously. The range of bearing types available is extensive, although they can be broadly split into two categories: sliding bearings (see Figure 6.14), where the motion is facilitated by a thin layer or film of lubricant, and rolling element bearings, where the motion is aided by a combination of rolling motion and lubrication. Lubrication is often required in a bearing to reduce friction between surfaces and to remove heat. Here consideration will be limited to the fluid flow associated with one particular type of sliding bearings, rotating journal bearings. An introduction to rolling element bearings is given in Childs (2004).

The term *sliding bearing* refers to bearings where two surfaces move relative to each other without the benefit of rolling contact. The two surfaces slide over each other, and this motion can be facilitated by means of a lubricant that gets squeezed by the motion of the components and can under certain conditions generate sufficient pressure to separate the surfaces, thereby reducing frictional contact and wear. A typical application of sliding bearings is to allow rotation of a load-carrying shaft. The portion of the shaft at the bearing is referred to as the journal, and the stationary part, which supports the load, is called the bearing (see Figure 6.15). For this reason, sliding bearings are

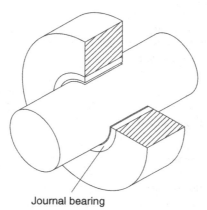

Journal bearing

FIGURE 6.14   A journal bearing.

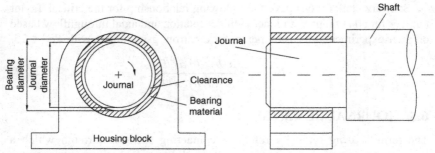

FIGURE 6.15   A plain surface, sliding, or journal bearing.

often collectively referred to as journal bearings, although this term ignores the existence of sliding bearings that support linear translation of components. Another common term is *plain surface bearings*. This section is concerned principally with bearings for rotary motion, and the terms *journal* and *sliding bearing* are used interchangeably.

There are three principal regimes of lubrication for sliding bearings:

1. boundary lubrication
2. mixed film lubrication
3. full film lubrication

Boundary lubrication typically occurs at low relative velocities between the journal and the bearing surfaces and is characterized by actual physical contact. The surfaces, even if ground to a very low value of surface roughness, will still consist of a series of peaks and troughs as illustrated schematically in Figure 6.16. Although some lubricant may be present, the pressures generated within it are not significant, and the relative motion of the surfaces brings the corresponding peaks periodically into contact. Mixed film lubrication occurs when the relative motion between the surfaces is sufficient to generate high enough

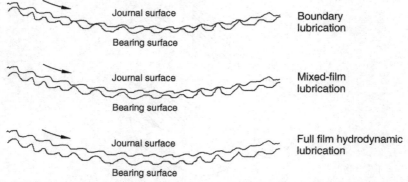

FIGURE 6.16   Schematic representation of the surface roughness for sliding bearings and the relative position depending on the type of lubrication occurring.

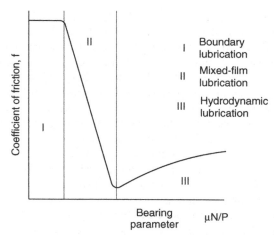

FIGURE 6.17   Schematic representation of the variation of bearing performance with lubrication.

pressures in the lubricant film, which can partially separate the surfaces for periods of time. There is still contact in places around the circumference between the two components. Full film lubrication occurs at higher relative velocities. Here the motion of the surface generates high pressures in the lubricant, which separate the two components and the journal can "ride" on a wedge of fluid. All of these types of lubrication can be encountered in a bearing without external pressure of the bearing. If lubricant under high enough pressure is supplied to the bearing to separate the two surfaces, it is called a hydrostatic bearing.

The performance of a sliding bearing differs markedly, depending on which type of lubrication is physically occurring. This is illustrated in Figure 6.17, which shows the characteristic variation of the coefficient of friction with a group of variables called the bearing parameter, which is defined by

$$\frac{\mu N}{P} \tag{6.68}$$

where

μ = viscosity of lubricant (Pa s)
N = speed (for this definition normally given in rpm)
P = load capacity (N/m^2) given in Equation 6.69

$$P = \frac{W}{LD} \tag{6.69}$$

where

W = applied load (N)
L = bearing length (m)
D = journal diameter (m)

The bearing parameter, $\mu N/P$, groups several of the bearing design variables into one number. Normally, of course, a low coefficient of friction is desirable in order to reduce the power lost in the overcoming friction. In general, boundary lubrication is used for slow speed applications where the surface speed is less than approximately 1.5 m/s. Mixed film lubrication is rarely used because it is difficult to quantify the actual value of the coefficient of friction (note the steep gradient in Figure 6.17 for this zone). The design of boundary-lubricated bearings is outlined in Section 6.5.1, and full film hydrodynamic bearings are described in Section 6.5.2.

As can be seen from Figure 6.17, bearing performance is dependent on the type of lubrication occurring and the viscosity of the lubricant. The viscosity is a measure of a fluid's resistance to shear. Lubricants can be solid, liquid, or gaseous, although the most commonly known are oils and greases. The principal classes of liquid lubricants are mineral oils and synthetic oils. Their viscosity is highly dependent on temperature and pressure as indicated for the case of temperature in Figure 6.18. They are typically solid at –35 °C, thin as paraffin at 100 °C, and burn above 240 °C. Many additives are used to affect their performance. For example, EP (extreme pressure) additives add fatty acids and other compounds to the oil, which attack the metal surfaces to form "contaminant" layers, which protect the surfaces and reduce friction even when the oil film is squeezed out by high-contact loads. Greases are oils mixed with soaps to form a thicker lubricant that can be retained on surfaces.

The viscosity variation with temperature of oils has been standardized, and oils are available with a class number—for example, SAE 10, SAE 20, SAE 30,

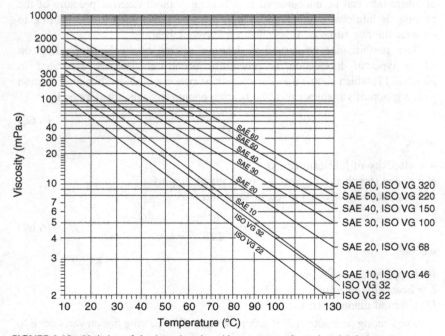

FIGURE 6.18 Variation of absolute viscosity with temperature for various lubricants.

SAE 40, SAE 5W, and SAE 10W. The Society of Automotive Engineers developed this identification system in order to class oils for general-purpose use and winter use, the "W" signifying the latter. The lower the numerical value, the thinner or less viscous the oil. Multigrade oil (e.g., SAE 10W/40) is formulated to meet the viscosity requirements of two oils, giving some of the benefits of the constituent parts. An equivalent identification system is also available from the International Organization for Standardization (ISO 3448).

6.5.1 Boundary-Lubricated Bearings

As described in Section 6.5, the journal and bearing surfaces in a boundary lubricated bearing are in direct contact in places. These bearings are typically used for very-low-speed applications such as bushes and linkages where their simplicity and compact nature are advantageous. Examples include shafts for gear wheels in electric tools, lawnmower wheels, garden hand tools such as shears, ratchet wrenches, and domestic and automotive door hinges incorporating the journal as a solid pin riding inside a cylindrical outer member.

General considerations in the design of a boundary lubricated bearing are:

- the coefficient of friction (both static and dynamic)
- the load capacity
- the relative velocity between the stationary and moving components
- the operating temperature
- wear limitations
- the production capability

A useful measure in the design of boundary-lubricated bearings is the PV factor (load capacity × peripheral speed), which indicates the ability of the bearing material to accommodate the frictional energy generated in the bearing. At the limiting PV value the temperature will be unstable, and failure will occur rapidly. A practical value for PV is half the limiting PV value. Values for PV depend on the material combinations concerned and typically vary between 0.035 in the case of thermoplastic with filler to 1.75 for PTFE with filler, bonded to steel backing. The preliminary design of a boundary lubricated bearing essentially consists of setting the bearing proportions, its length and its diameter, and selecting the bearing material such that an acceptable PV value is obtained. This approach is set out as a step-by-step procedure as follows.

1. Determine the speed of rotation of the bearing and the load to be supported.
2. Set the bearing proportions. Common practice is to set the length to diameter ratio between 0.5 and 1.5. If the diameter D is known as an initial trial, set L equal to D.
3. Calculate the load capacity, $P = W/(LD)$.
4. Determine the maximum tangential speed of the journal.
5. Calculate the PV factor ($P \times u_\phi$).
6. Multiply the PV value obtained by a factor of safety of 2.

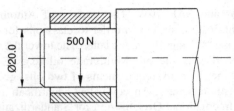

FIGURE 6.19 Boundary lubricated bearing design example.

7. Interrogate manufacturer's data or the limited example data in Table 6.2 to identify an appropriate bearing material with a value for PV factor greater than that obtained in (6).

Example 6.9.

A bearing is to be designed to carry a radial load of 500 N for a shaft of diameter 20 mm running at a speed of 100 rpm (see Figure 6.19). Calculate the PV factor, and by comparison with the available materials listed in Table 6.2 determine a suitable bearing material.

Solution

The primary data are $W = 500$ N, $D = 20$ mm, and $N = 100$ rpm.

Use $L/D = 1$ as an initial suggestion for the length to diameter ratio for the bearing. $L = 20$ mm.

Calculating the load capacity, P

$$P = \frac{W}{LD} = \frac{500}{0.02 \times 0.02} = 1.25 \text{ MN/m}^2$$

$$u_\phi = \Omega a = \left(\frac{2\pi}{60}\right) N \frac{D}{2} = 0.1047 \times 100 \times \frac{0.02}{2} = 0.1047 \text{ m/s}$$

$$PV \approx 0.13 \quad (\text{MN/m}^2)(\text{m/s})$$

Multiplying this by a safety of factor of 2 gives a PV factor of 0.26 (MN/m s)

A material with PV factor greater than this, such as filled PTFE (limiting PV factor up to 0.35), or PTFE with filler bonded to a steel backing (limiting PV factor up to 1.75), would give acceptable performance.

6.5.2 Design of Full Film Hydrodynamic Bearings

In a full film hydrodynamic bearing, the load on the bearing is supported on a continuous film of lubricant so that no contact between the bearing and the rotating journal occurs. The motion of the journal inside the bearing creates the

TABLE 6.2 Characteristics of some rubbing materials. Reproduced courtesy of Neale (1995)

MATERIAL	MAXIMUM LOAD CAPACITY P (MN/m²)	LIMITING PV FACTOR (MN/m s)	MAXIMUM OPERATING TEMPERATURE (°C)	COEFFICIENT OF FRICTION	COEFFICIENT OF EXPANSION (×10⁻⁶/°C)	COMMENTS	TYPICAL APPLICATION
Carbon/ graphite	1.4 – 2	0.11	350–500	0.1–0.25 dry	2.5–5.0	For continuous dry operation	Food and textile machinery
Carbon/ graphite with metal	3.4	0.145	130–350	0.1–0.35 dry	4.2–5		
Graphite impregnated metal	70	0.28–0.35	350–600	0.1–0.15 dry	12–13		
Graphite/ thermo-setting resin	2	0.35	250	0.13–0.5 dry	3.5–5	Suitable for sea water operation	
Reinforced thermo-setting plastics	35	0.35	200	0.1–0.4 dry	25–80		Roll-neck bearings
Thermo-plastic material without filler	10	0.035	100	0.1–0.45 dry	100		Bushes and thrust washers
Thermo-plastic with filler or metal backed	10–14	0.035–0.11	100	0.15–0.4 dry	80–100		Bushes and thrust washers

(Continued)

Table 6.2 (*Continued*)

MATERIAL	MAXIMUM LOAD CAPACITY P (MN/m²)	LIMITING PV FACTOR (MN/m s)	MAXIMUM OPERATING TEMPERATURE (°C)	COEFFICIENT OF FRICTION	COEFFICIENT OF EXPANSION (×10⁻⁶/°C)	COMMENTS	TYPICAL APPLICATION
Thermo-plastic material with filler bonded to metal back	140	0.35	105	0.2–0.35 dry	27	With initial lubrication only	For conditions of intermittent operation or boundary lubrication e.g. ball joints, suspension, steering
Filled PTFE	7	up to 0.35	250	0.05–0.35 dry	60–80	Glass, mica, bronze, graphite	For dry operations where low friction and wear required
PTFE with filler, bonded to steel backing	140	up to 1.75	280	0.05–0.3 dry	20	Sintered bronze bonded to steel backing impregnated with PTFE/lead	Aircraft controls, linkages, gearbox, clutch, conveyors, bridges
Woven PTFE reinforced and bonded to metal backing	420	up to 1.6	250	0.03–0.3	–	Reinforcement may be interwoven glass fibre or rayon	Aircraft and engine controls, linkages, engine mountings, bridge bearings

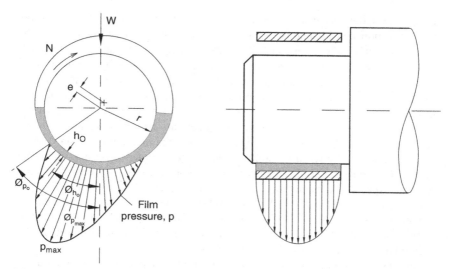

FIGURE 6.20 Motion of the journal generates pressure in the lubricant separating the two surfaces. Beyond h_o, the minimum film thickness, the pressure terms go negative and the film is ruptured.

FIGURE 6.21 Partial surface journal bearings.

necessary pressure to support the load (Figure 6.20). Hydrodynamic bearings are commonly found in internal combustion engines for supporting the crankshaft and in turbocharger applications. Hydrodynamic bearings can consist of a full circumferential surface or a partial surface around the journal (Figure 6.21).

Bearing design involves a significant number of often conflicting parameters, with improvement in one feature resulting in deterioration in another. One approach to designing a bearing system would be to assign attribute points to each aspect of the design and undertake an optimization exercise. This, however, would be time consuming if the software was not already in a developed state and would not necessarily produce an optimum result due to inadequacies in modeling and incorrect assignment of attribute weightings. An alternative approach, which is sensible as a starting point and outlined here, is to develop one or a number of feasible designs and use judgments to select the best, or combine the best features of the proposed designs.

The design procedure for a journal bearing, recommended here as a starting point, includes the specification of the journal radius a, the radial clearance c, the axial length of the bearing surface, L, the type of lubricant and its viscosity, μ, the journal speed, N, and the load, W. Values for the speed, the load, and possibly the journal radius are usually specified by the machine requirements and stress and deflection considerations. As such, journal bearing design consists of the determination of the radial clearance, the bearing length, and the lubricant viscosity. The design process for a journal bearing is usually iterative. Trial values for the clearance, the length and the viscosity are chosen, various performance criteria calculated, and the process repeated until a satisfactory or optimized design is achieved. Criteria for optimization may be minimizing of the frictional loss, minimizing the lubricant temperature rise, minimizing the lubricant supply, maximizing the load capability, and minimizing production costs.

The clearance between the journal and the bearing depends on the nominal diameter of the journal, the precision of the machine, surface roughness, and thermal expansion considerations. An overall guideline is for the radial clearance, c, to be in the range $0.001a < c < 0.002a$, where a is the nominal bearing radius ($0.001D < 2c < 0.002D$). Figure 6.22 shows values for the recommended diametral clearance ($2 \times c$) as a function of the journal diameter and rotational speed for steadily loaded bearings.

For a given combination of a, c, L, μ, N, and W, the performance of a journal bearing can be calculated. This requires determining the pressure distribution in the bearing, the minimum film thickness h_o, the location of the minimum film

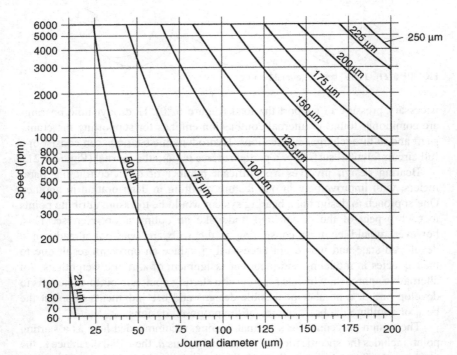

FIGURE 6.22 Minimum recommended values for the diametral clearance ($2 \times c$) for steadily loaded journal bearings (reproduced from Welsh, 1983).

thickness $\phi_{p_{max}}$, the coefficient of friction f, the lubricant flow Q, the maximum film pressure p_{max}, and the temperature rise ΔT of the lubricant.

The pressure distribution in a journal bearing (see Figure 6.20) can be determined by solving the relevant form of the Navier-Stokes fluid flow equations, which in the reduced form for journal bearings is called the Reynolds equation and was first derived by Reynolds (1886) and further developed by Harrison (1913) to include the effects of compressibility. Here the derivation of the general Reynolds equation is developed following the general outline given by Hamrock (1994).

The Navier-Stokes equations in a Cartesian coordinate system, for compressible flow assuming constant viscosity, are given by Equations 2.46–2.48 and repeated here for convenience, Equations 6.70–6.72. Here x is taken as the coordinate in the direction of sliding, y is the coordinate in the direction of side leakage and z is the coordinate across the lubricant film.

$$\rho \left(\frac{\partial u_x}{\partial t} + u_x \frac{\partial u_x}{\partial x} + u_y \frac{\partial u_x}{\partial y} + u_z \frac{\partial u_x}{\partial z} \right) = -\frac{\partial p}{\partial x} + \mu \left(\frac{\partial^2 u_x}{\partial x^2} + \frac{\partial^2 u_x}{\partial y^2} + \frac{\partial^2 u_x}{\partial z^2} \right)$$
$$+ \frac{\mu}{3} \frac{\partial}{\partial x} \left(\frac{\partial u_x}{\partial x} + \frac{\partial u_y}{\partial y} + \frac{\partial u_z}{\partial z} \right) + F_x \tag{6.70}$$

$$\rho \left(\frac{\partial u_y}{\partial t} + u_x \frac{\partial u_y}{\partial x} + u_y \frac{\partial u_y}{\partial y} + u_z \frac{\partial u_y}{\partial z} \right) = -\frac{\partial p}{\partial y} + \mu \left(\frac{\partial^2 u_y}{\partial x^2} + \frac{\partial^2 u_y}{\partial y^2} + \frac{\partial^2 u_y}{\partial z^2} \right)$$
$$+ \frac{\mu}{3} \frac{\partial}{\partial y} \left(\frac{\partial u_x}{\partial x} + \frac{\partial u_y}{\partial y} + \frac{\partial u_z}{\partial z} \right) + F_y \tag{6.71}$$

$$\rho \left(\frac{\partial u_z}{\partial t} + u_x \frac{\partial u_z}{\partial x} + u_y \frac{\partial u_z}{\partial y} + u_z \frac{\partial u_z}{\partial z} \right) = -\frac{\partial p}{\partial z} + \mu \left(\frac{\partial^2 u_z}{\partial x^2} + \frac{\partial^2 u_z}{\partial y^2} + \frac{\partial^2 u_z}{\partial z^2} \right)$$
$$+ \frac{\mu}{3} \frac{\partial}{\partial z} \left(\frac{\partial u_x}{\partial x} + \frac{\partial u_y}{\partial y} + \frac{\partial u_z}{\partial z} \right) + F_z \tag{6.72}$$

If the density is assumed constant, then the dilation, Equation 6.73, is zero.

$$\frac{\partial u_x}{\partial x} + \frac{\partial u_y}{\partial y} + \frac{\partial u_z}{\partial z} = 0 \tag{6.73}$$

The Navier-Stokes equations for constant density and constant viscosity reduce to

$$\rho \left(\frac{\partial u_x}{\partial t} + u_x \frac{\partial u_x}{\partial x} + u_y \frac{\partial u_x}{\partial y} + u_z \frac{\partial u_x}{\partial z} \right) = -\frac{\partial p}{\partial x} + \mu \left(\frac{\partial^2 u_x}{\partial x^2} + \frac{\partial^2 u_x}{\partial y^2} + \frac{\partial^2 u_x}{\partial z^2} \right) + F_x \tag{6.74}$$

$$\rho\left(\frac{\partial u_y}{\partial t}+u_x\frac{\partial u_y}{\partial x}+u_y\frac{\partial u_y}{\partial y}+u_z\frac{\partial u_y}{\partial z}\right)=-\frac{\partial p}{\partial y}+\mu\left(\frac{\partial^2 u_y}{\partial x^2}+\frac{\partial^2 u_y}{\partial y^2}+\frac{\partial^2 u_y}{\partial z^2}\right)+F_y$$

(6.75)

$$\rho\left(\frac{\partial u_z}{\partial t}+u_x\frac{\partial u_z}{\partial x}+u_y\frac{\partial u_z}{\partial y}+u_z\frac{\partial u_z}{\partial z}\right)=-\frac{\partial p}{\partial z}+\mu\left(\frac{\partial^2 u_z}{\partial x^2}+\frac{\partial^2 u_z}{\partial y^2}+\frac{\partial^2 u_z}{\partial z^2}\right)+F_z$$

(6.76)

For conditions known as slow viscous motion, where pressure and viscous terms predominate, simplifications are possible for the Navier-Stokes equations, making their solution more amenable to analytical and numerical techniques.

The Navier-Stokes equations can be nondimensionalized to enable a generalized solution, using the same procedure as described in Section 2.6. Here the process is repeated to give dimensionless groups of specific relevance to journal bearings, using the characteristic parameters given in Equations 6.77–6.86.

$$x_* = \frac{x}{l_o}$$

(6.77)

where l_o is a characteristic length in the x direction.

$$y_* = \frac{y}{b_o}$$

(6.78)

where b_o is a characteristic length in the y direction.

$$z_* = \frac{z}{h_o}$$

(6.79)

where h_o is a characteristic length in the z direction.

$$t_* = \frac{t}{t_o}$$

(6.80)

where t_o is a characteristic time.

$$u_{x*} = \frac{u_x}{u_{x,o}}$$

(6.81)

where $u_{x,o}$ is a characteristic velocity in the x direction.

$$u_{y*} = \frac{u_y}{u_{y,o}}$$

(6.82)

where $u_{y,o}$ is a characteristic velocity in the y direction.

$$u_{z*} = \frac{u_z}{u_{z,o}}$$

(6.83)

where $u_{z,o}$ is a characteristic velocity in the z direction.

$$\rho_* = \frac{\rho}{\rho_o} \tag{6.84}$$

where ρ_o is a characteristic density.

$$\mu_* = \frac{\mu}{\mu_o} \tag{6.85}$$

where μ_o is a characteristic viscosity.

$$p_* = \frac{h_o^2 p}{\mu_o u_{x,o} l_o} \tag{6.86}$$

Substitution of the above dimensionless parameters, Equations 6.77–6.86 in Equation 6.70, gives

$$\frac{l_o}{u_{x,o} t_o}\frac{\partial u_{x*}}{\partial t_*} + u_{x*}\frac{\partial u_{x*}}{\partial x_*} + \frac{l_o u_{y,o}}{b_o u_{x,o}}u_{y*}\frac{\partial u_{x*}}{\partial y_*} + \frac{l_o u_{z,o}}{h_o u_{x,o}}u_{z*}\frac{\partial u_{x*}}{\partial z_*} = \frac{l_o g}{u_{x,o}^2}$$

$$-\frac{\mu_o}{\rho_o u_{x,o} l_o}\left(\frac{l_o}{h_o}\right)^2\frac{1}{\rho_*}\frac{\partial p_*}{\partial x_*} - \frac{2}{3}\frac{\mu_o}{\rho_o u_{x,o} l_o}\frac{1}{\rho_*}\frac{\partial}{\partial x_*}\left[\mu_*\left(\frac{\partial u_{x*}}{\partial x_*} + \frac{u_{y,o}l_o}{u_{x,o}b_o}\frac{\partial u_{y*}}{\partial y_*} + \frac{u_{z,o}l_o}{u_{x,o}h_o}\frac{\partial u_{z*}}{\partial z_*}\right)\right]$$

$$+\frac{2\mu_o}{\rho_o u_{x,o} l_o}\frac{1}{\rho_*}\frac{\partial}{\partial x_*}\left(\mu_*\frac{\partial u_{x*}}{\partial x_*}\right) + \frac{\mu_o}{\rho_o u_{x,o} l_o}\left(\frac{l_o}{b_o}\right)^2\frac{1}{\rho_*}\frac{\partial}{\partial y_*}\left[\mu_*\left(\frac{\partial u_{x*}}{\partial y_*} + \frac{u_{y,o}b_o}{u_{x,o}l_o}\frac{\partial u_{y*}}{\partial x_*}\right)\right]$$

$$+\frac{\mu_o}{\rho_o u_{x,o} l_o}\left(\frac{l_o}{h_o}\right)^2\frac{1}{\rho_*}\frac{\partial}{\partial z_*}\left[\mu_*\left(\frac{\partial u_{x*}}{\partial z_*} + \frac{u_{z,o}h_o}{u_{x,o}l_o}\frac{\partial u_{z*}}{\partial x_*}\right)\right]$$

$$\tag{6.87}$$

The relative importance of inertia and viscous forces can be determined by examining the value of the Reynolds number

$$\text{Re} = \frac{\rho_o u_{x,o} l_o}{\mu_o} \tag{6.88}$$

The inverse of the Reynolds number occurs throughout Equation 6.87.

In fluid film lubrication, because of the dominance of the viscous term $\partial^2 u_{x*}/\partial^2 z_*$, a modified form of the Reynolds number is used, defined for the x component of velocity by

$$\text{Re}_x = \frac{\rho_o u_{x,o} h_o^2}{\mu_o l_o} \tag{6.89}$$

and similarly for the y and z directions

$$\text{Re}_y = \frac{\rho_o u_{y,o} h_o^2}{\mu_o b_o} \tag{6.90}$$

$$\mathrm{Re}_z = \frac{\rho_o u_{z,o} h_o}{\mu_o} \tag{6.91}$$

The squeeze number is defined by

$$\sigma_s = \frac{\rho_o h_o^2}{\mu_o t_o} \tag{6.92}$$

Typically, in hydrodynamically lubricated journal bearings, viscous forces are much greater than inertia forces, and a typical Reynolds number, using Equation 6.89, might be of the order of 1×10^{-4}.

Substitution of the Reynolds and squeeze numbers into Equation 6.87 gives

$$\sigma_s \frac{\partial u_{x*}}{\partial t_*} + \mathrm{Re}_x u_{x*} \frac{\partial u_{x*}}{\partial x_*} + \mathrm{Re}_y u_{y*} \frac{\partial u_{x*}}{\partial y_*} + \mathrm{Re}_z u_{z*} \frac{\partial u_{x*}}{\partial z_*} = \frac{l_o g}{u_{x,o}^2} \mathrm{Re}_x - \frac{1}{\rho_*} \frac{\partial p_*}{\partial x_*}$$

$$+ \frac{1}{\rho_*} \frac{\partial}{\partial z_*} \left(\mu_* \frac{\partial u_{x*}}{\partial z_*} \right) - \frac{2}{3} \left(\frac{h_o}{l_o} \right)^2 \frac{1}{\rho_*} \frac{\partial}{\partial x_*} \left[\mu_* \left(\frac{\partial u_{x*}}{\partial x_*} + \frac{u_{y,o} l_o}{u_{x,o} b_o} \frac{\partial u_{y*}}{\partial y_*} + \frac{u_{z,o} l_o}{u_{x,o} h_o} \frac{\partial u_{z*}}{\partial z_*} \right) \right]$$

$$+ \left(\frac{h_o}{b_o} \right)^2 \frac{1}{\rho_*} \frac{\partial}{\partial y_*} \left[\mu_* \left(\frac{\partial u_{x*}}{\partial y_*} + \frac{u_{y,o} b_o}{u_{x,o} l_o} \frac{\partial u_{y*}}{\partial x_*} \right) \right] + 2 \left(\frac{h_o}{l_o} \right)^2 \frac{1}{\rho_*} \frac{\partial}{\partial x_*} \left(\mu_* \frac{\partial u_{x*}}{\partial x_*} \right)$$

$$+ \frac{1}{\rho_*} \frac{\partial}{\partial z_*} \left(\mu_* \frac{u_{z,o} h_o}{u_{x,o} l_o} \frac{\partial u_{z*}}{\partial x_*} \right)$$

$$\tag{6.93}$$

Examination of the order of magnitude of the terms in Equation 6.93 provides an indication of the relative significance of each of these and therefore which need to considered. The inertia terms, gravity term, and $u_{z,o}/u_{x,o}$ are of order h_o/l_o. The pressure gradient term and the first viscous term are of order unity. The remaining viscous terms are of order $(h_o/l_o)^2$ or $(h_o/b_o)^2$ and therefore very small in comparison to the other terms. Neglecting terms of the order of $(h_o/l_o)^2$ and $(h_o/b_o)^2$ gives

$$\sigma_s \frac{\partial u_{x*}}{\partial t_*} + \mathrm{Re}_x u_{x*} \frac{\partial u_{x*}}{\partial x_*} + \mathrm{Re}_y u_{y*} \frac{\partial u_{x*}}{\partial y_*} + \mathrm{Re}_z u_{z*} \frac{\partial u_{x*}}{\partial z_*} = \frac{l_o g}{u_{x,o}^2} \mathrm{Re}_x - \frac{1}{\rho_*} \frac{\partial p_*}{\partial x_*}$$

$$+ \frac{1}{\rho_*} \frac{\partial}{\partial z_*} \left(\mu_* \frac{\partial u_{x*}}{\partial z_*} \right) \tag{6.94}$$

Similarly, for the y and z components

$$\sigma_s \frac{\partial u_{y*}}{\partial t_*} + \mathrm{Re}_x u_{x*} \frac{\partial u_{y*}}{\partial x_*} + \mathrm{Re}_y u_{y*} \frac{\partial u_{y*}}{\partial y_*} + \mathrm{Re}_z u_{z*} \frac{\partial u_{y*}}{\partial z_*} = \frac{b_o g}{u_{y,o}^2} \mathrm{Re}_y - \frac{1}{\rho_*} \frac{\partial p_*}{\partial y_*}$$
$$+ \frac{1}{\rho_*} \frac{\partial}{\partial z_*} \left(\mu_* \frac{\partial u_{y*}}{\partial z_*} \right) \tag{6.95}$$

$$\frac{\partial p}{\partial z_*} = 0 \tag{6.96}$$

Examination of Equations 6.94–6.96 shows that the pressure is a function of x_*, y_*, and t_*,

$$p = f(x_*, y_*, t_*) \tag{6.97}$$

The continuity equation can be expressed as

$$\sigma_s \frac{\partial p_*}{\partial t_*} + \mathrm{Re}_x \frac{\partial}{\partial x_*}(\rho_* u_{x*}) + \mathrm{Re}_y \frac{\partial}{\partial y_*}(\rho_* u_{y*}) + \mathrm{Re}_z \frac{\partial}{\partial z_*}(\rho_* u_{z*}) = 0 \tag{6.98}$$

A check can be made to identify if Taylor vortices are likely to occur. If the Taylor number, Equation 6.12, is greater than approximately 41.2, then Taylor vortices may form, and laminar flow conditions may no longer hold, invalidating the use of the equations developed in this section.

The Froude number, Equation 2.142, indicates the relative importance of inertia and gravity forces. A typical Froude number for a journal bearing might be of the order of 10, providing an indication that gravity forces can be neglected in comparison with viscous forces.

The importance of pressure relative to inertia can be judged by examining the Euler number, Equation 2.143. For a typical journal bearing, the Euler number may be of the order of 100, indicating that the pressure term is much larger than the inertia term.

If in addition to neglecting terms of order $(h_o/l_o)^2$ or $(h_o/b_o)^2$, terms of the order of h_o/l_o and h_o/b_o are neglected and only terms of the order of unity are considered, the Navier-Stokes equations reduce to

$$\frac{\partial p}{\partial x} = \frac{\partial}{\partial z} \left(\mu \frac{\partial u_x}{\partial z} \right) \tag{6.99}$$

$$\frac{\partial p}{\partial y} = \frac{\partial}{\partial z} \left(\mu \frac{\partial u_y}{\partial z} \right) \tag{6.100}$$

Equation 6.97 shows that the pressure is only a function of x and y for steady-state conditions. Equations 6.99 and 6.100 can therefore be integrated directly to give general expressions for the velocity gradients as follows.

$$\frac{\partial u_x}{\partial z} = \frac{z}{\mu} \frac{\partial p}{\partial x} + \frac{A}{\mu} \tag{6.101}$$

$$\frac{\partial u_y}{\partial z} = \frac{z}{\mu}\frac{\partial p}{\partial y} + \frac{C}{\mu} \tag{6.102}$$

where A and C are constants of integration.

The temperature across the thin layer of layer lubricant in a journal bearing may vary significantly. As viscosity is highly dependent on temperature, this leads to increased complexity in obtaining a solution to Equations 6.101 and 6.102. In many fluid film applications, however, it has been found acceptable to model the viscosity of a fluid film using an average value for the viscosity across the film. With μ taken as the average value of viscosity across the fluid film, integration of Equations 6.101 and 6.102 gives the velocity components as follows.

$$u_x = \frac{z^2}{2\mu}\frac{\partial p}{\partial x} + A\frac{z}{\mu} + B \tag{6.103}$$

$$u_y = \frac{z^2}{2\mu}\frac{\partial p}{\partial y} + C\frac{z}{\mu} + D \tag{6.104}$$

where B and D are constants of integration.

If the no-slip condition is assumed at the fluid solid interface, then the boundary conditions are as follows and illustrated in Figure 6.23.

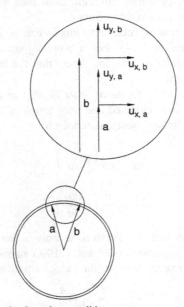

FIGURE 6.23 Journal bearing boundary conditions.

$$u_x = u_{x,a},\ u_y = u_{y,a} \text{ at } z = 0$$
$$u_x = u_{x,b},\ u_y = u_{y,b} \text{ at } z = h$$

Application of the boundary conditions to the equations for the velocity gradients and velocity components, Equations 6.103–6.104, gives

$$\frac{\partial u_x}{\partial z} = \frac{2z - h}{2\mu}\frac{\partial p}{\partial x} - \frac{u_{x,a} - u_{x,b}}{h} \tag{6.105}$$

$$\frac{\partial u_y}{\partial z} = \frac{2z - h}{2\mu}\frac{\partial p}{\partial y} - \frac{u_{y,a} - u_{y,b}}{h} \tag{6.106}$$

$$u_x = \frac{z(z - h)}{2\mu}\frac{\partial p}{\partial x} + \frac{h - z}{h}u_{x,a} + \frac{z}{h}u_{x,b} \tag{6.107}$$

$$u_y = \frac{z(z - h)}{2\mu}\frac{\partial p}{\partial y} + \frac{h - z}{h}u_{y,a} + \frac{z}{h}u_{y,b} \tag{6.108}$$

The viscous shear stresses are defined by

$$\tau_{zx} = \mu\left(\frac{\partial u_z}{\partial x} + \frac{\partial u_x}{\partial z}\right) \tag{6.109}$$

$$\tau_{zy} = \mu\left(\frac{\partial u_z}{\partial y} + \frac{\partial u_y}{\partial z}\right) \tag{6.110}$$

The order of magnitude of $\partial u_z/\partial x$ and $\partial u_z/\partial y$ are much smaller than $\partial u_x/\partial z$ and $\partial u_y/\partial z$, so the viscous shear stresses can be approximated by

$$\tau_{zx} = \mu\frac{\partial u_x}{\partial z} \tag{6.111}$$

$$\tau_{zy} = \mu\frac{\partial u_y}{\partial z} \tag{6.112}$$

From Equations 6.105 and 6.106, the viscous shear stresses acting on the solid surfaces can be expressed by

$$(\tau_{zx})_{z=0} = \left(\mu\frac{\partial u_x}{\partial z}\right)_{z=0} = -\frac{h}{2}\frac{\partial p}{\partial x} - \frac{\mu(u_{x,a} - u_{x,b})}{h} \tag{6.113}$$

$$(-\tau_{zx})_{z=h} = -\left(\mu\frac{\partial u_x}{\partial z}\right)_{z=h} = -\frac{h}{2}\frac{\partial p}{\partial x} + \frac{\mu(u_{x,a} - u_{x,b})}{h} \tag{6.114}$$

$$(\tau_{zy})_{z=0} = \left(\mu\frac{\partial u_y}{\partial z}\right)_{z=0} = -\frac{h}{2}\frac{\partial p}{\partial y} - \frac{\mu(u_{y,a} - u_{y,b})}{h} \tag{6.115}$$

$$(-\tau_{zy})_{z=h} = -\left(\mu\frac{\partial u_y}{\partial z}\right)_{z=h} = -\frac{h}{2}\frac{\partial p}{\partial y} + \frac{\mu(u_{y,a}-u_{y,b})}{h} \qquad (6.116)$$

The negative signs for the viscous shear stresses, in Equations 6.113 to 6.116, indicate that the stress acts in a direction opposite to the motion.

The volumetric flow rates per unit width in the x and y directions are defined by

$$q_x = \int_0^h u_x dz \qquad (6.117)$$

$$q_y = \int_0^h u_y dz \qquad (6.118)$$

Substituting for the velocity components using Equations 6.107 and 6.108 in Equations 6.117 and 6.118 gives

$$q_x = -\frac{h^3}{12\mu}\frac{\partial p}{\partial x} + \frac{u_{x,a}+u_{x,b}}{2}h \qquad (6.119)$$

$$q_y = -\frac{h^3}{12\mu}\frac{\partial p}{\partial y} + \frac{u_{y,a}+u_{y,b}}{2}h \qquad (6.120)$$

The Reynolds equation is formed by substituting the expressions for the volumetric flow rate into the continuity equation.

Integrating the continuity equation gives

$$\int_0^h\left[\frac{\partial \rho}{\partial t} + \frac{\partial}{\partial x}(\rho u_x) + \frac{\partial}{\partial y}(\rho u_y) + \frac{\partial}{\partial z}(\rho u_z)\right]dz = 0 \qquad (6.121)$$

The integral

$$\int_0^h\frac{\partial}{\partial x}[f(x,y,z)]dz = -f(x,y,h)\frac{\partial h}{\partial x} + \frac{\partial}{\partial x}\left[\int_0^h f(x,y,z)dz\right] \qquad (6.122)$$

If the density is assumed to be the mean density of the fluid across the film, then the u_x term in Equation 6.121 is

$$\int_0^h\frac{\partial}{\partial x}(\rho u_x)dz = -(\rho u_x)_{z=h}\frac{\partial h}{\partial x} + \frac{\partial}{\partial x}\left(\int_0^h \rho u_x dz\right) = -\rho u_{x,b}\frac{\partial h}{\partial x} + \frac{\partial}{\partial x}\left(\int_0^h \rho u_x dz\right) \qquad (6.123)$$

Similarly, for the u_y term in Equation 6.121,

$$\int_0^h\frac{\partial}{\partial y}(\rho u_y)dz = -\rho u_{y,b}\frac{\partial h}{\partial y} + \frac{\partial}{\partial y}\left(\int_0^h \rho u_y dz\right) \qquad (6.124)$$

The u_z term can be integrated directly giving

$$\int_0^h\frac{\partial}{\partial z}(\rho u_z)dz = \rho(u_{z,b}-u_{z,a}) \qquad (6.125)$$

The integral form of the continuity equation, Equation 6.121, on substitution of Equations 6.123-6.125, can be stated as

$$h\frac{\partial \rho}{\partial t} - \rho u_{x,b}\frac{\partial h}{\partial x} + \frac{\partial}{\partial x}\left(\rho \int_0^h u_x dz\right) - \rho u_{y,b}\frac{\partial h}{\partial y} + \frac{\partial}{\partial y}\left(\rho \int_0^h u_y dz\right) + \rho(u_{z,b} - u_{z,a}) = 0$$

(6.126)

The integrals in Equation 6.126 represent the volumetric flow rates per unit width. Substitution of the values for these integrals from Equations 6.119 and 6.120 gives the general Reynolds equation.

$$\frac{\partial}{\partial x}\left(-\frac{\rho h^3}{12\mu}\frac{\partial p}{\partial x}\right) + \frac{\partial}{\partial y}\left(-\frac{\rho h^3}{12\mu}\frac{\partial p}{\partial y}\right) + \frac{\partial}{\partial x}\left[\frac{\rho h(u_{x,a}+u_{x,b})}{2}\right] + \frac{\partial}{\partial y}\left[\frac{\rho h(u_{y,a}+u_{y,b})}{2}\right] +$$

$$\rho(u_{z,b}-u_{z,a}) - \rho u_{x,b}\frac{\partial h}{\partial x} - \rho u_{y,b}\frac{\partial h}{\partial y} + h\frac{\partial p}{\partial t} = 0$$

(6.127)

The first two terms of Equation 6.127 are the Poiseuille terms and describe the net flow rates due to pressure gradients within the lubricated area. The third and fourth terms are the Couette terms and describe the net entrained flow rates due to surface velocities. The fifth to the seventh terms are due to a squeezing motion. The eighth term describes the net flow rate due to local expansion as a result of density changes with time. Equation 6.127 is repeated below, with the identification of the various terms emphasized.

$$\underbrace{\frac{\partial}{\partial x}\left(-\frac{\rho h^3}{12\mu}\frac{\partial p}{\partial x}\right) + \frac{\partial}{\partial y}\left(-\frac{\rho h^3}{12\mu}\frac{\partial p}{\partial y}\right)}_{\text{Poiseuille}} + \underbrace{\frac{\partial}{\partial x}\left[\frac{\rho h(u_{x,a}+u_{x,b})}{2}\right] + \frac{\partial}{\partial y}\left[\frac{\rho h(u_{y,a}+u_{y,b})}{2}\right]}_{\text{Couette}} +$$

$$\underbrace{\rho(u_{z,b}-u_{z,a}) - \rho u_{x,b}\frac{\partial h}{\partial x} - \rho u_{y,b}\frac{\partial h}{\partial y}}_{\text{Squeeze}} + \underbrace{h\frac{\partial p}{\partial t}}_{\substack{\text{Net flow due to}\\\text{local expansion}}} = 0$$

For tangential only motion, where

$$u_{z,b} = u_{x,b}\frac{\partial h}{\partial x} + u_{y,b}\frac{\partial h}{\partial y}$$

(6.128)

and $u_{z,a} = 0$, Equation 6.127 reduces to

$$\frac{\partial}{\partial x}\left(\frac{\rho h^3}{\mu}\frac{\partial p}{\partial x}\right) + \frac{\partial}{\partial y}\left(\frac{\rho h^3}{\mu}\frac{\partial p}{\partial y}\right) = 12\bar{u}_x\frac{\partial(\rho h)}{\partial x} + 12\bar{u}_y\frac{\partial(\rho h)}{\partial y} \qquad (6.129)$$

where

$$\bar{u}_x = \frac{u_{x,a} + u_{x,b}}{2} = \text{constant} \qquad (6.130)$$

$$\bar{u}_y = \frac{u_{y,a} + u_{y,b}}{2} = \text{constant} \qquad (6.131)$$

For hydrodynamic lubrication, the fluid properties do not vary significantly through the bearing and can be considered constant. In addition for hydrodynamic lubrication, the motion is pure sliding and $\bar{u}_y = 0$.

The Reynolds equation can therefore be simplified to

$$\frac{\partial}{\partial x}\left(h^3\frac{\partial p}{\partial x}\right) + \frac{\partial}{\partial y}\left(h^3\frac{\partial p}{\partial y}\right) = 12\bar{u}_x\mu_o\frac{\partial h}{\partial x} \qquad (6.132)$$

In some lubrication applications, side leakage can be neglected, and Equation 6.129 can be restated as

$$\frac{\partial}{\partial x}\left(\frac{\rho h^3}{\mu}\frac{\partial p}{\partial x}\right) = 12\bar{u}_x\frac{\partial(\rho h)}{\partial x} \qquad (6.133)$$

Equation 6.133 can be integrated giving

$$\frac{1}{\mu}\frac{dp}{dx} = \frac{12\bar{u}_x}{h^2} + \frac{A}{\rho h^3} \qquad (6.134)$$

With boundary conditions $dp/dx = 0$ when $x = x_m$, $\rho = \rho_m$, $h = h_m$ gives $A = -12\bar{u}_x\rho_m h_m$. The subscript m refers to the condition for which $dp/dx = 0$ such as the point of maximum pressure. Substituting for A in Equation 6.134 gives

$$\frac{dp}{dx} = 12\bar{u}_x\mu\frac{\rho h - \rho_m h_m}{\rho h^3} \qquad (6.135)$$

If the density can be considered constant, then Equation 6.135 becomes

$$\frac{dp}{dx} = 12\bar{u}_x\mu\frac{h - h_m}{h^3} \qquad (6.136)$$

For the case of a gas-lubricated bearing, the density, using the ideal gas law, is given by

$$\rho = \frac{p}{RT} \qquad (6.137)$$

From Equation 6.129, and taking the viscosity as constant,

$$\frac{\partial}{\partial x}\left(\rho h^3\frac{\partial p}{\partial x}\right) + \frac{\partial}{\partial y}\left(\rho h^3\frac{\partial p}{\partial y}\right) = 12\bar{u}_x\mu_o\frac{\partial(\rho h)}{\partial x} \qquad (6.138)$$

Expressing Equation 6.129 in cylindrical coordinates

$$\frac{\partial}{\partial r}\left(\frac{r\rho h^3}{\mu}\frac{\partial p}{\partial r}\right) + \frac{1}{r}\frac{\partial}{\partial \phi}\left(\frac{\rho h^3}{\mu}\frac{\partial p}{\partial \phi}\right) = 12\left[\bar{u}_r\frac{\partial(\rho rh)}{\partial r} + \bar{u}_\phi\frac{\partial(\rho h)}{\partial \phi}\right] \qquad (6.139)$$

where

$$\bar{u}_r = \frac{u_{r,a} + u_{r,b}}{2} \qquad (6.140)$$

$$\bar{u}_\phi = \frac{u_{\phi,a} + u_{\phi,b}}{2} \qquad (6.141)$$

If the viscosity and density can be assumed to be constant, then Equation 6.139 becomes

$$\frac{\partial}{\partial r}\left(rh^3\frac{\partial p}{\partial r}\right) + \frac{1}{r}\frac{\partial}{\partial \phi}\left(h^3\frac{\partial p}{\partial \phi}\right) = 12\mu_o\left[\bar{u}_r\frac{\partial(rh)}{\partial r} + \bar{u}_\phi\frac{\partial h}{\partial \phi}\right] \qquad (6.142)$$

Equation 6.142 applies to a thrust bearing where the fluid film is in the z direction and the bearing dimensions are in the r, ϕ directions.

For turbulent flow, the Reynolds equation is

$$\frac{\partial}{\partial x}\left(\frac{h^3}{\mu k_x}\frac{\partial \bar{p}}{\partial x}\right) + \frac{\partial}{\partial y}\left(\frac{h^3}{\mu k_y}\frac{\partial \bar{p}}{\partial y}\right) = \frac{\bar{u}_x}{2}\frac{\partial h}{\partial x} \qquad (6.143)$$

where (from Constantinescu, 1962)

$$k_x = 12 + 0.026\left(\frac{\rho \Omega rh}{\mu}\right)^{0.102} \qquad (6.144)$$

$$k_y = 12 + 0.0198\left(\frac{\rho \Omega rh}{\mu}\right)^{0.091} \qquad (6.145)$$

For an infinitely wide journal bearing, the pressure in the axial direction can be assumed to be constant. This approach is valid for a length to diameter ratio of $L/D > 2$. The integrated form of the Reynolds Equation, Equation 6.136, on substitution for $\bar{u}_x = (u_a + u_b)/2 = \Omega a/2$, as $u_b = 0$ for a stationary bearing surface, gives

$$\frac{dp}{dx} = \frac{6\mu_o a\Omega(h - h_m)}{h^3} \qquad (6.146)$$

$$dx = ad\phi \qquad (6.147)$$

$$\frac{dp}{d\phi} = \frac{6\mu_o a^2\Omega(h - h_m)}{h^3} \qquad (6.148)$$

$$\cos\alpha = \frac{1}{b}[h + a + e\cos(\pi - \phi)] \qquad (6.149)$$

$$h = b \cos \alpha - a + e \cos \phi \qquad (6.150)$$

Hence, using trigonometric relationships,

$$h = b \left[1 - \left(\frac{e}{b}\right)^2 \sin^2 \phi\right]^{0.5} - a + e \cos \phi \qquad (6.151)$$

Expanding the series

$$\left[1 - \left(\frac{e}{b}\right)^2 \sin^2 \phi\right]^{0.5} = 1 - \frac{1}{2}\left(\frac{e}{b}\right)^2 \sin^2 \phi - \frac{1}{8}\left(\frac{e}{b}\right)^4 \sin^4 \phi - \dots \qquad (6.152)$$

Hence

$$h = b\left[1 - \frac{1}{2}\left(\frac{e}{b}\right)^2 \sin^2 \phi - \frac{1}{8}\left(\frac{e}{b}\right)^4 \sin^4 \phi - \dots\right] - a + e \cos \phi \qquad (6.153)$$

$$b - a = c \qquad (6.154)$$

$$h = c + e\left[\cos \phi - \frac{1}{2}\left(\frac{e}{b}\right) \sin^2 \phi - \frac{1}{8}\left(\frac{e}{b}\right)^3 \sin^4 \phi - \dots\right] \qquad (6.155)$$

The ratio e/b is typically of the order of 1×10^{-3}. As a result, terms in Equation 6.155 with this ratio can readily be neglected, giving

$$h = c + e \cos \phi = c\left(1 + \frac{e}{c}\cos \phi\right) = c(1 + \varepsilon \cos \phi) \qquad (6.156)$$

where ε is the eccentricity ratio defined by

$$\varepsilon = \frac{e}{c} \qquad (6.157)$$

Substitution of Equation 6.156 in Equation 6.148 gives

$$\frac{dp}{d\phi} = 6\mu\Omega\left(\frac{a}{c}\right)^2 \left[\frac{1}{(1 + \varepsilon \cos \phi)^2} - \frac{h_m}{c(1 + \varepsilon \cos \phi)^3}\right] \qquad (6.158)$$

Integration of Equation 6.158 gives the following expression for the pressure distribution.

$$p = 6\mu\Omega\left(\frac{a}{c}\right)^2 \int \left[\frac{1}{(1 + \varepsilon \cos \phi)^2} - \frac{h_m}{c(1 + \varepsilon \cos \phi)^3}\right] d\phi + A \qquad (6.159)$$

where A is a constant of integration.
For a journal bearing,

$$x = a\phi \qquad (6.160)$$

and

$$\bar{u} = \frac{u_{x,a}}{2} = \frac{a\Omega}{2} \tag{6.161}$$

From Equation 6.132,

$$\frac{\partial}{\partial \phi}\left(h^3 \frac{\partial p}{\partial \phi}\right) + a^2 \frac{\partial}{\partial y}\left(h^3 \frac{\partial p}{\partial y}\right) = 6\mu_o \Omega a^2 \frac{\partial h}{\partial \phi} \tag{6.162}$$

Substituting for the film thickness gives

$$\frac{\partial}{\partial \phi}\left(h^3 \frac{\partial p}{\partial \phi}\right) + a^2 h^3 \frac{\partial^2 p}{\partial y^2} = -6\mu_o \Omega a^2 e\sin\phi \tag{6.163}$$

The Reynolds equation, for example, Equation 6.163 or 6.129, can be solved by approximate mathematical methods or numerically. Once the pressure distribution has been established, the journal performance can be determined in terms of the bearing load capacity, frictional losses, lubricant flow requirements, and the lubricant temperature rise. Use can be made of a series of design charts that relate numerical solutions to key characteristic parameters. The solutions and accompanying charts presented here were originally produced by Raimondi and Boyd (1958a, b, c) who used an iterative technique to solve the Reynolds equations. These charts give the film thickness, coefficient of friction, lubricant flow, lubricant side flow ratio, minimum film thickness location, maximum pressure ratio, maximum pressure ratio position, and film termination angle versus the Sommerfield number (see Figures 6.24 to 6.31).

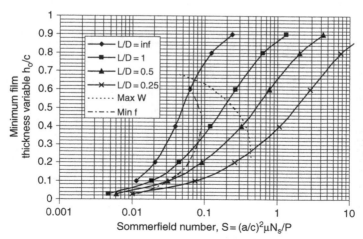

FIGURE 6.24 Chart for the minimum film thickness variable, (h_o/c) versus the Sommerfield number (data from Raimondi and Boyd, 1958c).

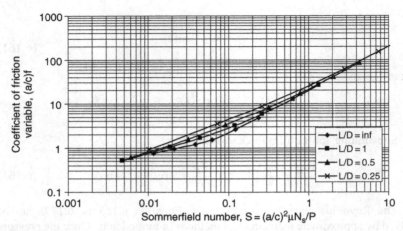

FIGURE 6.25 Chart for determining the coefficient of friction variable, $(a/c)f$ (data from Raimondi and Boyd, 1958c).

The Sommerfield number, which is also known as the bearing characteristic number, is defined in Equation 6.164. It is used as it encapsulates all the parameters usually defined by the designer, such as the journal radius, clearance, viscosity, rotational speed, and load. Great care needs to be taken with units in the design of journal bearings. Many design charts have been produced using English units (psi, reyn, Btu, etc.). As long as a consistent set of units is maintained, use of the charts will yield sensible results. In particular, note the use of revolutions per second in the definition of speed in the Sommerfield number given in Equation 6.164.

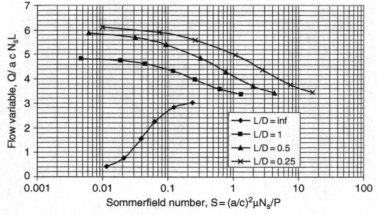

FIGURE 6.26 Chart for determining the flow variable, $Q/(acN_sL)$ (data from Raimondi and Boyd, 1958c).

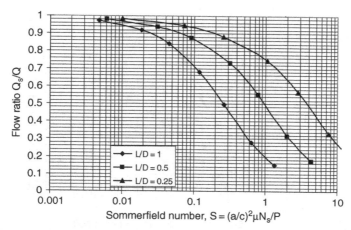

FIGURE 6.27   Chart for determining the ratio of side flow, Q_s, to total flow, Q (data from Raimondi and Boyd, 1958c).

$$S = \left(\frac{a}{c}\right)^2 \frac{\mu N_s}{P} \tag{6.164}$$

where: S is the bearing characteristic number

a is the journal radius, (m)
c is the radial clearance (m)
μ is the absolute viscosity (Pa s)
N_s is the journal speed (revolutions per second, rps)

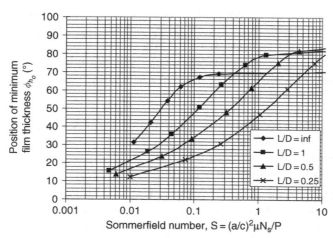

FIGURE 6.28   Chart for determining the position of the minimum film thickness ϕ_{h_o} (data from Raimondi and Boyd, 1958c).

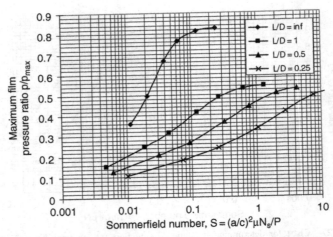

FIGURE 6.29 Chart for determining the maximum film pressure ratio, p/p_{max} (data from Raimondi and Boyd, 1958c).

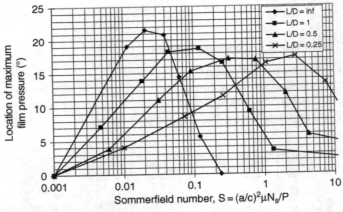

FIGURE 6.30 Chart for determining the position of maximum pressure, $\phi_{p_{max}}$ (data from Raimondi and Boyd, 1958c).

$P = W/LD$ is the load per unit of projected bearing area (N/m^2)

W is the load on the bearing (N),

D is the journal diameter (m),

L is the journal bearing length (m).

Consider the journal shown in Figure 6.32. As the journal rotates, it will pump lubricant in a clockwise direction. The lubricant is pumped into a wedge-shaped space, and the journal is forced over to the opposite side. The angular position where the lubricant film is at its minimum thickness h_o is called the attitude angle, ϕ_{h_o}.

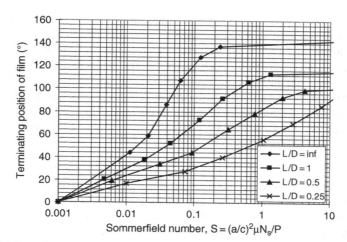

FIGURE 6.31   Chart for determining the film termination angle, ϕ_{p_o} (data from Raimondi and Boyd, 1958c).

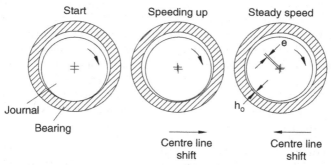

FIGURE 6.32   Full film hydrodynamic bearing motion from start-up.

The center of the journal is displaced from the center of the bearing by a distance e called the eccentricity. The ratio of the eccentricity e to the radial clearance c is called the eccentricity ratio ($\varepsilon = e/c$), Equation 6.157. The relationship between the film thickness, clearance, eccentricity, and eccentricity ratio are defined in Equations 6.165 and 6.166.

$$h_o = c - e \tag{6.165}$$

$$\frac{h_o}{c} = 1 - \varepsilon \tag{6.166}$$

One assumption made in the Raimondi and Boyd analysis is that the viscosity of the lubricant is constant as it passes through the bearing. However, work is done on the lubricant in the bearing, and the temperature of the lubricant leaving the bearing zone will be higher than the entrance value. Figure 6.18 shows the variation of viscosity with temperature for some of the SAE and ISO defined lubricants, and it can be seen that the value of the viscosity for a particular lubricant is highly dependent on the temperature. Some of the lubricant entrained into the bearing film emerges as side flow that carries away some of the heat. The remainder flows through the load-bearing zone and carries away the remainder of the heat generated. The temperature used for determining the viscosity can be taken as the average of the inlet and exit lubricant temperatures as given by Equation 6.167.

$$T_{av} = T_1 + \frac{T_2 - T_1}{2} = T_1 + \frac{\Delta T}{2} \tag{6.167}$$

where T_1 is the temperature of the lubricant supply and T_2 is the temperature of the lubricant leaving the bearing. Generally, petroleum lubrication oils should be limited to a maximum temperature of approximately 70 °C in order to prevent excessive oxidation.

One of the parameters that needs to be determined is the bearing lubricant exit temperature T_2. This is a trial-and-error or iterative process. A value of the temperature rise, ΔT, is guessed, the viscosity for a standard oil, corresponding to this value, determined, and the analysis performed. If the temperature rise calculated by the analysis does not correspond closely to the guessed value, the process should be repeated using an updated value for the temperature rise until the two match. If the temperature rise is unacceptable, it may be necessary to try a different lubricant or to modify the bearing configuration.

Given the length, the journal radius, the radial clearance, the lubricant type, and its supply temperature, the steps for determining the various bearing operating parameters are listed below.

1. Determine the speed of the journal and the load to be supported.
2. If L and D are not already determined, set the proportions of the bearing so that the load capacity, $P = W/LD$, is somewhere between 0.34 MN/m² for light machinery and 13.4 MN/m² for heavy machinery.
3. Determine a value for the radial clearance of the bearing using the data presented in Figure 6.22.
4. If not already specified, select a lubricant. Lubricant oil selection is a function of speed or compatibility with other lubricant requirements. Generally, as the design speed rises, oils with a lower viscosity should be selected.
5. Estimate a value for the temperature rise ΔT across the bearing. The value taken for the initial estimate is relatively unimportant. As a guide, a value of $\Delta T = 10\,°C$ is generally a good starting guess. This value can be increased for high-speed bearings and for low-bearing clearances.
6. Determine the average lubricant temperature $T_{av} = T_1 + \Delta T/2$ and find the corresponding value for the viscosity for the chosen lubricant.

7. Calculate the Sommerfield number, $S = (a/c)^2 \mu N_s /P$ and the length to diameter ratio (L/D).

8. Use the charts (Figures 6.25, 6.26, and 6.27) to determine values for the coefficient of friction variable, the total lubricant flow variable, and the ratio of the side flow to the total lubricant flow with the values for the Sommerfield number and the L/D ratio.

9. Calculate the temperature rise of the lubricant through the bearing using

$$\Delta T = \frac{8.30 \times 10^{-6} P}{1 - \frac{1}{2}(Q_s/Q)} \times \frac{(a/c)f}{Q/acN_s L} \tag{6.168}$$

10. If this calculated value does not match the estimated value for ΔT to within, say, 1 °C, repeat the procedure from step 6 using the updated value of the temperature rise to determine the average lubricant temperature.

11. Check that the values for the Sommerfield number and the length to diameter ratio give a design that is in the optimal operating region for minimal friction and maximum load capability on the chart for the minimum film thickness variable (between the dashed lines in Figure 6.24). If the operating Sommerfield number and L/D ratio combination do not fall within this zone, then it is likely that the bearing design can be improved by altering the values for c, L, D, the lubricant type, and the operating temperature as appropriate.

12. If the value for the temperature rise across the bearing has converged, values for the total lubricant flow rate, Q, the side flow rate Q_s and the coefficient of friction f can be calculated. The journal bearing must be supplied with the value of the total lubricant calculated in order for it to perform as predicted by the charts.

13. The charts given in Figures 6.24 to 6.31 can be used to determine values for the maximum film thickness, the maximum film pressure ratio, the location of the maximum film pressure, and its terminating angular location as required.

14. The torque required to overcome friction in the bearing can be calculated by

$$T_q = fWa \tag{6.169}$$

15. The power lost in the bearing is given by

$$\text{Power} = \Omega \times T_q = 2\pi N_s \times T_q \tag{6.170}$$

where Ω is the angular velocity (rad/s).

16. Specify the surface roughness for the journal and bearing surfaces. A ground journal with an arithmetic average surface roughness of 0.4 to 0.8 μm is recommended for bearings of good quality. For high-precision equipment both surfaces can be lapped or polished to give a surface

roughness of 0.2 to 0.4 μm. The specified roughness should be less than the minimum film thickness.

17. Design the recirculation and sealing system for the bearing lubricant.

For values of the length to diameter (L/D) ratio other than those shown in the charts given in Figures 6.24–6.31, values for the various parameters can be found by interpolation using Equation 6.171 (see Raimondi and Boyd, 1958b).

$$
y = \frac{1}{(L/D)^3}
\begin{bmatrix}
-\dfrac{1}{8}\left(1-\dfrac{L}{D}\right)\left(1-2\dfrac{L}{D}\right)\left(1-4\dfrac{L}{D}\right)y_{\infty} \\[2ex]
+\dfrac{1}{3}\left(1-2\dfrac{L}{D}\right)\left(1-4\dfrac{L}{D}\right)y_1 - \dfrac{1}{4}\left(1-\dfrac{L}{D}\right)\left(1-4\dfrac{L}{D}\right)y_{1/2} \\[2ex]
+\dfrac{1}{24}\left(1-\dfrac{L}{D}\right)\left(1-2\dfrac{L}{D}\right)y_{1/4}
\end{bmatrix}
$$

$$(6.171)$$

Here y is the desired variable and y_{∞}, y_1, $y_{1/2}$, and $y_{1/4}$ are the variables at $L/D = \infty$, 1, ½, and ¼, respectively.

Example 6.10.

A full journal bearing has a nominal diameter of 30.0 mm and a bearing length of 30.0 mm (see Figure 6.33). The bearing supports a load of 900 N, and the journal design speed is 10000 rpm. The radial clearance has been specified as 0.04 mm. An ISO VG22 oil, with a density of 900 kg/m³, has been chosen, and the lubricant supply temperature is 60 °C.

Find the temperature rise of the lubricant, the lubricant flow rate, the minimum film thickness, the torque required to overcome friction, and the heat generated in the bearing.

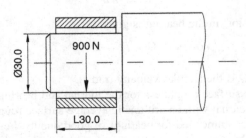

FIGURE 6.33 Bearing design example.

Solution

The primary data are $D = 30.0$ mm, $L = 30$ mm, $W = 900$ N, $N = 10000$ rpm, $c = 0.04$ mm, ISO VG 22, $T_1 = 60\ °C$.

Guess a value for the lubricant temperature rise ΔT across the bearing to be, say, $\Delta T = 15\ °C$.

$$T_{av} = T_1 + \frac{\Delta T}{2} = 60 + \frac{15}{2} = 67.5\ °C$$

From Figure 6.18 for ISO VG22 at 67.5 °C, $\mu = 0.0068$ Pa s.
The mean radius is given by

$$r_m = \frac{a+b}{2} = 15.02 \text{ mm}$$

$$b - a = 0.04 \text{ mm}$$

$$\nu = \mu/\rho = 0.0068/900 = 7.556 \times 10^{-6} \text{m}^2/\text{s}$$

So the Taylor number using this approximate value for the kinematic viscosity is given by

$$Ta_m = \frac{\Omega r_m^{0.5}(b-a)^{1.5}}{\nu} = 4.296$$

As this is much less than the critical Taylor number, 41.19, laminar flow can be assumed.

$$N_s = 10000/60 = 166.7\text{rps}, L/D = 30/30 = 1$$

$$P = \frac{W}{LD} = \frac{900}{0.03 \times 0.03} = 1 \times 10^6 \text{ N/m}^2$$

$$S = \left(\frac{a}{c}\right)^2 \frac{\mu N_s}{P} = \left(\frac{15 \times 10^{-3}}{0.04 \times 10^{-3}}\right)^2 \frac{0.0068 \times 166.7}{1 \times 10^6} = 0.1594$$

From Figure 6.25 with $S = 0.1594$ and $L/D = 1$, $(a/c)f = 4.0$
From Figure 6.26 with $S = 0.1594$ and $L/D = 1$, $Q/(acN_sL) = 4.2$
From Figure 6.27 with $S = 0.1594$ and $L/D = 1$, $Q_s/Q = 0.62$

The value of the temperature rise of the lubricant can now be calculated using Equation 6.168:

$$\Delta T = \frac{8.3 \times 10^{-6}P}{1-0.5\frac{Q_s}{Q}} \times \frac{\frac{r}{c}f}{\frac{Q}{rcN_sL}} = \frac{8.3 \times 10^{-6} \times 1 \times 10^6}{1-(0.5 \times 0.62)} \times \frac{4.0}{4.2} \approx 11.5\ °C$$

As the value calculated for the lubricant temperature rise is significantly different from the estimated value, it is necessary to repeat the above calculation but using the new improved estimate for determining the average lubricant temperature.

Using $\Delta T = 11.5\,°C$ to calculate T_{av} gives:

$$T_{av} = 60 + \frac{11.5}{2} = 65.7\,°C$$

Repeating the procedure using the new value for T_{av} gives:

$\mu = 0.007$ Pa s

$S = 0.1641$

From Figure 6.25 with $S = 0.1641$ and $L/D = 1$, $(a/c)f = 4.0$
From Figure 6.26 with $S = 0.1641$ and $L/D = 1$, $Q/(acN_sL) = 4.2$
From Figure 6.27 with $S = 0.1641$ and $L/D = 1$, $Q_s/Q = 0.61$

$$\Delta T = \frac{8.3 \times 10^{-6} \times 2.4 \times 10^{6}}{1-(0.5 \times 0.61)} \times \frac{4.0}{4.2} \approx 11.4\,°C,$$

$$T_{av} = 60 + \frac{11.4}{2} = 65.7\,°C$$

This value for T_{av} is the same as the previous calculated value, suggesting that the solution has converged. For $T_{av} = 65.7\,°C$, $\mu = 0.007$ Pa s, and $S = 0.1641$. The other parameters can now be found.

$$Q = acN_sL \times 4.2 = 15 \times 0.04 \times 166.7 \times 30 \times 4.2 = 12600 \text{ mm}^3/\text{s}$$

From Figure 6.24 $h_o/c = 0.48$. $h_o = 0.0192$ mm.
An indication of the Reynolds number can be obtained from

$$Re_x = \frac{\rho_o u_{x,o} h_o^2}{\mu_o l_o} = \frac{900 \times (1047 \times 0.01502) \times (0.0192 \times 10^{-3})^2}{0.007 \times 0.01502} = 0.04963$$

$$f = 4.0 \times (c/r) = 4.0 \times (0.04/15) = 0.01067$$

The torque is given by $T_q = fWa = 0.01067 \times 900 \times 0.015 = 0.144$ N m.
The power dissipated in the bearing is given by Power $= 2\pi \times T_q \times N_s$
$= 150.8$ W.

Example 6.11.

A full journal bearing is required for a 30-mm-diameter shaft, rotating at 2750 rpm supporting a load of 2000 N (see Figure 6.34). From previous experience, the lubrication system available is capable of delivering the lubricant to the bearing at 50 °C. Select an appropriate radial clearance, length, and lubricant for the bearing and determine the temperature rise of the lubricant, the total lubricant flow rate required, and the power absorbed in the bearing. Check that your chosen design operates within the optimum zone indicated by the dotted lines on the minimum film thickness ratio chart (Figure 6.24).

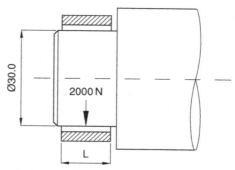

FIGURE 6.34   Bearing design example.

Solution

The primary design information is: $D = 30.0$ mm, $W = 2000$ N, $N = 2750$ rpm, $T_1 = 50\ ^\circ$C.

From Figure 6.22 for a speed of 2750 rpm and a nominal journal diameter of 30 mm, a suitable value for the diametral clearance is 0.05 mm. The radial clearance is therefore 0.025 mm. The next step is to set the length of the bearing. Typical values for L/D ratios are between 0.5 and 1.5. Here a value for the ratio is arbitrarily set as $L/D = 0.5$ to give a compact bearing, that is, L =15 mm. If this value is found to be unsuitable, it can be adjusted.

Given the speed of the bearing, 2750 rpm, as an initial proposal an SAE 10 lubricant is selected.

The procedure for determining the average lubricant temperature is the same as that for the previous example. As a first estimate, the temperature rise of the lubricant is taken as 10 °C, and the results of the iterative procedure are given in Table 6.3.

The mean radius is given by

$$r_m = \frac{a+b}{2} = \frac{15 + 15.025}{2} = 15.01 \text{ mm}$$

$$b - a = 0.025 \text{ mm}$$

An estimate for the kinematic viscosity can be found, taking the average temperature of the oil as 55 °C and assuming a density of 900 kg/m³. At 55 °C, the viscosity for SAE 10 is approximately 0.016 Pa.s

$$\nu = \mu/\rho = 0.016/900 = 1.78 \times 10^{-5}\,\text{m}^2/\text{s}$$

So the Taylor number using this approximate value for the kinematic viscosity is given by

$$Ta_m = \frac{\Omega r_m^{0.5}(b-a)^{1.5}}{\nu} = 0.2481$$

As this is much less than the critical Taylor number, 41.19, laminar flow can be assumed.

$$P = \frac{W}{LD} = \frac{2000}{0.015 \times 0.03} = 4.444 \times 10^6 \text{ N/m}^2$$

$$S = \left(\frac{a}{c}\right)^2 \frac{\mu N_s}{P} = \left(\frac{15 \times 10^{-3}}{0.025 \times 10^{-3}}\right)^2 \frac{\mu 45.83}{4.444 \times 10^6} = 3.713 \, \mu.$$

$$T_{av} = T_1 + \frac{\Delta T}{2} = 50 + \frac{10}{2} = 55 \, ^\circ C$$

Values for ΔT, T_{av}, μ, S, $(a/c)f$, Q/acN_sL, Q_s/Q from successive iterations using steps 6–10 of the calculation procedure are given in Table 6.3.

The total lubricant flow rate required is given by $Q = acN_sL \times 5.62 = 15 \times 0.025 \times 45.83 \times 15 \times 5.62 = 1449 \text{ mm}^3/\text{s}$.

With $S = 0.04084$ and $L/D = 0.5$, the design selected is within the optimum operating zone for minimum friction and optimum load capacity indicated in Figure 6.24.

The friction factor, $f = 1.95 \times (c/a) = 1.95 \times 0.025/15 = 0.00325$.

$T_q = fWa = 0.00325 \times 2000 \times 0.015 = 0.0975 \text{ N m}$
$Power = 2\pi N_s \times T_q = 28.08 \text{ W}$

TABLE 6.3 Tabular data for Example 6.11

Temperature rise, ΔT (°C)	10	29.11	24.40	23.81
Average lubricant temperature T_{av} (°C)	55	64.6	62.2	61.9
Average lubricant viscosity μ (Pa s)	0.016	0.012	0.011	0.011
Sommerfield number, S	0.0594	0.04455	0.04084	0.04084
Coefficient of friction variable $(a/c)f$	2.4	2.0	1.95	1.95
Flow variable Q/acN_sL	5.58	5.6	5.62	5.62
Side flow to total flow ratio Q_s/Q	0.91	0.92	0.925	0.925

6.6. ROTATING CYLINDERS AND SPHERES WITH CROSS-FLOW

A cylinder that is rotating about its axis and moving through a fluid in a direction perpendicular to its axis will experience a force perpendicular to both the direction of motion and the axis as illustrated in Figures 6.35 and 6.36. The production of lift forces by a rotating cylinder placed in a fluid stream is known as the Magnus effect, after Heinrich Magnus who investigated inaccuracies when firing cannonballs (Magnus, 1852).

An explanation of the Magnus effect can be developed by considering the schematic diagrams in Figures 6.35, 6.36, and 6.37. At the surface of the cylinder, fluid will rotate at the tangential velocity component of the cylinder's outer radius due to the no slip condition, and the action of viscosity will transmit that rotation to adjacent layers of fluid. This boosts the velocity of the fluid stream in the direction of rotation and opposes it in the reverse direction;

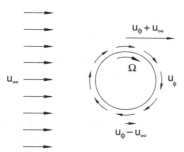

FIGURE 6.35   A rotating cylinder in a fluid stream cross-flow.

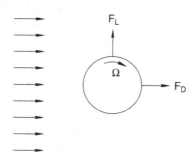

FIGURE 6.36   Lift and drag forces on a rotating cylinder in a fluid stream.

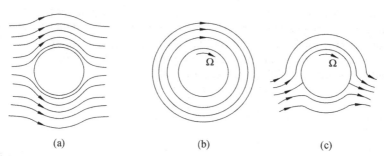

FIGURE 6.37   Flow patterns for (a) a circular cylinder placed in a parallel flow, (b) pure circulatory flow, and (c) combined viscous and circulatory flow.

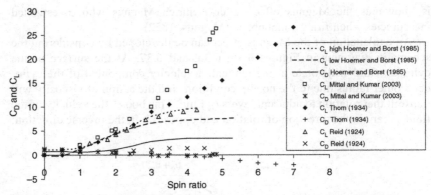

FIGURE 6.38 Lift and drag coefficients, C_L and C_D, for rotating cylinder in a cross-flow as a function of spin ratio, $\Omega r/u\infty$ (data from Hoerner and Borst, 1985; Mittal and Kumar, 2003; Thom, 1934; Reid, 1924.

see Figure 6.35. For a nonrotating cylinder in fluid stream, the flow field will be as illustrated in Figure 6.37a. Pure circulatory flow, resulting from rotation, will give a flow pattern of concentric circles as indicated in Figure 6.37b. Superimposing these two flow fields and plotting the streamlines by vector addition gives the flow field shown in Figure 6.37c. If the flow has been deflected to the right, then an acceleration has occurred, and therefore an associated force has caused this. The reaction to the force will therefore cause a force on the cylinder as shown in Figure 6.36 at right-angles to the fluid stream.

Lift and drag coefficients for a spinning cylinder in a cross-flow are presented in Figure 6.38 as a function of the spin ratio, the ratio of the surface speed to the free-stream flow speed, using data from Reid (1924), Thom (1934), Hoerner and Borst (1985), and Mittal and Kumar (2003). The definitions used for the lift and drag coefficients for the data presented are given in Equations 6.172 and 6.173. From potential flow theory, the lift per unit length of a rotor is 2π times the product of the tangential velocity component of the rotor and the free-stream flow speed. Discs or fences can be added to a cylinder to reduce tip vortices, in a manner that is comparable to the use of fences and the end of wing tips on aircraft. The data from Thom (1934) presented in Figure 6.38 is for a rotating cylinder with multiple disc fences along the length of the cylinder placed at intervals of 0.75 times the rotor diameter.

$$C_L = \frac{Lift}{0.5\rho U^2 DL} \qquad (6.172)$$

$$C_D = \frac{Drag}{0.5\rho U^2 DL} \qquad (6.173)$$

FIGURE 6.39   The Buckau with two 18-m-high and 2.7-m-diameter rotors.

Anton Flettner in the 1920s proposed the use of rotating cylinders, instead of sails, to propel ships, with motors driving large cylinders mounted on the ship's deck and the wind providing the parallel flow necessary to induce a propulsive force as a result of the interaction between the wind and the rotating cylinders. Anton Flettner built two ships based on this principle. In 1926, the three-masted schooner rig on the 52-m *Buckau* was replaced by two rotors, shown in Figure 6.39, and the ship was subsequently renamed the *Baden-Baden*. Each rotor was 18 m high and 2.7 m in diameter. The ship's performance exceeded all expectations, being faster than before and able to sail much closer to the wind.

The *Baden-Baden* successfully crossed the Atlantic in 1926 (see Seufert and Seufert, 1983). As a result of this success, the Transportation Department of the German Navy ordered the construction of another rotor ship, the *Barbara*, shown in Figure 6.40. The *Barbara* was 92 m long and was fitted with three rotors, each 17 m high, 4 m in diameter, and driven at 150 rpm by a 27-kW electric motor. The *Barbara* carried 3000 tons of cargo and a few passengers and operated between Hamburg and Italy for six years, further proving the technical viability of the concept. Of course, if there is no wind, then there is no cross-flow and a ship becomes becalmed with such a system.

Example 6.12.

Determine the force applied to each rotor by the wind for the *Barbara* rotor ship. Take the diameter of the rotors to be 4 m, length 17 m, and a rotor speed of 150 rpm. Assume that the speed of the wind relative to the rotor is 7 m/s.

FIGURE 6.40 The Barbara with three, 17-m-high, 4-m-diameter rotors.

Solution

The angular velocity of each rotor is

$$\Omega = 150 \times \frac{2\pi}{60} = 15.71 \text{ rad/s}$$

The spin ratio is

$$\frac{\Omega r}{u_{z=\infty}} = \frac{15.71 \times (4/2)}{7} = 4.488$$

From Figure 6.38, $C_L = 8.7$ and $C_D = 3.45$.
Assuming a density of 1.2047 kg/m³, the lift and drag forces are

$$\text{Lift} = \frac{1}{2}\rho U^2 DLC_L = 0.5 \times 1.2047 \times 7^2 \times 4 \times 17 \times 8.7 = 17460 \text{ N}$$

$$\text{Drag} = \frac{1}{2}\rho U^2 DLC_D = 0.5 \times 1.2047 \times 7^2 \times 4 \times 17 \times 3.45 = 6924 \text{ N}$$

The resultant force applied to each rotor is therefore

$$\sqrt{17460^2 + 7025^2} = 18780 \text{ N}$$

This acts at an angle of

$$tan^{-1}\left(\frac{6924}{17460}\right) = 21.63° \text{ to the left of the relative wind direction.}$$

Salter, Sortino, and Latham (2008) have proposed the use of rotor ships to propel a fleet of 1500 ships for the purpose of spraying a fine mist of seawater spray worldwide to alter the Earth's albedo and as a result affect the energy balance due to insolation and radiation to space. The concept is to offset the effects of the presumed 3.7 W/m^2 heating rise apportioned to worldwide industrial activity. A concept design for a spray vessel is shown in Figure 6.41. For the vessel shown, the wind would be blowing from the right-hand side of the image, the rotor angular velocity would be clockwise as viewed from above, and the resulting rotor thrust would be to the left.

A rotating sphere moving through a stream of fluid also experiences a force perpendicular to both the axis of rotation and the motion. The flow structure is considerably more complex than that of the equivalent two-dimensional cross-sectional rotating cylinder. Experimental data for the case of a smooth spinning sphere in a uniform flow is given in Figure 6.42 as a function of the spin ratio. The Reynolds number is of secondary importance. At low values of the spin ratio, the lift is negative. Above a spin ratio of approximately 0.5, the lift becomes positive and increases, with the spin ratio leveling off to an approximately constant value of lift coefficient equal to about 0.35 for spin ratios above about 1.5.

Spin has little effect on the drag coefficient, which ranges from about 0.5 to 0.65 for spin ratios up to about 5. For large values of the spin ratio the flow behavior can be understood using concepts similar to those described for the rotating cylinder case. Any quantitative data, however, needs to be derived from experiment. At low values of the spin ratio, the force reverses in direction. This effect is probably due to transition occurring in the boundary layer on the side where the relative velocity between the sphere and the fluid is higher. As transition delays separation, this will produce a flow of the form shown in Figure 6.43, with the flow deflected to the side where separation has occurred earliest. The deflection of the wake in this way produces a resultant force that is opposite to that of the conventional Magnus effect.

The phenomenon of variable lift and drag with spin rate is familiar in sports. In tennis and ping-pong, the spin rate is used to control the trajectory and bounce of a shot. A golf ball can be driven off a tee with a speed of the order of 90 m/s with a backspin of 9000 rpm. The spin results in an increase in the lift coefficient that substantially increases the range of the shot. In baseball the pitcher uses spin to throw a curve ball. Considerable interest in the science associated with sports has resulted in a substantial quantity of data becoming available. The general aerodynamics of sports balls is reviewed by Mehta (1985). Specific studies, for example, on cricket balls are reported by Mehta and Wood (1980) and Mehta et al. (1983); baseballs by Briggs (1959) and Watts and Sawyer (1975); and golf balls by Beerman and Harvey (1976), Smits (1994), and Beasley and Camp (2002).

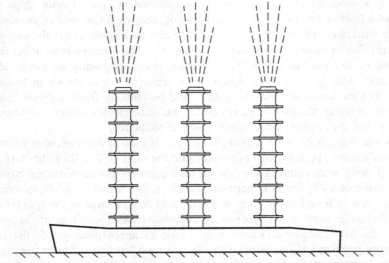

FIGURE 6.41 Concept design for a Flettner rotor spray vessel to alter the Earth's albedo. After J MacNeill (2006).

Example 6.13.

Determine the radius of curvature at maximum elevation in the vertical plane for a smooth tennis ball that is hit at 24 m/s with a topspin of 7800 rpm. Determine the comparable radius of curvature if the ball is hit with no spin. The radius of the tennis ball is 32.8 mm, and its mass is 57 g. A smooth tennis ball has been assumed for simplicity.

Solution

The Reynolds number based on the forward speed and ball diameter is given by

$$Re_D = \frac{\rho u D}{\mu} = \frac{1.2047 \times 24 \times 0.0656}{1.819 \times 10^{-5}} = 1.043 \times 10^5$$

The spin ratio is

$$\frac{\Omega r}{u_{z=\infty}} = \frac{7800 \times (2\pi/60) \times 0.0328}{24} = 1.116$$

From Figure 6.42, $C_L \approx 0.3$
The lift force is given by

$$F_L = 0.5\rho U^2 A C_L = 0.5\rho U^2 \pi r^2 C_L = 0.5 \times 1.205 \times 24^2 \pi \times 0.0328^2 \times 0.3 \approx 0.35 \text{ N}$$

As the ball is hit with topspin, this force will act downward, causing the ball to drop faster than it would otherwise.

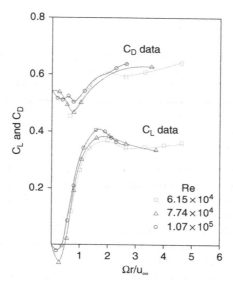

FIGURE 6.42   Lift and drag coefficients for a smooth spinning sphere in a uniform flow (data from Maccoll, 1928).

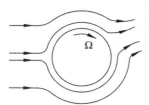

FIGURE 6.43   Flow pattern for a smooth spinning sphere in a uniform flow at low values of spin ratio.

From Newton's second law,

$$\sum F = -F_L - mg = ma_z = -m\frac{u^2}{R}$$

Hence the radius of curvature is given by

$$R = \frac{u^2}{(F_L/m) + g} = \frac{24^2}{(0.35/0.057) + 9.81} = 36.1 \text{ m}$$

In the absence of spin,

$$\sum F = -mg = ma_z = -m\frac{u^2}{R}$$

Hence the radius of curvature is given by

$$R = \frac{u^2}{g} = \frac{24^2}{9.81} = 58.7 \text{ m}$$

Comparison of these figures shows that topspin makes the ball drop faster, creating many tactical advantages, such as allowing a player to hit harder and higher over the net or to make the ball dip at an oncoming net-rusher's feet.

6.7. CONCLUSIONS

The flow phenomenon in rotating cylinder applications depends on the orientation of the cylinder to the flow and the axis of rotation. Correlations are available to determine the drag associated with rotating shaft applications. For an annulus where the inner or outer cylinder may be rotating, consideration can be given to the stability of the flow in order to identify whether vortices are likely to form. Under certain conditions of inner cylinder rotation or combined inner and outer cylinder rotation, toroidal vortices can form in an annulus known as Taylor vortices, with alternate vortices rotating in opposite directions.

Rotation of a cylinder or sphere in a cross-flow causes a deflection of the fluid stream and as a result a force due to reaction on the cylinder or sphere concerned. This phenomenon can be exploited to provide a propulsive mechanism for ships and boats using winds to provide the cross-flow; in sports, the variability in lift and drag with spin rate is exploited to control the trajectory and bounce of a shot and in some cases to provide ball or projectile flight paths that are more difficult to predict.

REFERENCES

Andereck, C.D., and Hayot, F. (eds.). *Ordered and turbulent patterns in Taylor Couette flow.* Plenum, 1992.

Beasley, D., and Camp, T. Effects of dimple design on the aerodynamic performance of a golf ball. In E. Thain (ed.), *Science and Golf* IV, pp. 328–340. Routledge, 2002.

Beerman, P.W., and Harvey, J.K. Golf ball aerodynamics. *Aeronaut. Quarterly,* 27 (1976), pp. 112–122.

Bilgen, E., and Boulos, R. Functional dependence of torque coefficient of coaxial cylinders on gap width and Reynolds numbers. *Trans. ASME, Journal of Fluids Engineering* (1973), pp. 122–126.

Briggs, L.J. Effect of spin and speed on the lateral deflection of a baseball and the Magnus effect for smooth spheres. *American Journal of Physics,* 27 (1959), pp. 589–596.

Chandrasekhar, S. Hydrodynamic stability of viscid flow between coaxial cylinders. *Proceedings of the National Academy of Sciences,* 46 (1960b), pp. 141–143.

Chandrasekhar, S. The hydrodynamic stability of inviscid flow between coaxial cylinders. *Proceedings of the National Academy of Sciences,* 46 (1960a), pp. 137–141.

Childs, P.R.N., and Noronha, M.B. The impact of machining techniques on centrifugal compressor performance. ASME Paper 97-GT-456, 1997.

Childs, P.R.N. *Mechanical design.* 2nd Edition. Elsevier, 2004.

Coles, D. Transition in circular Couette flow. *Journal of Fluid Mechanics*, 21 (1965), pp. 385–425.

Consantinescu, V.N.Analysis of bearings operating in turbulent regime. Transactions of the ASME, Journal of Basic Engineering, 84(1962), pp. 139–151.

Couette, M. Etudes sur le frottement des liquids. *Ann. Chim. Phys.* Ser. 6, 21 (1890), pp. 433–510.

Di Prima, R.C., and Swinney, H. Instabilities and transition in flow between concentric rotating cylinders. In Swinney, H. and Gollub, J.P. (eds.), *Hydrodynamic instabilities and the transition to turbulence*, pp. 139–180. Springer 1981.

Donnelly, R.J. Experiments on the stability of viscous flow between rotating cylinders. I. Torque measurements. *Proc. Royal Society*, London, Ser. A, 246 (1958), pp. 312–325.

Donnelly, R.J., and Fultz, D. Experiments on the stability of viscous flow between rotating cylinders. II. Visual observations. *Proc. Royal Society*, London, Ser. A (1960), 258, pp. 101–123.

Donnelly, R.J., and Schwarz, K.W. Experiments on the stability of viscous flow between rotating cylinders. VI. Finite amplitude experiments. *Proc. Royal Society*, London, Ser. A, 283 (1965), pp. 531–556.

Donnelly, R.J. and Simon, N.J. An empirical torque relation for supercritical flow between rotating cylinders, *Journal of Fluid Mechanics*, 7 (1960), pp. 401–418.

Fenstermacher, P.R., Swinney, H.L., and Gollub, J.P. Dynamical instabilities and the transition to chaotic Taylor vortex flow. *Journal of Fluid Mechanics*, 94 (1979), pp. 103–129.

Gollub, J.B., and Swinney, H.L. Onset of turbulence in a rotating fluid. *Physical Review Letters*, 35 (1975), pp. 927–930.

Hamrock, B.J. *Fundamentals of fluid film lubrication*. McGraw-Hill, 1994.

Harrison, W.J. The hydrodynamical theory of lubrication with special reference to air as a lubricant. *Trans. Cambridge Philos. Soc.*, xxii (1912–1925), pp. 6–54, 1913.

Hendriks, F. On Taylor vortices and Ekman layers in flow-induced vibration of hard disk drives. Microsystem Technologies-Micro-and Nanosystems-Information Storage and Processing Systems, Vol. 16, 2010, pp. 93–101.

Hoerner, S.F., and Borst H.V. Fluid-dynamic lift. Information on lift and its derivatives, in air and in water. Hoerner Fluid Dynamics, 1985.

Hoffmann, C., Altmeyer, S., Pinter, A., and Lucke, M. Transitions between Taylor vortices and spirals via wavy Taylor vortices and wavy spirals. *New Journal of Physics*, 11 (2009).

International Organization for Standardization. ISO 3448 Industrial liquid lubricants—ISO viscosity classification. 1992.

Kaye, J., and Elgar, E.C. Modes of adiabatic and diabatic fluid flow in an annulus with an inner rotating cylinder. *Trans. ASME* (1958), pp. 753–765.

Lamb, H. *Hydrodynamics*. 6th Edition, Cambridge University Press, 1932.

Lewis, J.W. Experimental study of the motion of a viscous liquid contained between two coaxial cylinders. *Proc. Royal Society*, London, Series A, 117 (1927), pp. 388–407.

Maccoll, J.W.. Aerodynamics of a spinning sphere. *Journal of the Royal Aeronautical Society*, 32 (1928), pp. 777–798.

MacNeill, J. www.johnmacneill.com Last accessed 17th July 2010.

Magnus, G. *Abhandlung der Akademie der Wissenschaftern*. Berlin, 1852.

Mallock, A. Determination of the viscosity of water. *Proc. Royal Society*, London, Ser. A, 45 (1888), pp. 126–132.

Mallock, A. Experiments on fluid viscosity. *Philos. Trans. Royal Society*, London, Ser. A, 187 (1896), p. 41–56.

Mehta, R.D. The aerodynamics of sports balls. *Annual Review of Fluid Mechanics*, 17 (1985), pp. 151–189.

Mehta, R.D., Bentley, K., Proudlove, M., and Varty, P. Factors affecting cricket ball swing. *Nature*, 303 (1983), pp. 787–788.

Mehta, R.D., and Wood, D.H. Aerodynamics of the cricket ball. *New Scientist*, 87 (1980), pp. 442–447.

Mittal, S., and Kumar, B. Flow past a rotating cylinder. *Journal of Fluid Mechanics*, 476 (2003), pp. 303–334.

Moalem, D., and Cohen, S. Hydrodynamics and heat/mass transfer near rotating surfaces. *Advances in Heat Transfer*, 21, pp. 141–183. Academic Press, 1991.

Neale, M.J. (ed.). *The tribology handbook*. Butterworth Heinemann, 1995.

Nissan, A.H., Nardacci, J.L., and Ho, L.Y. The onset of different modes of instability for flow between rotating cylinders. *American Institute of Chemical Engineers Journal*, 9 (1963), pp. 620–624.

Raimondi, A.A., and Boyd, J. A solution for the finite journal bearing and its application to analysis and design: I. *ASLE Transactions*, 1 (1958a), pp. 159–174.

Raimondi, A.A., and Boyd, J. A solution for the finite journal bearing and its application to analysis and design: II. *ASLE Transactions*, 1 (1958b), pp. 175–193.

Raimondi, A.A., and Boyd, J. A solution for the finite journal bearing and its application to analysis and design: III. *ASLE Transactions*, 1 (1958c), pp. 194–209.

Rayleigh, Lord. On the dynamics of revolving fluids. *Proc. Royal Society*, London, Ser. A, 93 (1916), pp. 148–154.

Reid, E.G. Tests of rotating cylinders. NACA Technical Note, NACA-TN-209, 1924.

Reynolds, O. On the theory of lubrication and its application to Mr Beauchamp Tower's experiments including an experimental determination of the viscosity of olive oil. *Philos. Trans. Royal Society*, 177 (1886), pp. 157–234.

Roberts, P.H. Appendix in experiments on the stability of viscous flow between rotating cylinders, VI. Finite amplitude experiments. *Proc. Royal Society*, London, Series A, 238 (1965), pp. 531–556.

Salter, S., Sortino, G., and Latham, J. Sea-going hardware for the cloud albedo method of reversing global warming. *Philos. Trans. Royal Society* A, 366 (2008), pp. 3989–4006.

Schultz-Grunow, F. Stabilitat einer rotierended flussigkeit. *Zeitschrift für angewandte Mathematik Mechanik* (1963), pp. 411–415.

Schwarz, K.W., Springnett, B.E., and Donnelly, R.J. Modes of instability in spiral flow between rotating cylinders. *Journal of Fluid Mechanics*, 20 (1964), pp. 281–289.

Seufert, W., and Seufert, S. Critics in a spin over Flettner's ship. *New Scientist* (1983), pp. 656–659.

Smits, A.J. A new aerodynamic model of a golf ball in flight. In A.J. Cochran and M.R. Farrally, (eds.), *Science and Golf* II, pp. 340–347. E&FN Spon, 1994.

Snyder, H.A. Change in wave form and mean flow associated with wave length variations in rotating Couette flow. *Journal of Fluid Mechanics*, Part L, 35 (1969), pp. 337–352.

Snyder, H.A. Waveforms in rotating Couette flow. *International Journal of Heat and Non-linear Mechanics*, 5 (1970), pp. 659–685.

Sparrow, E.M., Munro, W.D., and Johnson, V.K. Instability of the flow between rotating cylinders: the wide gap problem. *Journal of Fluid Mechanics*, 20 (1964), pp. 35–46.

Steinheuer, J. Three dimensional boundary layers on rotating bodies and in corners. AGARD No. 97 (1965), pp. 567–611.

Taylor, G.I. Stability of a viscous liquid contained between two rotating cylinders. *Proc. Royal Society*, London, Ser. A, 223 (1923), pp. 239–343.

Theodorsen, T., and Regier, A. Experiments on drag of revolving discs, cylinders and streamline rods at high speeds. NACA Report 793, 1944.

Thom, A. Effects of discs on the air forces on a rotating cylinder. Aeronautical Research Committee Reports and Memoranda 1623, 1934.

Walowit, J., Tsao, S., and Di-Prima, R.C. Stability of flow between arbitrarily spaced concentric cylindrical surfaces including the effect of a radial temperature gradient. *Trans. Journal of Applied Mechanics*, 31 (1964), pp. 585–593.

Watts, R.G., and Sawyer, E. Aerodynamics of a knuckleball. *American Journal of Physics*, 43 (1975), pp. 960–963.

Welsh, R.J. *Plain bearing design handbook*. Butterworth, 1983.

Chapter 7

Rotating Cavities

A cavity formed between two co-rotating co-axial discs with an outer shroud is a common feature in rotating machinery, for example, in compressor and turbine rotor assemblies. The behavior of flow contained between the discs is strongly dependent on whether there is a throughflow of fluid and on the relative temperature of the cavity surfaces.

7.1. INTRODUCTION

Cavities formed by co-rotating discs are shown in Figures 7.1 and 7.2 illustrating an axial turbine assembly and an axial compressor, respectively. In a gas turbine engine, air extracted from the compression system for turbine cooling and rim sealing is often led through the annular passage formed between disc bores and a central shaft. A bulk flow through the cavity between the discs can be arranged by providing a slot or holes in the shroud and either supplying or drawing fluid through central holes at the disc bores. Depending on the boundary conditions, the bulk flow will as a result either be radially inward or outward. If the outer shroud is sealed, then, for the case of discs with holes at the disc bores on the axis, a throughflow of fluid through the bores is possible, and the interactions of this flow with the cavity flow can be significant. The bulk passage of this air through the bore can provide ventilation of the rotating cavities formed between the adjacent compressor discs.

The range of configurations possible for rotating cavities is extensive, and a selection of these is illustrated in Figure 7.3. In general, the cavity comprises two discs of outer radius, b, and inner radius, a, separated by an axial gap, s, sometimes with a shroud located at the outer radius. For a gas turbine engine, knowledge and understanding of the flow and heat transfer in rotating cavities is required for accurate disc stressing and component life predictions. The flow behavior in a rotating cavity is strongly dependent on whether there is a throughflow of air in the cavity, which can be radial or axial, or both of these, and on the relative temperature between the components.

Sections 7.1.1, 7.1.2, 7.1.3, and 7.1.4 introduce the flow characteristics for closed and annular cavities, cavities with outflow, cavities with inflow, and

Rotating Flow, DOI: 10.1016/B978-0-12-382098-3.00007-X
© 2011 Elsevier Inc. All rights reserved.

249

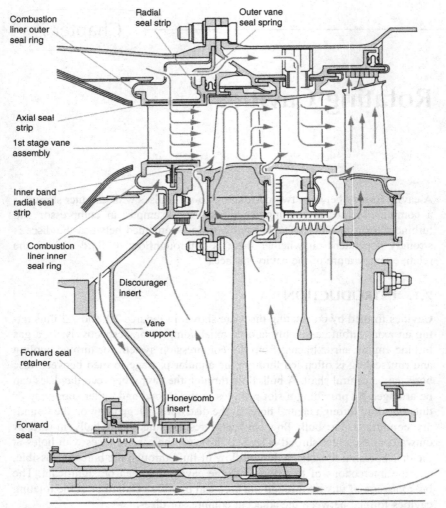

Combustion
liner outer
seal ring

Radial
seal strip

Outer vane
seal spring

Axial seal
strip

1st stage vane
assembly

Inner band
radial seal
strip

Combustion
liner inner
seal ring

Discourager
insert

Vane
support

Forward seal
retainer

Honeycomb
insert

Forward
seal

FIGURE 7.1 An axial turbine assembly, illustrating the high pressure turbine of a multi-spool engine. Image courtesy of Rolls-Royce plc.

cavities with axial throughflow, respectively. The bulk characterisation of rotating cavity flows is considered in Section 7.1.5, prior to more detailed development of modeling procedures in Sections 7.2 and 7.3 based on simplifications to the equations of motion and Section 7.4 presenting the relevant general equations of motion. Modeling of a variety of rotating cavity applications is considered in Section 7.5.

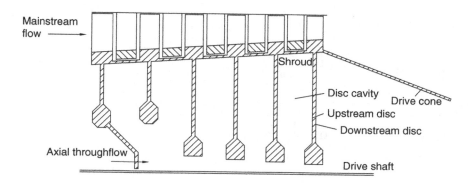

FIGURE 7.2   Schematic representation of the compressor discs of a high-pressure axial compressor.

7.1.1 Closed Rotating Cylindrical and Annular Cavities

A closed cylindrical cavity and an annular cavity are illustrated in Figure 7.4 where the discs and the outer shroud, and, in the case of the annular cavity the surface at the inner radius, are all co-rotating at the same angular velocity. Closed cylindrical and annular cavities are a relatively common feature in compressor and turbine rotor assemblies.

As an introduction to the range of flow characteristics, it is convenient to consider the three different geometric and heating configurations illustrated in Figure 7.5. For steady-state conditions in an isothermal cavity or annulus, where all the surfaces are at the same temperature as the fluid, there is no relative motion of the fluid inside the cavity and the fluid rotates as a solid body. For nonisothermal flow, buoyancy forces are generated from the centripetal acceleration and density variations in the fluid. The buoyancy forces result in fluid motion relative to the bounding surfaces that depend on the geometrical and heating configuration. For the case of an axially directed heat flux, where there is a temperature difference between the two discs, fluid moves radially inward adjacent to the hot disc and radially outward adjacent to the cooler disc as indicated in Figures 7.5(a) and 7.5(b). In a cavity with no central shaft, there is an additional recirculation zone close to the axis of rotation, but in an annular cavity this feature is not present.

Bohn et al. (1993) investigated the case of a rotating cavity with a radially directed heat flux, where there is a temperature difference between the inner and outer peripheral surfaces (see Figure 7.5(c)) and found a series of cyclonic (low-pressure) and anticyclonic (high-pressure) recirculations. A separate numerical study by King, Wilson, and Owen. (2007)—for the case of a closed rotating cavity where the outer radial surface is hotter than the inner—also showed that

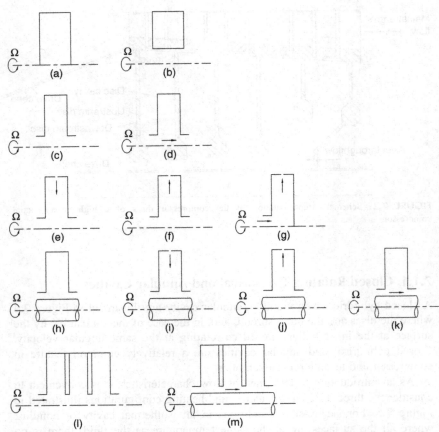

FIGURE 7.3 Selected rotating cavity geometric configurations: (a) Closed cylindrical cavity, (b) annular cavity, (c) open rotating cavity with no throughflow, (d) rotating cavity with axial throughflow, (e) rotating cavity with radial inflow, (f) rotating cavity with radial outflow, (g) rotating cavity with an axial inlet and radial outflow, (h) rotating cavity with a central shaft, (i) rotating cavity with a central shaft and radial inflow, (j) rotating cavity with a central shaft and radial outflow, (k) rotating cavity with a central shaft and axial throughflow, (l) multiple rotating cavities with axial throughflow, (m) multiple rotating cavities with a central shaft and axial throughflow.

the Rayleigh-Bénard convection type of flow structure could occur with counter-rotating cyclonic and anticyclonic vortices, as illustrated in Figure 7.6.

As a result of interest in the flow phenomena associated with the Earth's lower atmosphere and its oceans, a considerable number of experimental and modeling studies for the case of a rotating cylindrical container or annulus are available; see, for example, Hide (1968) and Greenspan (1968). Oceanic and atmospheric flows are considered in Chapter 8.

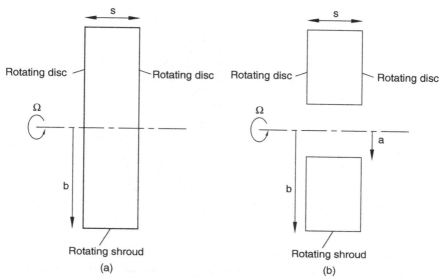

FIGURE 7.4   A closed cylindrical cavity and an annular cavity.

7.1.2 Rotating Cavity with Radial Outflow

A radial outflow is sometimes used to remove heat from a co-rotating turbine disc cavity, as illustrated schematically in Figure 7.7, and for the cavity formed between the mini-disc and right-hand turbine disc in Figure 7.1. The flow can be characterized in many cases by a rotating inviscid core and disc boundary layers.

Hide (1968) used the Taylor-Proudman theorem to give insight into the laminar source-sink flows in a rotating cavity. Owen, Pincombe, and Rogers (1985) and Owen and Rogers (1995) classify these regions as follows.

i. *The source region.* The source region defines the zone where the flow is supplied to the cavity and then feeds the two boundary layer flows on the disc faces. Hide (1968) for the flow structures he considered referred to this

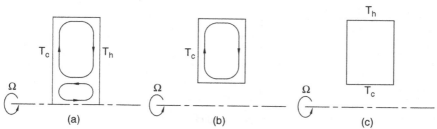

FIGURE 7.5   Sealed annular cavity configurations.

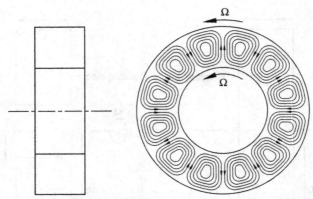

FIGURE 7.6 Schematic streamlines for Rayleigh-Bénard convection in a closed rotating annular cavity.

region as a source layer. For rotating cavities, the term *layer* is not necessarily appropriate as the incoming flow to the cavity can occupy a substantial proportion of the cavity before it divides to feed the two Ekman-type layers on the disc faces.

ii. *Ekman-type layers.* Flow entering the cavity is progressively entrained from the source region into entraining boundary layers on the discs. Once all of the fluid supplied to the cavity has been entrained into the boundary layers, the steady boundary layers become "nonentraining" and are referred to as

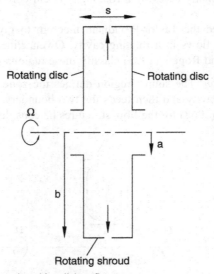

FIGURE 7.7 Rotating cavity with radial outflow.

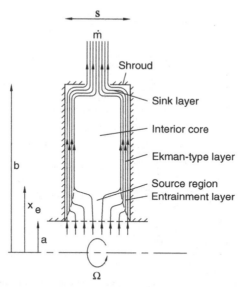

FIGURE 7.8 Characteristic streamlines for a rotating cavity with radial outflow with a radial inlet.

Ekman-type layers. The term *Ekman layer* is used to refer to the viscous region between a geostrophic zone, where viscous forces are negligible, and a solid boundary in a rotating fluid.

iii. *The sink layer.* The sink layer is the region of the cavity over which the two disc boundary layers combine to feed the cavity flow into the sink or exit.

iv. *The interior core.* For the case of an inviscid fluid between the two disc face boundary layer flows, an interior core of rotating fluid will be set up in a rotating cavity where the radial and axial velocity components are zero.

The regions are illustrated in Figures 7.8 and 7.9 for the cases of a radial inlet at the axis and also for an axial inlet where the radius of one of the discs extends to the axis. The temperature distribution on the discs can be highly significant in determining the flow structure in rotating cavities. Here the temperature distribution on the discs is assumed to be uniform.

7.1.3 Rotating Cavity with Radial Inflow

In order to provide the mass flow for cooling in a gas turbine engine, one option is to extract air from the axial compressor by drawing it radially inward through holes in the compressor rotor assembly and passing it through the cavity formed by the co-rotating discs supporting the compressor blades. The simplified geometric configuration associated with this scenario is illustrated in Figure 7.10. The shroud can have multiple holes in it to supply the bulk radial inward

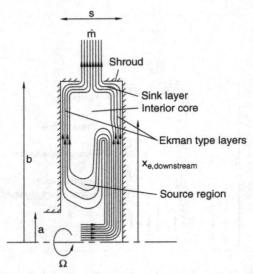

FIGURE 7.9 Characteristic streamlines for a rotating cavity with radial outflow with an axial inlet.

flow, a mesh, or slots. The flow in a rotating cavity with radial inflow can be characterized by a source region that feeds the flow to the disc boundary layers, an inviscid rotating core region, and a sink layer that is fed by the Ekman-type disc boundary layers, as illustrated in Figures 7.11 and 7.12.

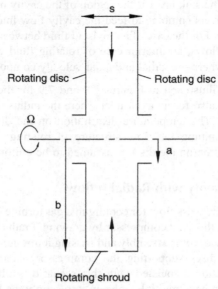

FIGURE 7.10 Rotating cavity with radial inflow.

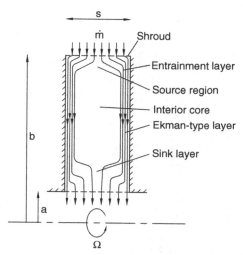

FIGURE 7.11   Characteristic streamlines for a rotating cavity with radial inflow ($c_{eff} = 1$).

Prior to entering the cavity, the flow may have a significant tangential component of velocity. This is characterized by the inlet swirl ratio c, Equation 7.1. As the flow enters the cavity, the tangential component may be significantly modified as it passes through inlet holes and slots and mixes with recirculating fluid; its swirl ratio as it enters the cavity is denoted by the effective swirl ratio, c_{eff}, Equation 7.2. The size and structure of the source region are dependent on the swirl ratio.

$$c = \frac{u_\phi}{\Omega r} \tag{7.1}$$

$$c_{eff} = \frac{u_{\phi,\, r=b}}{\Omega r} \tag{7.2}$$

For the case of an effective swirl ratio of unity, $c_{eff} = 1$, the flow structure in the cavity can be approximately modeled by a source region, Ekman-type layers, and a sink layer as illustrated in Figure 7.11, in a fashion similar to the case of radial outflow in Section 7.1.2.

For an effective swirl ratio of $c_{eff} < 1$, the source region is divided into two parts as illustrated in Figure 7.12. In the outer source region, $b > r > r_1$, where $u_\phi < \Omega r$; there is recirculation on both discs and inflow outside the boundary layers. The recirculation transfers angular momentum to the incoming fluid, and as a result the effective swirl ratio, c_{eff}, differs from the initial value, c. For the inner source region, the fluid flows radially inward across the whole width of the cavity and $u_\phi > \Omega r$, corresponding to the entrainment layer for the case of radial outflow (Section 7.1.2). The tangential component of velocity within the source region can be modeled as a free vortex (see Equation 3.35), with the

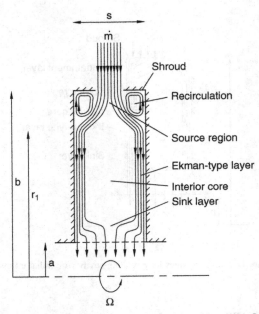

FIGURE 7.12 Characteristic streamlines for a rotating cavity with radial inflow ($c_{eff} < 1$).

tangential velocity increasing as the radius decreases. If the swirl ratio is less than unity, then the fluid rotates slower than the disc at the outer radii and faster than the disc at the inner radii. If angular momentum is conserved, then $u_{\phi,\infty}r =$ constant, where $u_{\phi,\infty}$ is the tangential velocity of the rotating core. If $u_{\phi,\infty}r < \Omega b$ at $r = b$, there will be a radius r' where $u_{\phi,\infty} = \Omega r'$. Hence $u_{\phi,\infty} < \Omega r$ for $r' < r < b$, and the flow in the boundary layers for this region will be radially outward. For $r < r'$ where $u_{\phi,c} > \Omega r$ the flow in the boundary layers for this region will be radially inward. For the case of radial inflow, the fluid in the interior core between the two Ekman-type layers rotates at an angular speed that is greater than that of the disc.

7.1.4 Rotating Cavities with Axial Throughflow

A rotating cavity with an axial throughflow of air can be used to model part of the stack formed by multiple compressor discs, supporting the blades in an axial flow compressor. The flow structure for the case of a rotating cavity with an axial throughflow of air at the inner bore is particularly complex. For isothermal conditions, where all the surfaces are at the same temperature as the fluid, vortex breakdown in the central jet of air can result in nonaxisymmetric flow inside the cavity. If the discs are at different temperatures, nonaxisymmetric flow predominates.

(a)   (b)   (c)

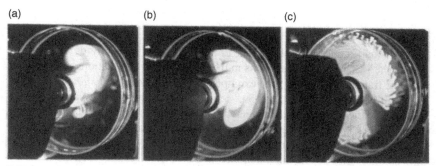

FIGURE 7.13   Flow visualization for a heated rotating cavity with axial throughflow. $G = 0.267$, $Re_z = 2\overline{w}a/\nu = 2180$, $Re_\phi = 13000$, $Ro = \overline{w}/\Omega a = 8.4$. The cavity is rotating anticlockwise. The time indicated represents the period after the smoke is introduced to the cavity (Farthing et al., 1992a).

For the case of a radial temperature gradient on the discs, experimental flow visualization data is available from Farthing et al. (1992a). A sequence of photographs of the cavity is shown in Figures 7.13(a)–(c). For these tests, air was fed axially into the cavity. The discs were heated by hot air blowers, giving a radial temperature distribution that rose to a peak and then reduced linearly. The flow visualization data from smoke introduced into the cavity for the sequence of photographs at $t = 2$ s, $t = 3$ s, and $t = 10.5$ s, shows the formation of a radial arm, Figure 7.13(a), that then bifurcates, Figure 7.13(b), forming two recirculating zones.

These two zones rotate in opposite directions. In Figure 7.13(b), the top recirculating zone is cyclonic, and rotates in the same direction as the cavity. The lower recirculating zone is anticyclonic and rotates in a direction opposite to that of the cavity. These recirculating zones continue to grow until they almost fill the cavity; see Figure 7.13(c). A dead zone, between the two fronts, can be seen in Figure 7.13(c), indicated by the dark segmental region into which smoke is not advected. Careful examination of the flow visualization data indicated a radial inflow of smoke on the heated discs.

7.1.5 Characterization of Cavity Flow and Modeling

As discussed in Chapter 2, rotating flows can sometimes be characterized by the Rossby number and the Ekman number. These are restated here for convenience as Equations 7.3 and 7.4, respectively. In Equation 7.3 the reference velocity used, U_o, is the tangential velocity component in the interior core of the cavity flow, $u_{\phi,c}$, and the relevant corresponding characteristic dimension, L, is sometimes taken to be the axial cavity gap, s, or the disc bore radius, a.

$$Ro = \frac{U_o}{\Omega L} = \frac{u_{\phi,c}}{\Omega s} \qquad (7.3)$$

$$Ek = \frac{\nu}{\Omega L^2} = \frac{\nu}{\Omega s^2} \qquad (7.4)$$

For small values of Rossby and Ekman numbers, typically defined by $Ro \ll Ek^{0.25} \ll 1$, the inertial terms in the boundary layer equations referred to a rotating frame of reference; for example, $u\partial u/\partial r$, $w\partial u/\partial z$, v^2/r, $u\partial v/\partial r$, $w\partial v/\partial z$, uv/r, can be neglected. This gives a set of linear equations that can be readily solved, and the assumptions associated with the simplifications to the governing equations are known as the boundary layer approximation.

The modeling of rotating cavity flows can require consideration of three-dimensional time-dependent phenomena. For certain cases, such as a cavity with radial outflow or inflow, approximations for the boundary layer and core flow can be made that significantly simplify the analysis. The boundary layer approximation for rotating cavity flows is considered in Section 7.2. For the case of a rotating cavity with radial outflow, it can be assumed that the boundary layers in the source region are similar to those for a rotating free disc, as considered in Chapter 4. For a rotating cavity with radial inflow, it can be assumed that the flow is characterized by a free vortex outside the boundary layer. Such assumptions make the flow amenable to relatively simple modeling techniques, in comparison to the full equations of motion; and solutions can be developed for flow in the source region and the extent of the source region for a cavity with either radial inflow or outflow.

When the velocity components relative to the rotating frame of reference are small in comparison with the local speed of the frame of reference, Ωr, the nonlinear terms in Equations 2.102 to 2.104 can be neglected. The resulting equations are referred to as the inviscid linear equations, or the linear Ekman layer equations, and the assumptions made in deriving them are called the linear approximation, as considered in Section 7.3.

The differential equations for continuity and momentum can be integrated across the boundary layer. The resulting integral equations provide useful insight into the flow phenomena occurring in a rotating cavity and are developed in Section 7.4.

Application of these methods to a variety of rotating cavities is considered in Section 7.5.

7.2. BOUNDARY LAYER APPROXIMATION FOR ROTATING CAVITY FLOWS

If the effects of viscous or turbulent stresses are only important near a solid boundary, then the governing equations can be simplified to give a set of equations that can be readily solved to model—for example, the flow in the boundary layer of a rotating disc. Outside the boundary layer the flow can be modeled as inviscid with accompanying ease of analysis. The assumptions associated with this approach are referred to as the boundary layer approximation. For a rotating cavity boundary, layers are formed on both rotating discs and on the shroud. The approximations made are:

that the rate of change of all quantities, except pressure, normal to surface are much greater in magnitude than that parallel to the surface;

that the rate of change of pressure normal to the surface is negligible;

that the component of velocity normal to the surface is much smaller in magnitude than parallel to the surface.

Pressure, as commonly assumed in boundary layer theory, is thought to be independent of the coordinate normal to the surface within the boundary layer.

For the case of a rotating disc, at $z = 0$, these approximations can be stated in the form:

$$\left|\frac{\partial p}{\partial z}\right| >> \left|\frac{\partial p}{\partial r}\right| \tag{7.5}$$

$$\left|\frac{\partial u_r}{\partial z}\right| >> \left|\frac{\partial u_r}{\partial r}\right| \tag{7.6}$$

$$\left|\frac{\partial u_\phi}{\partial z}\right| >> \left|\frac{\partial u_\phi}{\partial r}\right| \tag{7.7}$$

$$\left|\frac{\partial u_z}{\partial z}\right| >> \left|\frac{\partial u_z}{\partial r}\right| \tag{7.8}$$

$$\left|\frac{\partial p}{\partial z}\right| << \left|\frac{\partial p}{\partial r}\right| \tag{7.9}$$

$$|u_z| << |u_r| \tag{7.10}$$

$$|u_z| << |u_\phi| \tag{7.11}$$

The velocity gradients normal to the disc will have the same order of magnitude as the radial and tangential velocity gradients as indicated by the ~ symbol in the following equations.

$$\left|\frac{\partial u_z}{\partial z}\right| \sim \left|\frac{\partial u_r}{\partial r}\right| \tag{7.12}$$

$$\left|\frac{\partial u_z}{\partial z}\right| \sim \left|\frac{\partial u_\phi}{\partial r}\right| \tag{7.13}$$

As a result of these approximations, the following shear stresses can be taken as zero

$$\sigma_r = \sigma_\phi = \sigma_z = \tau_{r\phi} = \tau_{\phi r} = 0 \tag{7.14}$$

For convenience the nonzero shear stresses are given the symbols τ_r and τ_ϕ.

$$\tau_{rz} = \tau_{zr} = \tau_r \tag{7.15}$$

$$\tau_{\phi z} = \tau_{z\phi} = \tau_\phi \tag{7.16}$$

For steady axisymmetric flow, the continuity and momentum equations, 2.80 and 2.58–2.60, reduce to:

$$\frac{1}{r}\frac{\partial}{\partial r}(\rho r u_r) + \frac{\partial}{\partial z}(\rho u_z) = 0 \tag{7.17}$$

$$\frac{1}{r}\frac{\partial}{\partial r}(\rho r u_r^2) + \frac{\partial}{\partial z}(\rho u_r u_z) - \rho\left(\frac{u_\phi^2}{r}\right) = -\frac{\partial p}{\partial r} + \frac{\partial \tau_r}{\partial z} \tag{7.18}$$

$$\frac{1}{r^2}\frac{\partial}{\partial r}(\rho r^2 u_r u_\phi) + \frac{\partial}{\partial z}(\rho u_\phi u_z) = \frac{\partial \tau_\phi}{\partial z} \tag{7.19}$$

$$\frac{\partial p}{\partial z} = 0 \tag{7.20}$$

For laminar flow,

$$\tau_r = \mu\frac{\partial u_r}{\partial z} \tag{7.21}$$

$$\tau_\phi = \mu\frac{\partial u_\phi}{\partial z} \tag{7.22}$$

Substituting for the shear stress, Equations 7.18 to 7.20 therefore become

$$\frac{1}{r}\frac{\partial}{\partial r}(\rho r u_r^2) + \frac{\partial}{\partial z}(\rho u_r u_z) - \rho\left(\frac{u_\phi^2}{r}\right) = -\frac{\partial p}{\partial r} + \frac{\partial}{\partial z}\left(\mu\frac{\partial u_r}{\partial z}\right) \tag{7.23}$$

$$\frac{1}{r^2}\frac{\partial}{\partial r}(\rho r^2 u_r u_\phi) + \frac{\partial}{\partial z}(\rho u_\phi u_z) = \frac{\partial}{\partial z}\left(\mu\frac{\partial u_\phi}{\partial z}\right) \tag{7.24}$$

$$\frac{\partial p}{\partial z} = 0 \tag{7.25}$$

Equations 7.17 to 7.20 are valid for incompressible and compressible flow. The form of the pressure in the core, p_∞, which is a function of the radius only, must be determined by consideration of the inviscid equations. p_∞ is the value of the pressure in the core as $z \to 0$.

In a rotating frame of reference, the continuity and momentum equations, 2.110 and 2.102–2.104, respectively, reduce to

$$\frac{1}{r}\frac{\partial}{\partial r}(\rho r u) + \frac{\partial}{\partial z}(\rho w) = 0 \tag{7.26}$$

$$\frac{1}{r}\frac{\partial}{\partial r}(\rho r u^2) + \frac{\partial}{\partial z}(\rho u w) - \rho\left(\frac{v^2}{r} + 2\Omega v + \Omega^2 r\right) = -\frac{\partial p}{\partial r} + \frac{\partial \tau_r}{\partial z} \tag{7.27}$$

$$\frac{1}{r^2}\frac{\partial}{\partial r}(\rho r^2 u v) + \frac{\partial}{\partial z}(\rho v w) + 2\rho\Omega u = \frac{\partial \tau_\phi}{\partial z} \tag{7.28}$$

$$\frac{\partial p}{\partial z} = 0 \tag{7.29}$$

For laminar flow,

$$\tau_r = \mu\frac{\partial u}{\partial z} \tag{7.30}$$

$$\tau_\phi = \mu\frac{\partial v}{\partial z} \tag{7.31}$$

For the case of the stationary shroud, at $r = b$, the boundary layer approximations can be stated in the form:

$$\left|\frac{\partial p}{\partial r}\right| >> \left|\frac{\partial p}{\partial z}\right| \tag{7.32}$$

$$\left|\frac{\partial u_r}{\partial r}\right| >> \left|\frac{\partial u_r}{\partial z}\right| \tag{7.33}$$

$$\left|\frac{\partial u_\phi}{\partial r}\right| >> \left|\frac{\partial u_\phi}{\partial z}\right| \tag{7.34}$$

$$\left|\frac{\partial u_z}{\partial r}\right| >> \left|\frac{\partial u_z}{\partial z}\right| \tag{7.35}$$

$$\left|\frac{\partial p}{\partial r}\right| << \left|\frac{\partial p}{\partial z}\right| \tag{7.36}$$

$$|u_r| << |u_z| \tag{7.37}$$

$$|u_r| << |u_\phi| \tag{7.38}$$

$$\left|\frac{\partial u_r}{\partial z}\right| >> \left|\frac{u_r}{r}\right| \tag{7.39}$$

The radial velocity gradients have the same order of magnitude as the normal and tangential velocity gradients as indicated in Equations 7.40 and 7.41.

$$\left|\frac{\partial u_r}{\partial r}\right| \sim \left|\frac{\partial u_z}{\partial z}\right| \tag{7.40}$$

$$\left|\frac{\partial u_r}{\partial r}\right| \sim \left|\frac{\partial u_\phi}{\partial z}\right| \tag{7.41}$$

As a result of these approximations, the following shear stresses can be taken as zero

$$\sigma_r = \sigma_\phi = \sigma_z = \tau_{z\phi} = \tau_{\phi z} = 0 \tag{7.42}$$

For convenience, the nonzero shear stresses are given the symbols τ_r and τ_ϕ in Equations 7.43 and 7.44.

$$\tau_{rz} = \tau_{zr} = \tau_z \tag{7.43}$$

$$\tau_{\phi r} = \tau_{r\phi} = \tau_\phi \tag{7.44}$$

For steady axisymmetric flow, the continuity and momentum equations, 2.80 and 2.58–2.60, respectively, reduce to:

$$\frac{1}{r}\frac{\partial}{\partial r}(\rho r u_r) + \frac{\partial}{\partial z}(\rho u_z) = 0 \tag{7.45}$$

$$\rho\left(\frac{u_\phi^2}{r}\right) = \frac{\partial p}{\partial r} \tag{7.46}$$

$$\frac{1}{r^2}\frac{\partial}{\partial r}(\rho r^2 u_r u_\phi) + \frac{\partial}{\partial z}(\rho u_\phi u_z) = \frac{\partial \tau_\phi}{\partial z} \tag{7.47}$$

$$\frac{\partial}{\partial r}(\rho u_r u_z) + \frac{\partial}{\partial z}(\rho u_z^2) = \frac{\partial p}{\partial z} + \frac{\partial \tau_z}{\partial z} \tag{7.48}$$

For laminar flow,

$$\tau_z = \mu\frac{\partial u_z}{\partial r} \tag{7.49}$$

$$\tau_\phi = \mu\frac{\partial u_\phi}{\partial r} \tag{7.50}$$

In a rotating frame of reference the continuity and momentum equations, 2.110 and 2.102–2.104, respectively, reduce to Equations 7.51 and 7.52–7.54.

$$\frac{1}{r}\frac{\partial}{\partial r}(\rho r u) + \frac{\partial}{\partial z}(\rho w) = 0 \tag{7.51}$$

$$\rho\left(\frac{v^2}{r} + 2\Omega v + \Omega^2 r\right) = \frac{\partial p}{\partial r} \tag{7.52}$$

$$\frac{1}{r^2}\frac{\partial}{\partial r}(\rho r^2 uv)+\frac{\partial}{\partial z}(\rho vw) =\frac{\partial \tau_\phi}{\partial z} \tag{7.53}$$

The term $2\rho\Omega u$ has been omitted because it is of a smaller order of magnitude than the other terms on the left-hand side of the above equation.

$$\frac{\partial}{\partial r}(\rho vw) + \frac{\partial}{\partial z}(\rho w^2) = -\frac{\partial p}{\partial z}+\frac{\partial \tau_z}{\partial r} \tag{7.54}$$

For laminar flow,

$$\tau_z = \mu\frac{\partial w}{\partial r} \tag{7.55}$$

$$\tau_\phi = \mu\frac{\partial v}{\partial r} \tag{7.56}$$

The boundary layer equations developed are for the disc and shroud. The corresponding equations for $z = s$ can readily be developed from these.

The boundary layer system of partial differential equations, Equations 7.51–7.54, can be solved numerically. Alternatively, if they are converted to integral equations (see Section 7.4), solutions can be obtained by analytical or numerical methods. If the boundary-layer equations are expressed in a rotating frame of reference, then it is usually possible to obtain a simpler linear form referred to as the Ekman layer equations (see Section 7.3).

7.3. LINEAR EKMAN LAYER EQUATIONS

At high rotational speeds, where the Coriolis terms are much greater than the nonlinear inertial terms, the boundary layer equations can be simplified to give a set of linear equations called the Ekman layer equations. These can provide good insight into the flow field in a rotating cavity.

The nonlinear boundary equations for steady axisymmetric flow when no body forces act on the fluid in a rotating frame of reference, given in Equations 7.26–7.29, can be simplified when the magnitude of the velocity components u, v, w is small compared with Ωr.

$$-2\rho\Omega(v-v_\infty) = \frac{\partial \tau_r}{\partial z} \tag{7.57}$$

$$2\rho\Omega u = \frac{\partial \tau_\phi}{\partial z} \tag{7.58}$$

In practice, the radial component of velocity is zero throughout much of the core, $u_\infty = 0$.

For laminar flow

$$\tau_r = \mu \frac{\partial u_r}{\partial r} \tag{7.59}$$

$$\tau_\phi = \mu \frac{\partial u_\phi}{\partial r} \tag{7.60}$$

Hence,

$$-2\Omega(v - v_\infty) = \nu \frac{\partial^2 u}{\partial z^2} \tag{7.61}$$

and

$$2\Omega(u - u_\infty) = \nu \frac{\partial^2 v}{\partial z^2} \tag{7.62}$$

where the subscript ∞ has been used to denote the conditions outside the boundary layer. Equations 7.61 and 7.62 show the balance between Coriolis and viscous forces characteristic of Ekman-type boundary layers.

Alternatively, from Equations 2.106–2.108, for laminar flow, in a rotating frame of reference,

$$-\rho_\infty(2\Omega v_\infty + \Omega^2 r) = -\frac{\partial p_\infty}{\partial r} \tag{7.63}$$

$$2\rho_\infty \Omega u_\infty = -\frac{1}{r}\frac{\partial p_\infty}{\partial \phi'} \tag{7.64}$$

$$0 = F_z' - \frac{\partial p_\infty}{\partial z} \tag{7.65}$$

On rearranging and substitution,

$$-2\Omega v_\infty = -\frac{1}{\rho_\infty}\frac{\partial p_\infty}{\partial r} \tag{7.66}$$

$$2\Omega u_\infty = -\frac{1}{\rho_\infty r}\frac{\partial p_\infty}{\partial \phi'} \tag{7.67}$$

The Ekman layer equations are only valid for incompressible flow when

$$\frac{|v_\infty|}{\Omega r} << 1 \tag{7.68}$$

Although the Ekman layer equations involve a series of assumptions, their solution provides very valuable insight into flow in a rotating system. The linear equations are considerably easier to solve than the nonlinear equations, and the results in both cases are qualitatively similar.

The continuity equation for incompressible flow, in a rotating frame of reference, is given by Equation 2.109, which in component form is given by

$$\frac{\partial u}{\partial r} + \frac{u}{r} + \frac{1}{r}\frac{\partial v}{\partial \phi} + \frac{\partial w}{\partial z} = 0 \tag{7.69}$$

From Equations 7.66, 7.67, and 7.69 it can be deduced that the axial velocity gradient must be zero (as for Taylor-Proudman flow—see Section 3.3). For the case of an isothermal cavity, symmetry about the midplane of the cavity can also be assumed. In the region between the Ekman layers in the cavity, the axial velocity will be zero and the Ekman layers are nonentraining. This is a valid assumption for both laminar and turbulent incompressible flow but is not necessarily the case if the flow is not symmetrical about the cavity midplane.

The corresponding boundary conditions for the inviscid linear equations for an Ekman layer on a rotating disc are

$$u = 0, \quad v = 0 \quad \text{at } z = 0 \tag{7.70}$$

$$u \rightarrow u_\infty, \quad v \rightarrow v_\infty \quad \text{as } z \rightarrow \infty \tag{7.71}$$

For axisymmetric flow, from Equation 7.67, the radial velocity outside the boundary layer will be zero, $u_\infty = 0$. The radial velocity outside the boundary layers will also be zero for turbulent flow and for buoyancy-driven flows.

The swirl ratio, β, is taken as the ratio of the tangential velocity component in the rotating core to the disc tangential velocity component.

$$\beta = \frac{u_{\phi,\infty}}{\Omega r} \tag{7.72}$$

For the boundary conditions given in 7.70 and 7.71, the solution to Equations 7.61 and 7.62, for $u_\infty = 0$, is

$$u = -v_\infty e^{-z/\sqrt{\nu/\Omega}} \sin\left(\frac{z}{\sqrt{\nu/\Omega}}\right) \tag{7.73}$$

$$v = v_\infty \left[1 - e^{-z/\sqrt{\nu/\Omega}} \cos\left(\frac{z}{\sqrt{\nu/\Omega}}\right)\right] \tag{7.74}$$

or in a stationary frame of reference,

$$u_r = \Omega r(1-\beta) e^{-z/\sqrt{\nu/\Omega}} \sin\left(\frac{z}{\sqrt{\nu/\Omega}}\right) \tag{7.75}$$

$$u_\phi - u_{\phi,\infty} = (\Omega r - u_{\phi,\infty}) e^{-z/\sqrt{\nu/\Omega}} \cos\left(\frac{z}{\sqrt{\nu/\Omega}}\right) = \Omega r(1-\beta) e^{-z/\sqrt{\nu/\Omega}} \cos\left(\frac{z}{\sqrt{\nu/\Omega}}\right) \tag{7.76}$$

The local mass flow rate through the boundary layer is given by

$$\dot{m}_o = 2\pi\rho r \int_0^\infty u\,dz \tag{7.77}$$

Substituting for u using Equation 7.73 and integrating gives

$$\dot{m}_o = -\pi\mu b Re_\phi^{0.5}\frac{v_\infty}{\Omega r}\left(\frac{r}{b}\right)^2 = \pi\rho\Omega r^2\left(\frac{\mu}{\rho\Omega}\right)^{0.5}(1-\beta) \tag{7.78}$$

In nondimensional form, Equation 7.78 can be stated as

$$\frac{\dot{m}_o}{\mu r} = \pi(1-\beta)\,x\,Re_\phi^{0.5} \tag{7.79}$$

If it is assumed the fluid entering the cavity is divided equally between the flows in the two boundary layers,

$$\dot{m}_o = \frac{\dot{m}}{2} = \frac{\mu b Cw}{2} \tag{7.80}$$

where $\dot{m}$ is the mass flow rate entering the cavity. $\dot{m}$ is normally taken as positive if the flow is radially outward and negative for inflow.

Hence from Equation 7.78, the velocity outside the boundary layer can be related to the dimensionless mass and rotational Reynolds number by

$$\frac{v_\infty}{\Omega r} = -\frac{1}{2\pi}CwRe_\phi^{-0.5}x^{-2} \tag{7.81}$$

or from Equation 7.72,

$$(1-\beta) = \frac{1}{2\pi}CwRe_\phi^{-0.5}x^{-2} \tag{7.82}$$

If the local flow rate is known, then Equation 7.82 can be used to determine the core velocity. Alternatively, if the core velocity is known, then the local flow rate can be determined. If the core is rotating faster than the disc, then the local flow rate is negative and the fluid will flow radially inward. Conversely, if the core rotates more slowly than the discs, then the flow will be radially outward.

By differentiating Equations 7.75 and 7.76 with respect to ζ and setting $\zeta = \zeta_0$, where

$$\zeta = \frac{z}{\sqrt{\nu/\Omega}} \tag{7.83}$$

then

$$\frac{\tau_{r,o}}{\rho\Omega^2 b^2} = -\frac{\tau_{\phi,o}}{\rho\Omega^2 b^2} = (1-\beta)\,x\,Re_\phi^{-0.5} \tag{7.84}$$

From Equation 7.84, the ratio of the radial and tangential shear stresses is

$$\alpha = 1 \tag{7.85}$$

and

$$\left(\frac{\sqrt{\tau_0/\rho}}{\Omega r(1-\beta)}\right)^2 = \sqrt{\frac{2}{(1-\beta)^2 x^2 Re_\phi}} \tag{7.86}$$

The boundary layer thickness, δ, can be estimated by choosing δ to be the smallest positive vale of z for which u_r is zero. For values of $z > \delta$, viscous effects will be present, but small.

$$\delta = \pi b Re_\phi^{-0.5} = \pi\sqrt{\frac{\nu}{\Omega}} \tag{7.87}$$

For isothermal turbulent flow, it is possible to obtain a solution to Equations 7.61 and 7.62 using a mixing length model for the Reynolds stresses. However, a simple expression, comparable to Equation 7.81, relating the tangential velocity outside the boundary layer to Cw and the rotational Reynolds number, is not forthcoming.

Equations 7.66 and 7.67 are valid for the rotating core in turbulent flow as well as for laminar flow. The radial velocity in the core can be assumed to be zero, $u_\infty = 0$, so

$$-2\rho\Omega(v-v_\infty) = \frac{\partial \tau_r}{\partial z} \tag{7.88}$$

$$2\rho\Omega u = \frac{\partial \tau_\phi}{\partial z} \tag{7.89}$$

By assuming the flow in the disc boundary layer can be modeled by free disc theory with the velocity profile being given by Equation 4.96, that is,

$$u = \alpha r\Omega \left(\frac{z}{\delta}\right)^{1/7}\left(1-\frac{z}{\delta}\right) \quad \text{for } z < \delta \tag{7.90}$$

and

$$u = 0 \text{ for } z > \delta. \tag{7.91}$$

By analogy with Equation 4.97 for the tangential velocity component,

$$v = v_\infty \left(\frac{z}{\delta}\right)^{1/7} \quad \text{for } z \leq \delta \tag{7.92}$$

$$v = v_\infty \text{ for } z > \delta \tag{7.93}$$

Integration of Equations 7.88 and 7.89 with respect to z from 0 to ∞ gives

$$\frac{\tau_{r,o}}{\rho} = -\frac{1}{4}\Omega v_\infty \delta \tag{7.94}$$

and

$$\frac{\tau_{\phi,o}}{\rho} = -\frac{49}{60}\Omega u \delta \tag{7.95}$$

Equations 7.94 and 7.95 can be restated, in dimensionless form, as

$$\frac{\tau_{r,o}}{\rho\Omega^2 b^2} = \frac{1}{4}(1-\beta)x^2\frac{\delta}{r} \tag{7.96}$$

$$\frac{\tau_{\phi,o}}{\rho\Omega^2 b^2} = -\frac{49}{60}\alpha(1-\beta)x^2\frac{\delta}{r} \tag{7.97}$$

By using an approach analogous to that used by von Kármán to obtain the radial and tangential stresses for the rotor in the case of the free disc, Equations 4.115 and 4.116,

$$\frac{\tau_{r,o}}{\rho} = 0.0225\left(\frac{\nu}{\delta}\right)^{0.25}u_o(u_o^2 + v_\infty^2)^{3/8} \tag{7.98}$$

$$\frac{\tau_{\phi,o}}{\rho} = 0.0225\left(\frac{\nu}{\delta}\right)^{0.25}v_o(u_o^2 + v_\infty^2)^{3/8} \tag{7.99}$$

Equations 7.96 to 7.99 can be solved to give

$$\alpha = -\frac{\tau_{r,o}}{\tau_{\phi,o}} = -\frac{u_o}{v_\infty} = \frac{\sqrt{15}}{7} = 0.553 \tag{7.100}$$

and

$$\frac{\delta}{b} = 0.0983 Re_\phi^{-0.5}\left(\frac{|v_\infty|}{\Omega b}\right)^{0.6} \tag{7.101}$$

or

$$\frac{\delta}{b} = 0.0983 Re_\phi^{-0.5}[(1-\beta)^2 x^2 Re_\phi]^{0.3} \tag{7.102}$$

The negative sign in Equation 7.100 is necessary in order to ensure that δ is positive.

From Equations 7.90 and 7.77,

$$\frac{\dot{m}_o}{\mu r} = \frac{49}{60}\pi\alpha(1-\beta)x^2 Re_\phi \frac{\delta}{r} \tag{7.103}$$

From Equations 7.103 and 7.102,

$$\frac{\dot{m}_o}{\mu r} = 0.1395[(1-\beta)^2 x^2 Re_\phi]^{0.8} \tag{7.104}$$

or

$$\frac{\dot{m}_o}{\mu b} = 0.1395 Re_\phi^{0.8} \left(\frac{r}{b}\right)^{2.6} \left(\frac{|v_\infty|}{\Omega r}\right)^{1.6} \tag{7.105}$$

If the local flow is equal to half the mass flow rate entering the cavity, then,

$$\frac{v_\infty}{\Omega r} = 2.22 C w^{5/8} Re_\phi^{-0.5} \left(\frac{r}{b}\right)^{-13/8} \tag{7.106}$$

7.4. INTEGRAL EQUATIONS

As noted in Chapter 2, quantities of interest in engineering applications can typically be expressed in terms of integrals. An example is volumetric flow rate, which can be determined by the integral of the velocity over an area. In order to determine an integral quantity, the integrand must be known, or sufficient information must be available so that a realistic estimate for it can be made. Integration can then be performed to give the integral quantity. If the integrand is not known and cannot be estimated, then the appropriate differential equations can be solved to give the required integrands. The equations of fluid mechanics that relate the integral quantities are known as the integral equations or control-volume formulation of fluid mechanics. The integral equations for rotating cavities provide insight into the flow phenomena occurring in a rotating cavity. For example, under isothermal conditions, the tangential component of velocity for the flow in the central core of a rotating cavity with a radial source–sink flow has been shown to depend only on a parameter involving distance from the axis of rotation, volumetric flow rate, and the rotational speed of the cavity.

The partial differential equations for continuity and momentum can be integrated across the boundary layer, giving the so-called integral equations. A number of assumptions are made in order to simplify the analytical approach. Outside the boundary layer, the radial pressure gradient, assuming inviscid flow in this region, can be modeled by

$$\frac{dp}{dr} = \rho \frac{u_\phi^2}{r} \tag{7.107}$$

The ratio of the radial and tangential shear stresses on the disc is given by

$$\frac{\tau_{r,o}}{\tau_{\phi,o}} = \lim_{z \to 0} \left(\frac{u_r}{u_\phi - u_{\phi,o}}\right) \tag{7.108}$$

The following densities are defined

$$\int_0^\infty \rho u_r dz = \tilde{\rho}_1(r) \int_0^\infty u_r dz \tag{7.109}$$

$$\int_0^\infty \rho u_r^2 dz = \tilde{\rho}_2(r) \int_0^\infty u_r^2 dz \tag{7.110}$$

$$\int_0^\infty (\rho u_\phi^2 - \rho_\infty u_{\phi,\infty}^2) dz = \tilde{\rho}_3(r) \int_0^\infty (u_\phi^2 - u_{\phi,\infty}^2) dz \tag{7.111}$$

$$\int_0^\infty \rho u_r u_\phi dz = \tilde{\rho}_4(r) \int_0^\infty u_r u_\phi dz \tag{7.112}$$

$$\int_0^\infty \rho(u_\phi - u_{\phi,\infty}) dz = \tilde{\rho}_5(r) \int_0^\infty (u_\phi - u_{\phi,\infty}) dz \tag{7.113}$$

For incompressible flow, $\tilde{\rho}$ is a constant, and Equations 7.109–7.113 are identities. For a perfect gas the pressure is taken as p_∞ and $T = \tilde{T}(r)$. The temperature should be chosen to ensure that Equations 7.109–7.113 give a good approximation, and it is usual to assume that this is given by the mean of the disc and core temperatures:

$$\tilde{T} = \frac{T_0 + T_\infty}{2} \tag{7.114}$$

The assumption made is that

$$\tilde{\rho}_1(r) = \tilde{\rho}_2(r) = \tilde{\rho}_3(r) = \tilde{\rho}_4(r) = \tilde{\rho}_5(r) = \tilde{\rho}(r) \tag{7.115}$$

From Equations 7.107, 7.18, and 7.19

$$\frac{1}{r}\frac{\partial}{\partial r}(\rho r u_r^2) + \frac{\partial}{\partial z}(\rho u_r u_z) - \rho\left(\frac{u_\phi^2 - u_{\phi,\infty}^2}{r}\right) = \frac{\partial \tau_r}{\partial z} \tag{7.116}$$

$$\frac{1}{r^2}\frac{\partial}{\partial r}[\rho r^2 u_r(u_\phi - u_{\phi,\infty})] + \frac{\partial}{\partial z}[\rho u_z(u_\phi - u_{\phi,\infty})] = \frac{\partial \tau_\phi}{\partial z} \tag{7.117}$$

Equations 7.116 and 7.117 can be integrated with respect to z from the disc surface to the core (0 to ∞) to give:

$$\frac{d}{dr}(\tilde{\rho} r \int_0^\infty u_r^2 dz) - \tilde{\rho}\int_0^\infty [(u_\phi - u_{\phi,\infty})^2 + 2u_{\phi,\infty}(u_\phi - u_{\phi,\infty})] dz = -r\tau_{r,0} \tag{7.118}$$

$$\frac{d}{dr}[\tilde{\rho} r^2 \int_0^\infty u_r(u_\phi - u_{\phi,\infty}) dz] + \tilde{\rho} r(\int_0^\infty u_r dz)\frac{d}{dr}(r u_{\phi,\infty}) = -r^2 \tau_{\phi,0} \tag{7.119}$$

Independent variables are defined as follows.

$$\alpha = \lim_{z \to 0}\left(\frac{u_r}{u_\phi - u_{\phi,0}}\right) = \frac{\tau_{r,0}}{\tau_{\phi,0}} \tag{7.120}$$

$$\beta = \frac{u_{\phi,\infty}}{u_{\phi,o}} = \frac{u_{\phi,\infty}}{\Omega r} \tag{7.121}$$

$$\eta = \frac{\dot{m}_d}{\dot{m}_{ref}} \tag{7.122}$$

where $\dot{m}_d$ is the mass flow rate through the boundary layer and $\dot{m}_{ref}$ is a constant reference mass flow rate. For a rotating cavity with a superposed radial flow it is convenient to take $\dot{m}_{ref} = 0.5\dot{m}$, where $\dot{m}$ is the superposed flow rate, and for a free disc the value taken for $\dot{m}_{ref}$ is arbitrary.

$$\dot{m}_d = 2\pi\tilde{\rho}r\int_0^\infty u_r dz \tag{7.123}$$

It is convenient to define a number of ratios of integrals, called shape factors, as follows.

$$K_u = \frac{\int_0^\infty u_r^2 dz}{\alpha(u_{\phi,o}-u_{\phi,\infty})\int_0^\infty u_r dz} \tag{7.124}$$

$$K_v = \frac{\int_0^\infty u_r(u_\phi-u_{\phi,\infty})dz}{(u_{\phi,o}-u_{\phi,\infty})\int_0^\infty u_r dz} \tag{7.125}$$

$$K_1 = \frac{\int_0^\infty (u_\phi-u_{\phi,\infty})dz}{\int_0^\infty u_r dz} \tag{7.126}$$

$$K_2 = \frac{\alpha\int_0^\infty (u_\phi-u_{\phi,\infty})^2 dz}{(u_{\phi,o}-u_{\phi,\infty})\int_0^\infty u_r dz} \tag{7.127}$$

The factors α and $1/\alpha$ have been introduced in order to ensure that the shape factors are constant when the velocity profiles are similar.

Values for K_u, K_v, K_1, and K_2 are given in Table 7.1.

The stress flow rate parameter K_w is given by

$$K_w = -\pi\frac{r\tau_{\phi,o}}{\Omega\dot{m}_d} \tag{7.128}$$

The factor π is introduced in the stress flow rate parameter in order to ensure that $K_w = 1$ when the Ekman layer approximation is valid.

Using these definitions, the density relationships become

$$\tilde{\rho}\int_0^\infty u_r dz = \frac{\dot{m}_{ref}}{2\pi r}\eta \tag{7.129}$$

$$\tilde{\rho}\int_0^\infty u_r^2 dz = \frac{\Omega\dot{m}_{ref}}{2\pi}\alpha(1-\beta)\eta K_u \tag{7.130}$$

TABLE 7.1 Velocity shape factor and various constants relating to velocity profiles (Owen and Rogers, 1995)

	Laminar Flow Von Kármán model	Laminar Flow Ekman type model	Turbulent Flow 1/7th power law model
K_u	0.1477	0.25	35/69
K_v	0.3842	0.25	1/6
K_1	2.382	1	15/49
K_2	1.260	0.75	10/147
I_∞	0.5338	0.5	49/120
σ	0.5	0.5	0.8
C_1	0.6159	1	0.0225
C_2	6.4903	π^2	0.08946

$$\tilde{\rho}\int_0^\infty u_r(u_\phi - u_{\phi,\infty})dz = \frac{\Omega\dot{m}_{ref}}{2\pi}(1-\beta)\eta K_v \qquad (7.131)$$

$$\tilde{\rho}u_{\phi,\infty}\int_0^\infty (u_\phi - u_{\phi,\infty})dz = \frac{\Omega\dot{m}_{ref}}{2\pi}\frac{\beta\eta}{\alpha}K_1 \qquad (7.132)$$

$$\tilde{\rho}\int_0^\infty (u_\phi - u_{\phi,\infty})^2 dz = \frac{\Omega\dot{m}_{ref}}{2\pi}\frac{(1-\beta)\eta}{\alpha}K_2 \qquad (7.133)$$

$$\tau_{\phi,o} = \frac{\Omega\dot{m}_{ref}}{\pi}\frac{1}{r}\eta K_w \qquad (7.134)$$

$$\tau_{r,o} = \frac{\Omega\dot{m}_{ref}}{\pi}\frac{1}{r}\alpha\eta K_w \qquad (7.135)$$

Substitution from Equations 7.129 to 7.135 in Equations 7.118 and 7.119 gives

$$\frac{d}{dr}[r\alpha(1-\beta)\eta K_u]-[(1-\beta)K_2 + 2\beta K_1]\frac{\eta}{\alpha} = -2\alpha\eta K_w \qquad (7.136)$$

$$\frac{d}{dr}[r^2(1-\beta)\eta K_v] + \eta\frac{d}{dr}(\beta r^2) = 2r\eta K_w \qquad (7.137)$$

Substituting $x = a/b$, Equations 7.136 and 7.137 become

$$1+\frac{x}{a}\frac{da}{dx}-\frac{x}{1-\beta}\frac{d\beta}{dx}+\frac{x}{\eta}\frac{d\eta}{dx}+\frac{x}{K_u}\frac{dK_u}{dx}-\left(\frac{K_2}{K_u}+\frac{2\beta}{1-\beta}\frac{K_1}{K_u}\right)\frac{1}{\alpha^2}+\frac{2K_w}{(1-\beta)K_u} = 0 \quad (7.138)$$

$$-2\left(\frac{1}{K_v}-1\right)+\frac{2}{K_v}\frac{1}{1-\beta}+\left(\frac{1}{K_v}-1\right)\frac{x}{1-\beta}\frac{d\beta}{dx}+\frac{x}{\eta}\frac{d\eta}{dx}+\frac{x}{K_v}\frac{dK_v}{dx}-\frac{2K_w}{(1-\beta)K_v}=0$$

(7.139)

7.5. MODELING OF SELECTED ROTATING CAVITY APPLICATIONS

The analysis introduced in Sections 7.2, 7.3, and 7.4 can be applied to a rotor-rotor disc cavity with radial inflow or outflow to provide insight to the flow phenomena and approximate values for the locations of flow features and bulk quantities such as flow rates. The flow in a rotating cavity with radial outflow is considered in Sections 7.5.2 and 7.5.3 for a radial and an axial inlet, respectively. Radial inflow is considered in Section 7.5.4. Modeling of a rotating cavity with a throughflow at the bore is considered in Section 7.5.5. Flow in a closed cavity is considered in the following section.

7.5.1 Closed Rotating Cavities

Closed cylindrical cavities are illustrated in Figures 7.4 and 7.5. The flow in a closed rotating cavity is governed by convection, and in the case of isothermal boundary conditions there is no relative motion of the fluid inside a rotating cavity. For nonisothermal flow, where the surfaces are at different temperatures, buoyancy forces are generated from the centripetal acceleration and density variations in the fluid. The buoyancy forces result in fluid motion relative to the rotating surfaces that depends on the geometrical and heating configuration.

For the case of an axially directed heat flux, where there is a temperature difference between the two discs, fluid moves radially inward adjacent to the hot disc and radially outward adjacent to the cooler disc (Ostrach and Braun, 1958) as indicated in Figures 7.5(a) and 7.5(b). In a cavity with no central shaft, there is an additional recirculation zone close to the axis of rotation, but in an annular cavity this feature is not present.

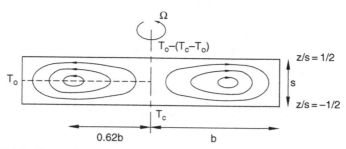

FIGURE 7.14   Characteristic streamlines for a rotating cavity with an axial heat flux (Ostrach and Braun, 1958).

The convection of flow inside a cavity with an axial heat flux, where one disc is at a temperature $T_o - (T_c - T_o)$ and the other at T_c was studied by Ostrach and Braun (1958). They developed an analytical solution to the governing equations of motion for the velocity components given in Equations 7.140–7.145. For higher rotational Reynolds numbers, this gives the flow profiles illustrated in Figure 7.14, with fluid moving radially inward adjacent to the hot disc and radially outward adjacent to the cooler disc. The velocity component profiles for a range of Reynolds numbers are illustrated in Figures 7.15–7.17, where $Re_\phi = \Omega s^2/\nu$ for this application.

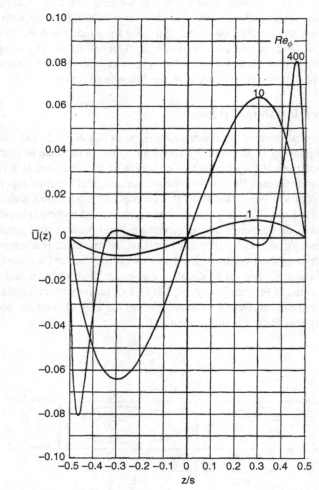

FIGURE 7.15 Dimensionless radial velocity component profiles $\bar{U}(z)$, for a rotating cavity with an axial heat flux (Ostrach and Braun, 1958).

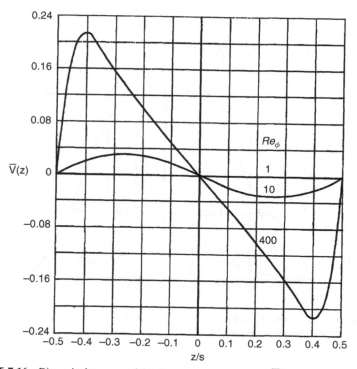

FIGURE 7.16 Dimensionless tangential velocity component profiles $\overline{V}(z)$, for a rotating cavity with an axial heat flux (Ostrach and Braun, 1958).

For a rotating cavity with no rotating shaft towards the inner radius, an additional recirculation zone is observed in practical applications, as indicated in Figure 7.5(a).

$$\frac{u}{\varepsilon\Omega b} = \frac{2}{\alpha}J_1\left(\alpha\frac{r}{b}\right)\left(C_1\sinh\sqrt{Re_\phi}\,\frac{z}{s}\cos\sqrt{Re_\phi}\,\frac{z}{s} - C_2\cosh\sqrt{Re_\phi}\,\frac{z}{s}\sin\sqrt{Re_\phi}\,\frac{z}{s}\right)$$

$$(7.140)$$

$$\frac{1}{\varepsilon}\left(\frac{v}{\Omega b} - \frac{r}{b}\right) = \frac{2}{\alpha}J_1\left(\alpha\frac{r}{b}\right)$$

$$\left(C_1\cosh\sqrt{Re_\phi}\,\frac{z}{s}\sin\sqrt{Re_\phi}\,\frac{z}{s} + C_2\sinh\sqrt{Re_\phi}\,\frac{z}{s}\cos\sqrt{Re_\phi}\,\frac{z}{s} - \frac{z}{2s}\right)$$

$$(7.141)$$

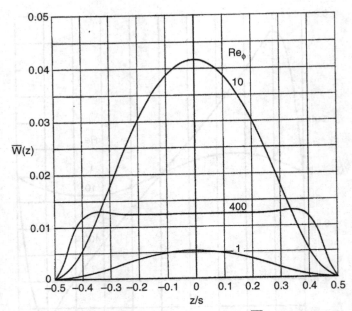

FIGURE 7.17 Dimensionless axial velocity component profiles $\overline{W}(z)$, for a rotating cavity with an axial heat flux (Ostrach and Braun, 1958).

$$\frac{w}{\varepsilon\Omega s} = \frac{1}{\mathrm{Re}_\phi} J_0\left(\alpha\frac{r}{b}\right) \begin{bmatrix} C_1\left(\sinh\dfrac{\sqrt{\mathrm{Re}_\phi}}{2}\sin\dfrac{\sqrt{\mathrm{Re}_\phi}}{2} + \cosh\dfrac{\sqrt{\mathrm{Re}_\phi}}{2}\cos\dfrac{\sqrt{\mathrm{Re}_\phi}}{2}\right) \\[2ex] +C_2\left(\cosh\dfrac{\sqrt{\mathrm{Re}_\phi}}{2}\cos\dfrac{\sqrt{\mathrm{Re}_\phi}}{2} - \sinh\dfrac{\sqrt{\mathrm{Re}_\phi}}{2}\sin\dfrac{\sqrt{\mathrm{Re}_\phi}}{2}\right) \\[2ex] -C_1\left(\sinh\sqrt{\mathrm{Re}_\phi}\dfrac{z}{s}\sin\sqrt{\mathrm{Re}_\phi}\dfrac{z}{s} + \cosh\sqrt{\mathrm{Re}_\phi}\dfrac{z}{s}\cos\sqrt{\mathrm{Re}_\phi}\dfrac{z}{s}\right) \\[2ex] -C_2\left(\cosh\sqrt{\mathrm{Re}_\phi}\dfrac{z}{s}\cos\sqrt{\mathrm{Re}_\phi}\dfrac{z}{s} - \sinh\sqrt{\mathrm{Re}_\phi}\dfrac{z}{s}\sin\sqrt{\mathrm{Re}_\phi}\dfrac{z}{s}\right) \end{bmatrix}$$

$$(7.142)$$

where

$$\varepsilon = 2\beta_v(T_c - T_o) \qquad (7.143)$$

β_v is the volumetric expansion coefficient.

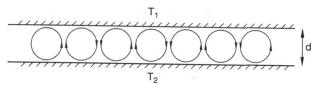

FIGURE 7.18   Rayleigh-Bénard convection between horizontal plates with the lower plate heated.

$$C_1 = \frac{\cosh \frac{\sqrt{\mathrm{Re}_\phi}}{2} \sin \frac{\sqrt{\mathrm{Re}_\phi}}{2}}{4 \left(\sinh^2 \frac{\sqrt{\mathrm{Re}_\phi}}{2} + \sin^2 \frac{\sqrt{\mathrm{Re}_\phi}}{2} \right)} \tag{7.144}$$

$$C_2 = \frac{\sinh \frac{\sqrt{\mathrm{Re}_\phi}}{2} \cos \frac{\sqrt{\mathrm{Re}_\phi}}{2}}{4 \left(\sinh^2 \frac{\sqrt{\mathrm{Re}_\phi}}{2} + \sin^2 \frac{\sqrt{\mathrm{Re}_\phi}}{2} \right)} \tag{7.145}$$

$J_0(\alpha r)$ and $J_1(\alpha r)$ are zero and first-order Bessel functions, and α is the first root of $J_1(\alpha) = 0$.

The flow in a cavity for the case of a radially directed heat flux, where there is a temperature difference between the inner and outer peripheral surfaces, has been studied by Bohn et al. (1993) and King et al.(2007) who found a series of cyclonic (low pressure) and anticyclonic (high-pressure) recirculations. The Rayleigh–Bénard convection flow structure shown in Figure 7.6 is normally associated with the flow structure resulting from fluid contained between two stationary horizontal plates with the lower plate heated, with the Rayleigh number, Equation 7.146, above the critical value. Instability tends to occur for Rayleigh numbers above about 1700, and a buoyancy-induced flow structure is produced as illustrated in Figure 7.18, comprising an array of contra-rotating vortices.

$$Ra = \frac{g\alpha(T_2 - T_1)d^3}{\nu\kappa} \tag{7.146}$$

where

   T_2 is the temperature of the lower plate (K);
   T_1 is the temperature of the upper plate (K);
   d is the distance between the plates (m);
   g is the local acceleration due to gravity or the centripetal acceleration (m/s^2)
   α is the coefficient of expansion (K^{-1});
   ν is the kinematic viscosity (m^2/s);
   κ is the thermal diffusivity (m^2/s).

For rotating cavity applications, the Rayleigh number is sometimes stated in terms of the product of the Prandtl and Grashof numbers, Equation 7.147. Definitions for the Prandtl and Grashof numbers are given in Equations 7.148 and 7.149, respectively. The Prandtl number provides an indication of the relative thicknesses of the momentum and thermal boundary layers, whereas the Grashof number indicates the ratio of buoyancy to viscous forces acting on a fluid.

$$Ra = \Pr Gr \tag{7.147}$$

$$Pr = \frac{\mu c_p}{k} = \frac{\nu}{\kappa} \tag{7.148}$$

$$Gr = \frac{\beta_v \Delta T g L^3}{\nu^2} \tag{7.149}$$

where
 Ra is the Rayleigh number
 Pr is the Prandtl number
 Gr is the Grashof number
 k is the thermal conductivity of the fluid (W/m K)
 κ is the thermal diffusivity (m^2/K)
 ΔT is the driving temperature gradient (K)
 β_v is the volumetric expansion coefficient (m^3/m^{3}K)
 L is a characteristic length scale, normally taken as the disc outer radius, b, in rotating cavity applications (m)

The continuity equation for an incompressible flow is given by Equation 7.150 (see Chapter 2, Equation 2.109). For steady rotating flow when both the Rossby number and Ekman number are small, indicating that the Coriolis force is significant in comparison to inertial forces and viscous forces, respectively, then the equation of motion for a rotating flow can be stated as Equation 7.151 (see Chapter 2, Equation 2.163).

$$\rho \nabla \cdot \boldsymbol{w} = 0 \tag{7.150}$$

$$2\boldsymbol{\Omega} \times \boldsymbol{w} = -\frac{1}{\rho} \nabla p \tag{7.151}$$

In component form these equations can be stated as

$$\frac{\partial(\rho r u)}{\partial r} + \frac{\partial(\rho v)}{\partial \phi} + \frac{\partial(\rho r w)}{\partial z} = 0 \tag{7.152}$$

$$2\Omega v = \frac{1}{\rho} \frac{\partial p}{\partial r} - \Omega^2 r \tag{7.153}$$

$$2\Omega u = -\frac{1}{\rho r} \frac{\partial p}{\partial \phi} \tag{7.154}$$

$$0 = \frac{1}{\rho} \frac{\partial p}{\partial z} \tag{7.155}$$

Equations 7.153 and 7.154 can be differentiated, giving

$$\frac{\partial(\rho r u)}{\partial r} + \frac{\partial(\rho v)}{\partial \phi} = \frac{\Omega r}{2}\frac{\partial \rho}{\partial \phi} \tag{7.156}$$

From comparison of Equations 7.152 and 7.156

$$\frac{\partial(\rho w)}{\partial z} = \frac{\Omega}{2}\frac{\partial \rho}{\partial \phi} \tag{7.157}$$

For incompressible or axisymmetric flow, $\partial \rho / \partial \phi = 0$, and therefore ρw is constant in the direction parallel with the axis of rotation, that is, ρw is invariant with z. This is the Taylor-Proudman theorem, previously introduced in Chapter 3, Section 3.3. The velocity components u and v are also invariant with z. The consequences of the Taylor-Proudman theorem are that rotation tends to produce stratified flow, with the suppression of circulation and that radial flow can only occur if the tangential pressure gradient, $\partial p / \partial \phi$, is not zero. For a rotating inviscid fluid, a radial velocity is possible only if there is a circumferential pressure gradient. Without a pressure gradient, the flow would be thermally stratified and convection would not take place. The counter-rotating cyclonic and anticyclonic vortices illustrated in Figure 7.6 generate regions of low and high pressure, respectively, that provide the circumferential pressure gradients enabling convection to occur.

Equation 7.153 can be written as

$$2\Omega v = \frac{1}{\rho}\frac{\partial p_{reduced}}{\partial r} \tag{7.158}$$

where the reduced pressure (see Section 2.6) is given by

$$p_{reduced} = p - \frac{1}{2}\rho\Omega^2 r^2 \tag{7.159}$$

For a cyclonic circulation, the radial velocity is greater than zero, and the pressure increases with radial location. At the center of the circulation, the pressure will be lower than that at the periphery of the circulation. As in atmospheric circulations, a cyclonic circulation in a rotating cavity is associated with low pressure at the center. An anticyclonic circulation is associated with high pressure at the center. King et al. (2007) identified that with Rayleigh-Bénard convection in a rotating cavity there will always be an even number of vortices, whereas in a stationary enclosure odd or even numbers of vortices are possible. For a viscous flow the effect of viscosity is to reduce the circulation of flow.

7.5.2 Rotating Cavity with Radial Outflow

For the case of a radial inlet, the flow can be taken as symmetrical about a midcavity plane, $z = s/2$. Fluid entering the cavity at $r = a$, is progressively entrained from the source region into the entraining boundary layers, which are also known as entrainment layers, on the rotating discs. When all of the fluid supplied by the source region has been entrained into the two boundary layers, the boundary layers become nonentraining and are referred to as nonentrained Ekman-type layers, by analogy with Ekman layers where the magnitude of the relative velocity of the fluid outside the layer is small in comparison with the disc speed, Ωr.

The term *Ekman layer* is strictly valid only when the nonlinear terms are negligible. In practice, the nonentraining boundary layers found on rotating discs are very similar to Ekman layers. The mass flow in each of the Ekman-type layers is $\dot{m}/2$. Near the shroud, the fluid flows from the Ekman-type layers into the sink layer where it then exits through a radial slit or circumferential holes around the shroud. The source region, the two Ekman-type layers and the sink region, form the boundaries of the annular-shaped interior core that rotates at an angular velocity below that of the discs. The size of the source region increases as the mass flow supplied to the cavity increases and decreases with rotational Reynolds number. For the case of radial outflow, the fluid in the interior core between the two Ekman-type layers rotates at an angular speed that is less than the disc.

For the case of an axial inlet, a significant proportion of the fluid impinges on the downstream disc and flows radially outward as a circular wall jet. The wall jet feeds the Ekman-type layer, and at a radius, x_e, a proportion of the jet flow reverses and flows radially inward to feed the upstream disc Ekman-type layer.

Flow visualization data to support the models described here and illustrated in Figures 7.8 and 7.9 was originally supplied for the case of laminar flow in a rotating cavity with radial outflow by Owen and Pincombe (1980) and data from CFD studies by Chew, Owen, and Pincombe (1984). Similar flow structures to those illustrated in Figures 7.8 and 7.9 exist for both laminar and turbulent flow and have been modeled by Ong and Owen (1989, 1991), Morse (1988), and Iacovides and Theofanopoulos (1991). Transition has been observed to occur at a rotational Reynolds number of $Re_\phi \approx 2 \times 10^5$ in the source region (Long and Owen, 1986). Owen, Pincombe, and Rogers (1985) found that transition in the Ekman-type layers occurred for a radial Reynolds number, $Re_r = Cw/2\pi x = 180$. It is possible for reverse transition to occur, with turbulent flow becoming laminar, in the Ekman-type layers.

In general, the radial thickness of the sink layer is an order of magnitude smaller than the source region. The size of the source region increases with the mass flow supplied to the cavity. If the mass flow is high enough, the source region can completely fill the cavity extending out to the shroud, and there will be no Ekman-type layers and no interior core. The extent of the source region can be modeled using linear solutions or an Ekman layer approximation (Owen, Pincombe, and Rogers (1985); Owen and Rogers, (1995)).

Asymmetric heating has a significant effect on the flow structure, and this has been investigated by Owen and Bilimoria (1977), Owen and Onur (1983) and Long, Morse and Zafiropoulos (1993). The following regimes of flow and heat transfer have been observed by Owen and Bilimoria (1977) and Owen and Onur (1983).

1. A core-dominated regime where the source region filled the entire cavity. The average Nusselt numbers increased with Cw but they were virtually independent of Re_ϕ;
2. A developing Ekman-layer regime where the average Nusselt numbers increased with both Cw and Re_ϕ.
3. A fully developed Ekman-layer regime where the rate of increase of Nusselt number with Re_ϕ was reduced.
4. Chaotic flow where the average Nusselt numbers were found to depend on the Grashof number, Gr_b, which is defined in Equation 7.160. Here the effect of rotation is incorporated into the Grashof number, and the length scale is defined as the disc outer radius.

$$Gr_b = \frac{\beta_v \Delta T \Omega^2 b^4}{\nu^2}. \tag{7.160}$$

Owen and Onur (1983) noted that for the third regime, (3), the source region appeared to oscillate at approximately 70% of the cavity speed and that hysteresis occurred in the transition from one regime to another, depending on whether the speed was increasing or decreasing. These features were also observed in the flow visualization work of Pincombe (1983) where it was also noted that the thickness of the Ekman layer on the heated downstream disc was greater than that of the unheated upstream one, and this effect was attributed to the axial movement of fluid from the hot to the cold disc. This observation was confirmed in the computations of Long, Morse, and Zafiropoulos (1993).

For radial outflow, with a radial inlet where the flow can be assumed to be symmetrical about the cavity midplane, $z = s/2$, the flow can be assumed to comprise a source region, two boundary layers, a central core region, and a sink layer. Fluid from the source is entrained equally between the boundary layers on each disc. At a certain radius, r_e, it is assumed that the boundary layer will become nonentraining, and the central region of the cavity is filled by the rotating core with no axial flow. For a radius $r < r_e$ it is assumed that Equations 4.30 and 4.23 are valid for the flow entrained into the boundary layer and the boundary thickness for the case of laminar flow and Equation 4.52 or 4.191 for turbulent flow.

The radial extent of the source region, x_e, which is defined in Equation 7.161 and illustrated in Figure 7.8, is based on the assumption that the entrainment layer persists until half of the incoming fluid to the cavity has been entrained.

$$x_e = \frac{r_e}{b} \tag{7.161}$$

For laminar flow, from Equations 4.30 and 7.161, the location at which the boundary layer becomes nonentraining is given by

$$x_e = \sqrt{\frac{Cw}{2 \times 2.779 Re_\phi^{0.5}}} = 0.424 Cw^{0.5} Re_\phi^{-0.25} \qquad (7.162)$$

For turbulent flow, from Equations 4.52 and 4.191, making a substitution for Cw, Equation 4.29, gives

$$x_e = \left(\frac{Cw Re_\phi^{4/5}}{2 \times 0.2186}\right)^{5/13} = 1.37 Cw^{5/13} Re_\phi^{-4/13} \qquad (7.163)$$

Equation 7.163 is in good agreement with the correlation given by Owen and Pincombe (1980).

7.5.3 Rotating Cavity with Radial Outflow and Axial Inlet

For the case of an axial inlet, where fluid enters the cavity through the center of the upstream disc, the flow structure depends on the magnitude of the mass flow into the cavity. For low-mass flows, vortex breakdown occurs, the central jet loses its axisymmetry, and fluid appears to be entrained equally by the two discs. In this case, Equations 7.162 and 7.163 provide a reasonable estimate for the extent of the source region.

For the case of an axial inlet, for higher mass flows where the flow impinges axisymmetrically on the downstream disc, an estimate for the radial extent of the source region on the downstream disc can be obtained by assuming that all of the fluid is entrained on the downstream disc.

For laminar flow for an axial inlet cavity, from Equations 4.30 and 7.161, the location at which the boundary layer becomes nonentraining on the downstream disc is therefore given by

$$x_e = \sqrt{\frac{Cw}{2.779 Re_\phi^{0.5}}} = 0.600 Cw^{0.5} Re_\phi^{-0.25} \qquad (7.164)$$

For turbulent flow, from Equations 4.52 and 4.191, making a substitution for Cw, Equation 4.29, gives

$$x_e = \left(\frac{Cw Re_\phi^{4/5}}{0.2186}\right)^{5/13} = 1.795 Cw^{5/13} Re_\phi^{-4/13} \qquad (7.165)$$

Equation 7.165 was found to be in good agreement with the size of the source region measured for $950 < Cw < 6600$ and $10^5 < Re_\phi < 10^6$ (Owen et al., 1985).

As indicated in Figure 7.9, the flow on the downstream divides at $x_{e,down-stream}$ in order to supply the Ekman-type layers on the downstream and upstream discs. Equations 7.162 and 7.163 can therefore be used for estimating the radial extent of the source region on the upstream disc.

7.5.4 Radial Inflow

As introduced in Section 7.1.3, the flow structure for a rotating cavity with a net radial inflow of fluid can be characterized by a source region feeding two Ekman layers flowing radially inward on the faces of the two discs, sandwiching an interior core of fluid. The two Ekman layers combine in the sink layer near the inner radius of the cavity, and the flow exits through the outlet near the axis of the cavity. In the rotating core, for the case of an inviscid fluid, the radial and axial velocity components will be zero, and the core rotates with a tangential velocity that is some fraction of the disc speed. Flow visualization data supporting this model characterizing radial inflow is given by Firouzian et al. (1985).

Modeling techniques based on the Ekman layer equations provide an effective means of estimating the size of the source region (Owen et al., 1985; Owen and Rogers, 1995). Firouzian et al. (1985, 1986) obtained a simple expression for the pressure distribution inside a rotating cavity based on the Ekman layer equations.

For radial inflow through holes, slots, or a mesh located in the outer shroud of a rotating cavity, the tangential velocity component of the incoming fluid will depend on the nature of the shroud and local conditions. The local tangential velocity can be defined by the swirl fraction given by

$$c = \frac{\bar{u}_\phi}{\Omega b} \tag{7.166}$$

Normally, the range for c in the absence of pre-swirl into the cavity will be in the range $0 < c < 1$. For flow in the cavity away from the boundary layers where angular momentum is conserved, it is assumed that the flow structure can be modeled by a free vortex. In which case, for the source region,

$$\frac{\bar{u}_\phi}{\Omega r} = \frac{c}{x^2} \tag{7.167}$$

Some fluid will be entrained into the boundary layers on the discs. If, $\bar{u}_\phi/\Omega b < 1$ then the flow will be radially outward in the boundary layer, and if, $\bar{u}_\phi/\Omega b > 1$ then the flow will be radially inward in the boundary layer.

The entrainment into the boundary layer can be quantified by assuming that the relationship developed based on the free disc, Equation 7.82, provides a relationship between the flow through the layer and the tangential component of velocity outside the boundary layer (Owen et al., 1985). From Equation 7.82 and $\bar{u}_\phi = \bar{v} + \Omega r$,

$$\frac{\dot{m}_1}{\mu b} = \pi Re_\phi^{0.5}(x^2 - c) \tag{7.168}$$

For turbulent flow, from Equation 7.104,

$$\frac{\dot{m}_1}{\mu b} = 0.140\left(1 - \frac{c}{x^2}\right)^{1.6} Re_\phi^{0.8} x^{2.6} \tag{7.169}$$

At a certain radial location the radial inflow meets the radial outflow and the flow stagnates. This occurs at

$$x = \sqrt{c} \tag{7.170}$$

For values of $x > \sqrt{c}$ there is outflow, and for $x < \sqrt{c}$ there is inflow in the entrainment layers.

The source region is taken between $r = b$ and $r = r_e$ where half of the incoming fluid has been entrained by each disc. Taking $\dot{m}_1 = \dot{m}/2$ at $r = r_e$, then for laminar flow,

$$x_e = \left(c - \frac{1}{2\pi}|Cw|Re_\phi^{-0.5}\right)^{0.5} \tag{7.171}$$

For turbulent flow,

$$x_e = (c - 2.22|Cw|^{5/8} Re_\phi^{-0.5} x_e^{3/8})^{0.5} \tag{7.172}$$

For a rotating cavity with radial inflow, the radial pressure gradient, in regions where viscous effects are negligible and where the radial and axial velocity components are small compared to the tangential velocity component, can be calculated from

$$\frac{dp}{dr} = \rho \frac{u_\phi^2}{r} \tag{7.173}$$

Equation 7.173 is valid for the source region and the interior core. If the sink layer can be neglected, then Equation 7.173 can be used to calculate the pressure drop that occurs as fluid moves from the source, $r = b$, to the sink at $r = a$.

A pressure coefficient can be defined, given by

$$C_p = \frac{p_s - p_a}{0.5\rho_{ref}\Omega^2 b^2} \tag{7.174}$$

where
   p_s = the static pressure in the fluid at the shroud, $r = b$
   p_a = the static pressure in the fluid at $r = a$
   ρ_{ref} = a reference density
   Assuming incompressible flow, Equation 7.173 can be written in integral form as

$$C_p = 2\int_{x_a}^{1} x \left(\frac{u_\phi}{\Omega r}\right)^2 dx \tag{7.175}$$

Equation 7.175 for laminar flow can be integrated, using 7.166 and 7.167, to give

$$C_p = (x_e^2 - x_a^2) + 4\lambda \ln\left(\frac{x_e}{x_a}\right) + \lambda_p^2 \left(\frac{1}{x_a^2} - \frac{1}{x_e^2}\right) + c^2 \left(\frac{1}{x_e^2} - 1\right) \tag{7.176}$$

where x_e can be determined from Equation 7.171 and

$$\lambda_p = \frac{1}{2\pi} CwRe_\phi^{-0.5} \tag{7.177}$$

For turbulent flow, using Equations 7.167 and 7.106,

$$C_p = (x_e^2 - x_a^2) + \frac{32}{3} \gamma_p (x_e^{3/8} - x_a^{3/8}) + \frac{8}{5} \gamma_p^2 (x_e^{-5/4} - x_a^{-5/4}) + c^2 \left(\frac{1}{x_e^2} - 1\right) \tag{7.178}$$

where x_e can be determined from Equation 7.172 and

$$\gamma_p = 2.22 Cw^{5/8} Re_\phi^{-0.5} \tag{7.179}$$

It should be noted that Equations 7.176 and 7.178 do not include the pressure drop across the shroud and are only valid for $x_e > x_a$ when the source region does not fill the entire cavity. If the source region does occupy the whole cavity, then free vortex flow can be assumed to occur throughout the cavity and for both laminar and turbulent flow,

$$C_p = c^2 \left(\frac{1}{x_a^2} - 1\right) \tag{7.180}$$

For the case when $Cw = 0$, $\lambda_p = \gamma_p = 0$ and the pressure coefficient for the cases of laminar and turbulent flow, from Equations 7.176 and 7.178, reduce to

$$C_p = 2c - c^2 - x_a^2 \tag{7.181}$$

7.5.5 Rotating Cavities with Axial Throughflow

In a jet engine, propulsive efficiency improves when a larger mass of air is ejected at lower speed. This principle is used in a turbofan or bypass engine where the bulk of air entering the front fan is ducted around the core engine. In such engines, the front fan needs to operate at much lower speeds than the core engine, and a multispool arrangement is typically used with a high-pressure turbine driving the high-pressure compressor at a high rotational speed, typically of the order of 15,000 rpm, and a low pressure turbine driving the front fan at a lower speed, of the order of 3000 rpm. In the case of some Rolls-Royce engines, an additional intermediate pressure turbine drives an intermediate pressure compressor. The engine architecture typically employed comprises the drive shaft between the high-pressure turbine and the high-pressure compressor running concentrically around the intermediate and low-pressure drive shafts. With blades supported on turbine and compressor discs, this gives rise to the geometric configuration of rotating cavities comprising a rotor-rotor disc cavity, with a shaft running through the inner bore of the discs. Cooling air is usually passed through this bore to reduce the effects of windage heating in the cavities and to convey coolant to other areas of the engine.

A simplified model of a rotating cavity with axial throughflow is shown in Figure 7.19. Two discs of outer radius, b, and inner radius, a, are separated by an axial gap, s. The rotational speed of the cavity is Ω, and the bulk average velocity of the axial throughflow is $\overline{w}$. The flow structure for the case of a cavity with an axial throughflow at the inner bore is particularly complex. For the case of isothermal conditions, vortex breakdown in the central jet of air can result in nonaxisymmetric flow inside the cavity. If the discs are at different temperatures, nonaxisymmetric flow predominates.

The relevant rotational and axial Reynolds numbers are defined in Equations 7.182 and 7.183.

$$Re_\phi = \frac{\rho \Omega b^2}{\mu} \tag{7.182}$$

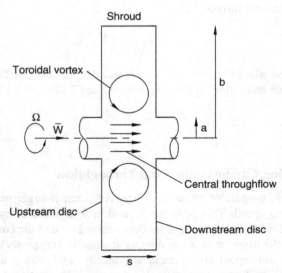

FIGURE 7.19 Nomenclature for a rotating cylindrical cavity with axial throughflow and the isothermal flow structure.

$$\frac{Re_z = \rho \overline{w} d_h}{\mu} \tag{7.183}$$

where d_h is the hydraulic diameter of the inlet.

For a cavity without an inner drive shaft,

$$d_h = 2a \tag{7.184}$$

For a cavity with an inner drive shaft of radius, r_s,

$$d_h = 2(a - r_s) \tag{7.185}$$

A further nondimensional parameter, the Rossby number, Ro links the effects of both rotation and the inertia of the throughflow. For the case of a rotor-rotor cavity with throughflow, this can be defined as the ratio of the mean velocity of the throughflow to the tangential velocity at the bore radius:

$$Ro = \frac{\overline{w}}{\Omega a} \tag{7.186}$$

The Rossby number can also be expressed in terms of the rotational and axial Reynolds numbers as

$$Ro = \left\{ \frac{b^2 \, Re_z}{2a(a - r_s)Re_\phi} \right\} \tag{7.187}$$

Laser illuminated flow visualization and LDA were used by Farthing et al. (1992) to study the flow structure in unheated (or isothermal) and heated cavities with $a/b \approx 0.1$. A series of schematic diagrams of the isothermal flow structure are shown in Figure 7.20.

The principal parameters to affect the resulting flow are the Rossby number, Ro, and the gap ratio, $G = s / b$. For no rotation, $Ro = \infty$, the throughflow generates one or more axisymmetric toroidal vortices. The number of vortices depends on the gap ratio. Rotation has the effect of suppressing the toroidal vortex and destabilizing the central throughflow, creating a change in the behavior of the central jet. This is characterized by a number of regimes of axisymmetric and nonaxisymmetric vortex breakdown. For turbulent flow ($Re_z >$ 2000), and for a constant gap ratio four separate regimes of vortex breakdown were identified as the Rossby number is decreased (from around 100 to a value of less than 1). These were given the following names (in order of decreasing Rossby number): Mode 1a, Mode 2a, Mode 1b, and Mode 2b. The characteristics of these vortex breakdown modes are broadly defined as follows.

Mode 1. Nonaxisymmetric vortex breakdown
Mode 2. Axisymmetric vortex breakdown
Mode 1a. Nonaxisymmetric vortex breakdown with precession of the jet for certain values of Rossby number

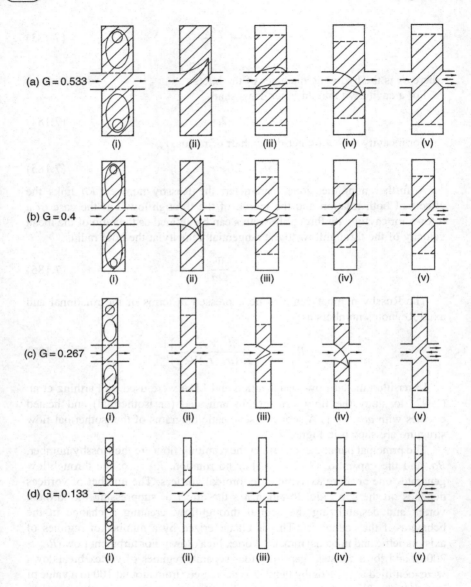

FIGURE 7.20 Visual impressions of smoke patterns in an isothermal rotating cavity with axial throughflow: $Re_z = 5000$ (Farthing et al,. 1992).

	(i)	(ii)	(iii)	(iv)	(v)
Ro	∞ (stationary)	25	4	2	1

(Shaded areas represent regions into which smoke is convected rapidly.)

Mode 2a. Axisymmetric vortex breakdown with occasional oscillations
Mode 1b. Nonaxisymmetric vortex breakdown with occasional excursions into
the cavity (referred to as the "flickering flame effect").
Mode 2b. Axisymmetric vortex breakdown with signs of reverse flow at the
downstream edge of the jet

For a gap ratio of $G = 0.533$, the respective boundaries of these regimes
occur at $Ro = 20$, 2.6, and 1.5. The Mode 1 regimes are associated with a
nonaxisymmetric behavior of the central throughflow; Mode 2 behavior is
associated with axisymmetric behavior. Decreasing the gap ratio appears to
suppress the formation of Mode 1a behavior.

In a cavity for which the rotating discs are hotter than the axial throughflow
of air, nonaxisymmetric buoyancy-driven flow occurs. When the discs are
heated, Farthing et al. (1992) found that significantly more of the throughflow
appears to penetrate further into the outer part of the cavity. Most observations
of the heated flow structure were made with a cavity ($G = 0.124$ and 0.267)
having a surface temperature distribution that decreased with radius. A schematic diagram, in the $r\phi$ plane, of the heated flow structure is shown in
Figure 7.21.

Flow is seen to enter the cavity through a radial arm, bifurcate near the outer
radius, and form a region of cyclonic flow (with the same sense of rotation as
the cavity) and anticyclonic flow (having the opposite sense of rotation). It was
not possible to obtain such clear visual or photographic evidence of the flow
structure in a cavity with a surface temperature distribution that increased with
radius. The overall impression was that significantly more of the central
throughflow penetrated the cavity, the higher velocities leading to an ill-defined
flow structure, especially in the region adjacent to the peripheral shroud. These

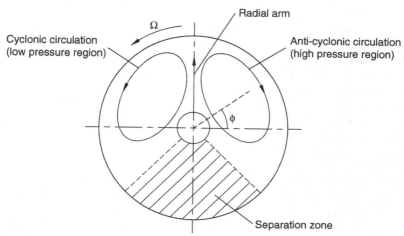

FIGURE 7.21   Schematic diagram of the heated flow structure in $r - \phi$ plane.

differences in flow structure may be qualitatively explained from consideration of thermal stratification in a centrifugal force field. A surface-to-fluid temperature difference that decreases with radius results in stable stratification. For a surface-to-fluid temperature difference that increases with radius there is unstable stratification, resulting in increased radial mixing. This suggests that heating the cavity creates buoyancy forces that act to destabilize the central throughflow. The resulting flow structure may or may not be dominated by buoyancy forces—it depends on cavity geometry as well as physical parameters such as the axial and rotational Reynolds numbers.

The role of buoyancy forces is confirmed in the various studies carried out to measure heat transfer in rotating cavities with axial throughflow. A more detailed review can be found in the thesis of Alexiou (2000) and Alexiou et al. (2000). Measurements appear to indicate a flow that can be dominated by either free convection or forced convection, or a combination of these. The heat transfer behavior depends not only on the values of rotational and axial Reynolds numbers but also on geometric parameters such as gap ratio, inlet radius ratio, the presence of a shaft, and its radius and sense of rotation relative to the cavity surfaces.

Owen and Powell (2004) made velocity and heat transfer measurements in a single cavity rig with an annular inlet, $a/b = 0.4$ and $s/b = 0.2$. Tests were carried out for $4 \times 10^5 < Re_\phi < 3.2 \times 10^6$ and $1.4 \times 10^3 < Re_z < 5 \times 10^4$. The radial velocity was found to be approximately two orders of magnitude smaller than the tangential velocity, although the radial velocity is of a similar magnitude to the relative tangential velocity. Further analysis of the velocity measurements revealed behavior that is consistent with one, two, or three pairs of cyclonic and anticyclonic vortices in the flow field. The heat transfer measurements add further support to the existence of two different regimes: a buoyancy-induced regime at large rotational speeds and small axial flow, and a throughflow-dominated regime at the smaller values of rotational speed and larger values of axial throughflow.

Various numerical studies have been carried out for rotating cavities with throughflow, including Long and Tucker (1994b), Long, Morse, and Tucker (1997), Tucker and Long (1995, 1996), Wong (2002), Sun, Kilfoil et al. (2004), and Tian et al. (2004). In general, these CFD studies have given results that are qualitatively similar to experimental observations and broadly similar to heat transfer and LDA measurements. However, huge computational resources (months of CPU time in 2004) are required for this time-dependent 3-D flow, and the accuracy of much of the work prior to circa 2000 is constrained by relatively coarse grid sizes and questionable turbulence models. The recent 3-D numerical study by Tian et al. (2004) supports the earlier qualitative flow visualization work of Farthing et al. (1992). Owen, Abrahamsson, and Linblad (2006) carried out a numerical simulation and made comparisons with experimental measurements of tangential velocity and heat transfer in a single cavity rig with only the downstream disc heated. The computed flow structure shows the features identified by Farthing et al., and with the exception of one test case,

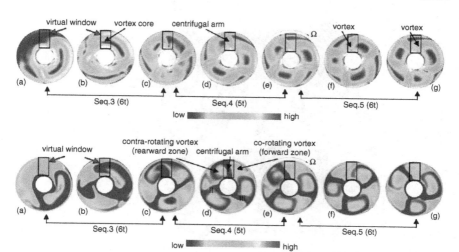

FIGURE 7.22   Investigation of the unstable flow structure in a rotating cavity (Bohn, Ren, and Tuemmers, 2006).

there is good agreement between the numerical results and the experimental measurements. Bohn, Ren, and Tuemmers (2006) make a comparison between experimental observations of flow structure and those computed using a 3-D time-dependent numerical code. The cyclone and anticyclone pairs observed in the experimental work are also found in the computations (see Figure 7.22).

It has also been found that the number of cyclone and anticyclone pairs can change with time. Bohn et al. (2006) suggest that the cyclone and anticyclone pairs are created by the buoyancy force and that their number and strength are due to the Coriolis force. What appears to be the first reported application of a Large Eddy Simulation to this flow is presented by Sun et al. (2006), who also make favorable comparisons with measurements of shroud heat transfer and tangential velocity reported by Long and Childs (2006, 2007) and Long et al. (2007).

In order to mitigate against the difficulties associated with numerical CFD modeling of the flow in a rotating cavity with axial throughflow, Owen (2007) proposed a scheme based on maximum entropy production where the flow field is modeled as a self-organizing system. The scheme was derived from far-from-equilibrium thermodynamics and states that a system will be stable in a state in which the entropy production is maximized. In the context of a rotating cavity, the maximum entropy production principle predicts that the flow is stable when the heat transfer out of the cavity is maximized; conversely, the system is unstable when the heat transfer is minimized.

Simultaneous measurements of the axial and tangential velocity components u_z and u_ϕ inside a cavity using Laser Doppler Anemometry for a rig with four rotating cavities (see Figure 7.23) are reported by Long and Childs (2006, 2007) and Long et al. (2007).

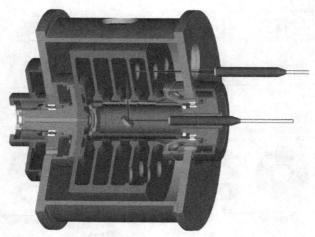

FIGURE 7.23 Multiple rotating cavity rig.

The axial component of velocity measured inside the cavity was found to be close to zero for all the conditions investigated. The measured values of axial velocity in the annular passage between disc bore and drive shaft were consistent with the value of the bulk mean velocity. Figure 7.24 shows the variation of the radial distribution of normalized tangential velocity ($u_\phi/\Omega r$) for the wide annular gap ($d_h / b = 0.164$) tests with $Ro = 0.27, 0.53, 0.94, 1.67,$ and 3.6. These measurements were made in the midplane of Cavity 3. However, measurements made in the midplane of Cavity 2 gave similar results. Measurements were also made close to the disc surfaces, and these too gave similar values. The outer radius of the shaft is located at $r / b = 0.236$ and the bore of the discs at $r/b = 0.318$. For $Ro = 0.53$ the tangential velocity exceeds solid body rotation in the region $r / b < 0.4$ and then decreases to $u_\phi/\Omega r = 1$ further into the cavity. As the Rossby number is reduced, the tangential velocity tends to behave less like a free vortex (where $u_\phi \propto 1 / r$) toward the inner radius. The behavior of the tangential velocity profile with Rossby number in these tests is consistent with the measurements of Farthing et al. (1992a), who found that reducing the Rossby number reduces the peak value of the normalized tangential velocity near the disc bore.

Data obtained from measurements made in the test rig with a narrower annular gap ($d_h / b = 0.092$) is also presented in Figure 7.24. It is clear that a significant difference exists between the results for the wide and the narrow annular gap. For the narrow annular gap, the values of $u_\phi/\Omega r$ are seen to be always lower than solid body rotation and to increase to $u_\phi/\Omega r \approx 1$, further into the cavity. Since the size of the annular gap tested differs by a factor of almost two, it is not unreasonable to expect that this does have an influence on the flow inside the cavity. In this case, a reduction of the annular gap between the shaft

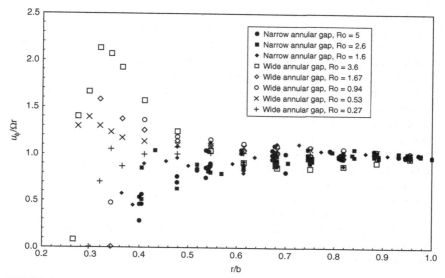

FIGURE 7.24   Variation of the radial distribution of the nondimensional tangential velocity with Rossby number for a wide ($d_h/b = 0.164$) and narrow ($d_h/b = 0.092$), annular gap.

and disc bore attenuates the influence of the throughflow on the tangential velocity inside the cavity. Some support for this can be found in the measurements of Owen and Powell (2004). These also show the behavior noted by Long and Childs (2006) for the narrow annular gap data. The detailed measurements of Pincombe (1983) display two different types of behavior corresponding to free and forced vortex flow. For $G = 0.53$, these differences were attributed to a change from Mode 1a to Mode 2a (nonaxisymmetric to axisymmetric) vortex breakdown. This could be of direct relevance to the measurements in Figure 7.24, with the geometry of the cavity affecting the mechanism of vortex breakdown, which in turn affects the tangential velocity distribution inside the cavity.

7.6.   CONCLUSIONS

The flow characteristics in the cavity formed between two co-rotating discs are dependent on whether there is a net radial throughflow, a net throughflow at the bore, and an axial or a radial temperature gradient between the discs or between inner and outer radial surfaces.

For the cases of a rotating cavity with net radial inflow or outflow, the flow is characterized by source and sink regions that feed the boundary layers that form on the disc faces. The boundary layers are separated by an interior rotating core of fluid where viscous effects are not significant. For small values of the Rossby and Ekman numbers, the inertial terms in the boundary layer equations can be neglected, giving a set of linear equations for modeling the flow in the boundary layers, which are known as Ekman-type boundary layers.

For the case of a closed annular rotating cavity with an axial temperature gradient, fluid tends to move radially inward adjacent to the hot disc and radially outward adjacent to the cooler disc. For a closed rotating cavity with a radial temperature gradient, a buoyancy-induced flow structure is produced, comprising an array of contra-rotating vortices. The counter-rotating vortices generate regions of low and high pressure, providing the circumferential pressure gradients that enable convection to occur.

The flow structure for the case of a cavity with an axial throughflow at the inner bore depends on the thermal boundary conditions. For the case of isothermal conditions, vortex breakdown in the central jet of air can result in nonaxisymmetric flow inside the cavity. If the discs are at different temperatures, nonaxisymmetric flow predominates with a buoyancy-induced flow regime at large rotational speeds and low axial flow rates and a throughflow-dominated flow regime, at small values of rotational speed and large values of axial throughflow.

REFERENCES

Alexiou, A. Flow and heat transfer in gas turbine H.P. compressor internal air systems. D.Phil. thesis, School of Engineering, University of Sussex, 2000.

Alexiou, A., Hills, N.J., Long, C.A., Turner, A.B., and Millward, J.A. Heat transfer in high pressure compressor internal air systems: A rotating disc-cone cavity with axial throughflow. *International Journal of Experimental Heat Transfer*, 13 (2000), pp. 299–328.

Bohn, D., Deuker, E., Emunds, R., and Gorzelitz, V. Experimental and theoretical investigations of heat transfer in closed gas-filled rotating annuli. ASME Paper 93-GT-292, 1993.

Bohn, D., Ren, J., and Tuemmers, C. Investigation of the unstable Flow Structure in a Rotating Cavity. Paper No. GT2006-90494, Presented at the ASME Turbo Expo, Barcelona, Spain, 2006.

Chew, J.W., Owen, J.M., and Pincombe, J.R. Numerical predictions for laminar source-sink flow in rotating cylindrical cavity. *Journal of Fluid Mechanics*, 143 (1984), pp. 451–466.

Farthing, P.R., Long, C.A., Owen, J.M., and Pincombe, J.R. Rotating cavity with axial throughflow of cooling air: Heat transfer. *ASME Journal of Turbomachinery*, 114 (1992b), pp. 229-236.

Farthing, P.R., Long, C.A., Owen, J.M., and Pincombe J.R. Rotating cavity with axial throughflow of cooling air: Flow structure. *ASME, Journal of Turbomachinery*, 114 (1992a), pp. 237–246.

Firouzian, M., Owen, J.M., Pincombe, J.R., and Rogers, R.H. Flow and heat transfer in a rotating cavity with a radial inflow of fluid. Part 1. The flow structure. *Int. J. Heat and Fluid Flow*, 6 (1985), pp. 228–1985.

Firouzian, M., Owen, J.M., Pincombe, J.R., and Rogers, R.H. Flow and heat transfer in a rotating cavity with a radial inflow of fluid. Part 2. Velocity, pressure and heat transfer measurements. *Int. J. Heat and Fluid Flow*, 7 (1986), 21–27.

Greenspan, H.P. *The theory of rotating fluids.* Cambridge, 1968.

Hide, R. On source-sink flows in a rotating fluid. *J. Fluid Mech.*, 32 (1968), pp. 737–764.

Iacovides, H., and Theofanopoulos, I.P. Turbulence modelling of axisymmetric flow inside rotating cavities. *Int. J. Heat and Fluid Flow*, 12 (1991), pp. 2–11.

King, M.P., Wilson, M., and Owen, J.M. Rayleigh-Bénard convection in open and closed rotating cavities. *Trans. ASME, Journal of Engineering for Gas Turbines and Power*, 129 (2007), pp. 305–311.

Long, C.A., and Childs, P.R.N. The effect of inlet conditions on the flow and heat transfer in a multiple rotating cavity with axial throughflow. Proceedings of 1st ISJPPE, the First International Symposium on Jet Propulsion and Power Engineering, Kunming, China, September 17–22, 2006.

Long, C.A., and Childs, P.R.N. Shroud heat transfer measurements inside a heated multiple rotating cavity with axial throughflow. *International Journal of Heat and Fluid Flow*, 28 (2007), pp. 1405–1417.

Long, C.A., and Tucker, P.G. Shroud heat transfer measurements from a rotating cavity with an axial throughflow of air. *ASME, Journal of Turbomachinery*, 116 (1994a), pp. 525–534.

Long, C.A., and Tucker, P.G. Numerical computation of laminar flow in a heated rotating cavity with an axial throughflow of air. *International Journal of Numerical Methods for Heat and Fluid Flow*, 4 (1994b), pp. 347–365.

Long, C.A., Miché, N.D.D., and Childs, P.R.N. Flow measurements inside a heated rotating multiple cavity with axial throughflow. *International Journal of Heat and Fluid Flow*, 28 (2007), pp. 1391–1404.

Long, C.A., Morse, A.P., and Tucker, P.G. Measurement and computation of heat transfer in high pressure compressor drum geometries with axial throughflow. *ASME, Journal of Turbomachinery*, 119 (1997), pp. 51–60.

Long, C.A., Morse, A.P., and Zafiropoulos, N. Buoyancy affected flow and heat transfer in asymmetrically heated rotating cavities. ASME Paper 93-GT-88, 1993.

Long, C.A., and Owen, J.M. The effect of inlet conditions on heat transfer in a rotating cavity with a radial outflow of fluid. *Trans. ASME, Journal of Turbomachinery*, 108 (1986), pp. 145–152.

Morse, A.P. Numerical prediction of turbulent flow in rotating cavities. *Trans. ASME, Journal of Turbomachinery*, 110 (1988), pp. 202–215.

Ong, C.L., and Owen, J.M. Boundary layer flows in rotating cavities. *Trans. ASME, Journal of Turbomachinery*, 111 (1989), pp. 341–348.

Ong, C.L., and Owen, J.M. Prediction of heat transfer in a rotating cavity with a radial outflow. *Trans. ASME, Journal of Turbomachinery*, 113 (1991), pp. 115–122.

Ostrach, S., and Braun, W.H. Natural convection inside a flat rotating container. NACA TN 4323, 1958.

Owen, J.M. Thermodynamic analysis of buoyancy-induced flow in rotating cavities, ASME Paper GT2007-27387, 2007.

Owen, J. M., Abrahamsson, H., and Linblad, K. Buoyancy induced flow in open rotating cavities. ASME Paper GT2006-91134, 2006.

Owen, J.M., and Bilimoria, E.D. Heat transfer in rotating cylindrical cavities. *Journal of Mechanical Engineering Science*, 19 (1977), pp. 175–187.

Owen, J.M., and Onur, H.S. Convective heat transfer in a rotating cylindrical cavity. *J. Eng. Power*, 105 (1983), pp. 265–271.

Owen, J.M., and Pincombe, J.R. Velocity measurements inside a rotating cylindrical cavity with a radial outflow of fluid. *Journal of Fluid Mechanics*, 99 (1980), pp. 111–127.

Owen, J.M., Pincombe, J.R., and Rogers, R.H. Source-sink flow inside a rotating cylindrical cavity. *Journal of Fluid Mechanics*, 155 (1985), pp. 233–265.

Owen, J.M., and Powell J. (2004) Buoyancy induced flow in a heated rotating cavity. Paper No. GT2004-53210, Presented at the ASME Turbo Expo, Vienna, Austria.

Owen, J.M., & Rogers, R.H. *Flow and heat transfer in rotating-disc systems*. Volume 2—*Rotating cavities*. Research Studies Press, 1995.

Pincombe, J.R.Optical measurements of the flow inside a rotating cylinder. D.Phil Thesis, School of Engineering and Applied Science, University of Sussex, 1983.

Sun, Z., Kilfoil, A., Chew, J.W., and Hills, N.J. Numerical simulation of natural convection in stationary and rotating cavities. Paper No. GT2004-53528, Presented at the ASME Turbo Expo, Vienna, Austria, 2004.

Sun, Z., Lindblad, K., Chew, J.W., and Young, C. LES and RANS investigations into buoyancy-affected convection in a rotating cavity with a central axial throughflow. Paper No. GT2006-90251, Presented at the ASME Turbo Expo, Barcelona, Spain, 2006.

Tian, S. Tao, Z, Ding, S. and Xu, G. Investigation of flow and heat transfer instabilities in a rotating cavity with axial throughflow of cooling air. Paper No. GT2004-53525. Presented at the ASME Turbo Expo, Vienna, Austria, 2004.

Tucker, P.G., and Long, C.A. CFD prediction of vortex breakdown in a rotating cavity with an axial throughflow of air. *International Communications in Heat and Mass Transfer*, 22, No. 5 (1995), pp. 639–648.

Tucker, P.G., and Long, C.A. Numerical investigation into influence of geometry on flow in a rotating cavity with an axial throughflow. *International Communications in Heat and Mass Transfer*, 23, No. 3 (1996), pp. 335–344.

Wong, L.S.Flow and heat transfer in rotationally induced buoyancy flow. D.Phil. Thesis, School of Engineering and Information Technology, University of Sussex, 2002.

Zysina,-Molozhen, L.M., and Salov, N.N. Heat exchange and flow regimes of liquid in a closed rotating annular cavity. *Aviatsionnaya Tekhnika*, 20, No. 1(1977), pp. 54–59.

Chapter 8

Atmospheric and Oceanic Circulations

Flow in the Earth's atmosphere and its oceans are significantly influenced by the Earth's rotation. Examples include the major circulations in the atmosphere and oceans, cyclones and anticyclones and large-scale vortices. This chapter provides a brief description of the atmosphere and oceans and the principal flow structures, with a specific focus on the associated flow circulations and the influence of the Earth's rotation.

8.1. INTRODUCTION

The circulation of flow in the atmosphere and oceans involves phase changes between ice, liquid, and vapor, interactions between phenomena with length scales from centimeters to thousands of kilometers, and timescales from seconds to millennia as well as the transfer of radiation through the semitransparent medium of variable composition that forms the atmosphere. The interactions of these factors give rise to the complex fluid motions that are familiar to us from our observations of the skies and oceans. In developing an understanding and modeling capability for geophysical flows, it is important to know about the composition of the atmosphere and oceans, the energy transfers involved, stability, and convection.

The Earth's atmosphere and the oceans occupy comparatively thin layers, relative to the Earth's radius, which are acted upon by gravity, rotation, and differential heating by incident solar radiation. The flow and heat transfer exchanges within this layer have a critical bearing on our lives. As introduced in Section 2.5, the Rossby number, Ro, can be used to determine whether the effect of rotation will be significant in governing the flow structure. If U_o represents the typical horizontal current in the flow and L the typical distance over which the flow varies, then the timescale of the motion is given by L/U_o. This timescale can be compared with the period of rotation or angular velocity, depending on the particular definition of Rossby number. Using Equation 2.160, if $Ro \gg 1$, then the timescale of motion is short relative to the rotation period and rotation will not significantly influence the fluid motion.

Rotating Flow, DOI: 10.1016/B978-0-12-382098-3.00008-1

In the case of the Earth's atmosphere, the relevant length scale is the Earth's mean radius, $a = 6371 \times 10^3$m. A typical velocity for atmospheric flows, representing say a wind speed, might be of the order of 10 m/s.

The planet Earth has three principal motions. The Earth spins on its axis at an angular velocity of one revolution per solar day, giving night and day. The Earth also revolves around the Sun, this revolution giving the two basic weather patterns, summer and winter. The Earth also moves through the Milky Way along with the rest of the solar system. The Earth's angular velocity, based on a 24-hour day, is approximated within about 0.3% of its actual value by

$$\Omega = \frac{1}{24 \times 60 \times 60} = 7.27 \times 10^{-5} \text{ rad/s} \qquad (8.1)$$

Rather than the 24-hour period normally taken for a day, it actually takes the Earth approximately 23 hours, 56 minutes, and 4.09 seconds to make one complete revolution about its axis. This length of time is known as a sidereal day, and the angular velocity corresponding to this period, Equation 8.2, is 7.2921159×10^{-5} rad/s.

$$\Omega = \frac{1}{(23 \times 60 \times 60) + (56 \times 60) + 4.09} = 7.2921159 \times 10^{-5} \text{ rad/s} \quad (8.2)$$

Using these figures in the definition of the Rossby number, Equation 2.160, with the length scale taken as the mean radius of the Earth,

$$Ro = \frac{U_o}{\Omega L} = \frac{10}{7.29 \times 10^{-5} \times 6371 \times 10^3} \approx 0.022 \qquad (8.3)$$

This Rossby number is much less than unity, indicating that rotation is decisive in determining the flow. It can therefore be assumed that flow in the atmosphere is significantly influenced by the Earth's rotation.

Similarly, taking a length scale in the ocean of the order of 1,000,000 m, a typical ocean current might have a velocity of 0.1 m/s, giving a Rossby number of

$$Ro = \frac{U_o}{\Omega L} = \frac{0.1}{7.29 \times 10^{-5} \times 10^6} \approx 0.0014 \qquad (8.4)$$

Again the Rossby number is much less than unity, and rotational effects will therefore be significant and indeed a controlling factor on the fluid motion.

As indicated by the low Rossby numbers involved, rotational effects are important in the fluid motion in both the atmosphere and the oceans. A further factor is energy transfer. The principal energy source for atmospheric flows is thermal radiation from the Sun. On a historical timescale, with a time average of a year, the Earth is neither warming nor cooling significantly, but instead is in equilibrium with the incoming energy balanced by outgoing energy lost by thermal radiation to space. The energy exchanges are not uniform, however, with the poles receiving less thermal radiation than equatorial regions. This imbalance results in global scale heat transfer processes and as a consequence associated fluid flows in the atmosphere and oceans. It should be noted that this model does not account for the effects of global warming caused by greenhouse gases, which has become a concern in recent years (see, for example, Inter-governmental Panel on Climate Change (2007)).

Convection is also important in atmospheric flows. In conventional fluid mechanics it is commonly assumed that either density is constant or density depends on pressure, $\rho = \rho(p)$. A fluid in which density depends on pressure is called a barotropic fluid. A fluid in which density depends on both pressure and temperature, $\rho = \rho(p, T)$, is called a baroclinic fluid. In a baroclinic fluid there can be gradients of density and temperature along a surface of constant pressure, and these can lead to instabilities in the flow that give rise to eddies superimposed on a bulk fluid motion. In a barotropic fluid, however, such gradients do not exist, and the fluid motion is comparatively less complex.

If the density in either the atmosphere or oceans was only dependent on pressure, then less dense fluid would be located on top of denser fluid, as illustrated in Figure 8.1. This represents a stable system, and one would not expect any significant fluid motions associated with such a system. However, our experience of the atmosphere and oceans suggests a much more complex situation. For a fluid where the density is a function of both pressure and temperature, it is possible for thermal energy to be converted into kinetic energy, thereby producing thermally induced bulk fluid motions where fluid that is heated locally by incident thermal radiation becomes less dense and rises in a convective current, as illustrated in Figure 8.2.

In the absence of rotation, a fluid can be characterized as a substance that flows freely and readily occupies and conforms to the shape of its container. A fluid undergoing rotation cannot, however, move in arbitrary paths. If the scales of motion are sufficiently large and sluggish, the motion will be significantly affected by rotation.

A turntable can be used to illustrate the effect of rotation on a fluid, and such experiments have provided significant insight into geophysical flows (see, for

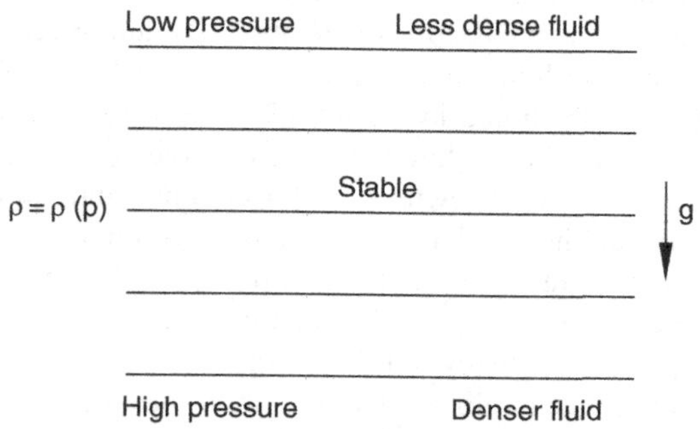

FIGURE 8.1 Stability associated with a barotropic fluid system where the density depends only on pressure.

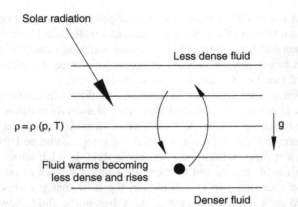

FIGURE 8.2 The convective flow of a parcel of less dense fluid characteristic of a baroclinic fluid where the fluid density is dependent on both pressure and temperature.

example, Hide, 1958, 1964, 1966, 1967; Greenspan, 1968; Marshall and Plumb, 2008). The series of experiments developed by Marshall and Plumb (2008) at the Massachusetts Institute of Technology to demonstrate selected aspects of geophysical flow and heat transfer phenomena provide a very accessible starting point; some of the experimental setups and visualizations developed are also presented here. Using a cylindrical container with a diameter of approximately 700 mm (the exact shape of the container is not critical, and a rectangular tank could readily be used), fill it with water to a depth of approximately 200 mm and then rotate it at about 20 rpm for about 15 minutes to allow all of the fluid to settle. After 15 minutes of rotation and while the container is still rotating at 20 rpm, using your hand or a ladle agitate the water by immersing it in the water and with your palm open pull it through the water up and down a few times. Remove your hand and pour in some food coloring to act as a dye and allow observation of the fluid motion. Repeat the process but this time without rotating the container.

In the case of the nonrotating container, the dye disperses in all directions. For the rotating container, streaks of dye are observed falling vertically, with the dye being drawn out by horizontal motion into vertical "curtains" that wrap around each other, as illustrated in Figure 8.3. If two different color dyes are used, the interleaving of fluid columns can be observed (see Figure 8.4).

The vertical columns are known as Taylor columns and are a result of the "rigidity" imparted to the fluid by rotation of the container. Under certain conditions, which are induced in this experiment, the fluid moves in vertical columns that are aligned parallel to the rotation vector, which in this case is coincident with the axis of the cylinder. It is in this sense,—that is, movement parallel to the rotation vector alone—that rotating fluids are "stiff" or relatively rigid. With increasing horizontal spatial scales and increasing timescales, the effect of rotation becomes stronger with an increasing impact on fluid motion.

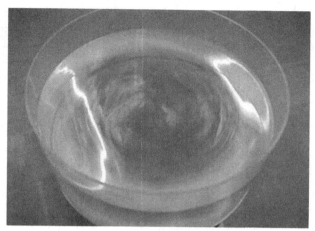

FIGURE 8.3 Taylor columns in a rotating tank (photograph Marshall and Plumb, 2008).

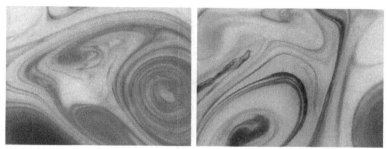

FIGURE 8.4 Dyes of different color being stirred around each other by rotationally constrained motion (photograph Marshall and Plumb, 2008).

In the case of the rotating container experiment, $U_o \approx 0.01$ m/s, $L = 0.3$ m and $\Omega \approx 2$ rad/s, giving a Rossby number of

$$Ro = \frac{U_o}{\Omega L} = \frac{0.01}{2 \times 0.3} \approx 0.017 \qquad (8.5)$$

As this value is much less than unity, it indicates, as observed, that rotational effects will be significant on the fluid motion.

The characteristic Rossby numbers for the atmosphere and oceans reveal the importance of rotation in influencing fluid motion in geophysical applications. The flow fields illustrated in Figures 1.11 and 1.14 as well as in Figure 8.4 reveal some of the complexity involved. The conventional grid coordinate system, providing a means to define the location on the planet, and the principal planetary dynamics, influencing the atmosphere and oceans, are introduced in Sections 8.1.1 and 8.1.2, respectively. The structure of the atmosphere is introduced in

Section 8.1.3, and a model for the exchange of energy in Section 8.1.4. The principal flow recirculations in the atmosphere are described in Section 8.2. An overview of some aspects of pressure systems and weather fronts is given in

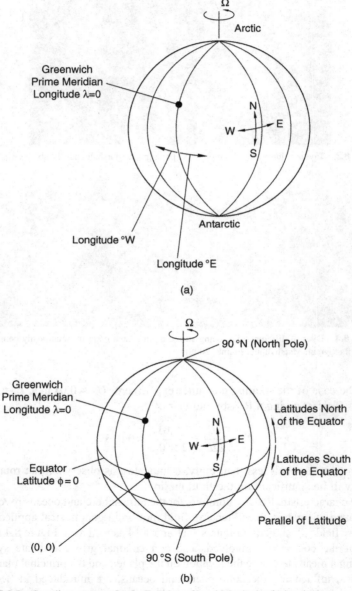

FIGURE 8.5 Coordinate system for the Earth's surface: (a) Latitude, (b) longitude.

Section 8.3. An introduction to intense atmospheric vortices such as typhoons and tornadoes is given in Section 8.4. The principal flow structure and associated flow circulations in the oceans are introduced in Section 8.5.

8.1.1 Coordinate System

The vast majority of flow phenomena related to the Earth's surface occur within the relatively thin (in comparison to the Earth's radius) atmosphere and oceans. While spherical polar coordinates offer a possible means for modeling the flow, it is conventional to relate flows to the Earth's surface and use the system of latitudes and longitudes, illustrated in Figure 8.5. This is a grid system of reference lines that cross at right angles over the Earth's surface. The grid lines are determined by internal Earth angle measurements and are called lines of latitude and longitude.

In this system lines of latitude, also called parallels, are defined relative to the equator. At the equator the latitude is 0°. Latitude lines encircle the Earth, parallel to the equator and extend to the north pole, defined as 90° north, and to the south pole, defined as 90° south, and are specified in terms of an angle, ϕ, in degrees. Each degree is divided into 60 minutes of arc and each minute into 60 seconds of arc. Lines of latitude must be designated as either south or north of the equator to avoid ambiguity. The circumferential location is defined by means of longitude, λ. The lines of longitude, also called meridians, are formed at right angles to the latitude grid. The longitude datum, $\lambda = 0$ or the prime meridian, begins at an arbitrarily chosen datum that is taken as a line passing from the north pole to the south pole through the Royal Naval Observatory in Greenwich, England. Lines of longitude extend 180° east and west of the datum around the Earth and meet at 180° east and 180° west longitude. As a result, a location on the Earth's surface can be defined uniquely by specification of the longitude and latitude. For example, Anchorage, in Alaska, has a longitude and latitude of approximately 150° and 61° north of the equator. Many geophysical flows can be modeled on a local scale using two-dimensional Cartesian coordinates.

Because of the thinness of the atmosphere, relative to the Earth's radius, it is possible to approximately model many geophysical flows using a two-dimensional Cartesian coordinate system, thereby neglecting the Earth's curvature. However, it is still necessary to account for the Coriolis force in such a system, and the Coriolis parameter is used for this purpose. For a latitude ϕ, the local coordinate is defined by (x, y, z) where x represents the eastward direction, y the northward direction, and z the vertical direction (radial distance with respect the Earth's surface). The components of the angular velocity vector are given by

$$\Omega = \begin{pmatrix} 0 \\ \Omega\cos\phi \\ \Omega\sin\phi \end{pmatrix} \qquad (8.6)$$

The cross product $\Omega \times w$ is given by

$$\Omega \times w = (0, \Omega \cos \phi, \Omega \sin \phi) \times (u, v, w) \tag{8.7}$$

$$= \Omega \cos \phi\, w - \Omega \sin \phi\, v, \Omega \sin \phi\, u, -\Omega \cos \phi\, u \tag{8.8}$$

The vertical component is acted on by gravity. If $\Omega u << g$,then we can be justified in neglecting terms with u. A typical value for the magnitude of velocity in the atmosphere is of the order of 10 m/s. The product of Ωu is of the order of 7×10^{-4} m/s^2. Comparing this with gravity ($g = 9.81$ m/s^2), gives $\Omega u << g$. In the atmosphere and oceans, vertical velocities are typically much less than horizontal velocities, and it is acceptable to neglect the terms involving w in the horizontal components of $\Omega \times w$. The cross product $\Omega \times w$ can therefore be approximated by

$$\Omega \times w = (-\Omega \sin \phi\, v, \Omega \sin \phi\, u, 0) \tag{8.9}$$

The Coriolis term is given by $2\Omega \times w$
that is,

$$2\Omega \times w = (-2\Omega \sin \phi\, v, 2\Omega \sin \phi\, u, 0) \tag{8.10}$$

or

$$2\Omega \times w = f z \times w \tag{8.11}$$

where, f is the Coriolis parameter given by

$$f = 2\Omega \sin \phi \tag{8.12}$$

$\Omega \sin \phi$ represents the vertical component of the Earth's rotational vector. It is the only component that has a significant effect on geophysical flow, and this is a

TABLE 8.1: Selected values for the Coriolis parameter as a function of latitude

Latitude (°)	f
90	1.4584×10^{-4}
80	1.4363×10^{-4}
60	1.2630×10^{-4}
45	1.0313×10^{-4}
30	7.2921×10^{-5}
10	2.5325×10^{-5}
0	0

consequence of the thin nature of the atmosphere and oceans relative to the Earth's radius. At the equator, $\phi = 0$, and the Coriolis parameter $f = 2\Omega sin\phi$ will also be zero; rotational effects are therefore negligible at the equator. Values for the Coriolis parameter are given in Table 8.1.

8.1.2 Earth, Sun, and Moon Dynamics

The Sun spins about a fixed axis with its equator nearly aligned with the Earth's orbital plane. At its equator the Sun rotates at an angular velocity of one revolution per 25 days with its poles spinning slightly more slowly. The direction of rotation is counterclockwise, when viewed from a position in space north of the Earth. The Earth's path around the Sun is approximately elliptical, and as the Earth rotates around the Sun the Earth also spins. The axis of the Earth's rotation is tilted at an angle of approximately 23.5° relative to the Sun. The motion of the Earth, its moon, and the Sun are illustrated in Figure 8.6.

Seasonal variations occur as a result of the Earth's orbit around the Sun, which is illustrated schematically in Figure 8.6. The combination of the Earth's tilt and elliptical path results in a variation of thermal radiation received by the Earth's surface. A comparatively short-term variation occurs by day and night due to the Earth's angular velocity. The variation of the distance between the Earth and the Sun and the Earth's angle of inclination results in the longer term variations associated with the seasons. The reason for seasonal changes can be described by reference to Figure 8.7. As the Earth moves with its axis tilted at an angle from the vertical, during the year the Earth's north pole is sometimes tilting toward the Sun and sometimes tilting away from it. The northern hemisphere receives its maximum hours of sunlight and hence thermal radiation from the Sun, when the north pole is tilted toward the Sun, position (a), defining the northern hemisphere's summer. In this period the south pole is tilted away from the Sun, so that the southern hemisphere receives less sunlight and less

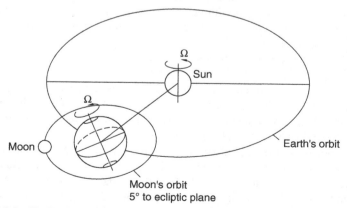

FIGURE 8.6   The Earth's rotation, inclination, and orbital motion. Not to scale.

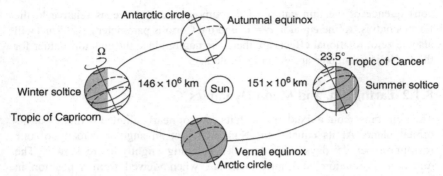

FIGURE 8.7 The Earth's seasons. Not to scale.

thermal radiation, defining the southern hemisphere's winter. At the other side of the orbit, position (c), when the north pole is tilted away from the Sun and the south pole is tilted toward the Sun, defines the northern hemisphere's winter and the southern hemisphere's summer. These extremes of orbit also correspond to the Earth being at its maximum distance from the Sun due to the elliptical orbit of the Earth around the Sun. During the northern hemisphere's summer, the Earth is further away from the Sun (151×10^6 km) than it is in winter (146×10^6 km).

As the Earth rotates around the Sun, the proportion of daylight to darkness on a given day varies because of the tilt of the Earth, except at the equator where the number of daylight hours is always the same. Between the extremes of summer and winter are the autumnal and vernal equinoxes when the period of daylight and darkness for the day is equal at the equator. These occur on approximately March 21 and September 21. When equinoxes occur at the Tropic of Cancer on June 21 and at the Tropic of Capricorn on December 21, it is called a solstice. The amount of sunlight incident on the Earth can be quantified by the term *insolation*. Insolation is defined as the thermal radiation received from the Sun per unit surface area of the Earth. Because of the Earth's tilt, there are several months with no insolation north or south of a latitude of 66.5°. Between the tropics of cancer and Capricorn, there is little seasonal variation in solar radiation levels, as the Sun does not move beyond these boundaries. In these regions the seasons tend to be characterized by just two periods, dry and wet.

Because the Earth spins and is not rigid, it has deformed into an oblate spheroidal shape with a bulge along the equator. This is illustrated in Figure 8.8. The equatorial radius is about 6378.4 km, and the polar radius is 6356.9 km. The mean radius is taken as 6371 km.

The Moon orbits the Earth in 29.5 days. Its plane is inclined to the Earth by 5°, and the same side always faces the Earth. The gradient of a gravity field causes an object in the field to be stretched in the radial direction. The Moon's gravitational force is stronger on particles on the side of the Earth closest to the Moon, while the centrifugal force, due to the Earth's rotation, is stronger on

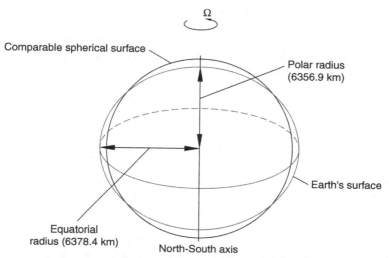

FIGURE 8.8   Schematic of the Earth's oblate spheroidal shape in comparison with a sphere. Not to scale.

particles furthest away from the center of the mass of the Earth–Moon system. This distribution of forces acts to pull surface particles away from the center of the Earth and generates the tide-raising force field on the Earth. As water is deformable, the Moon's gravity tends to move water to a point closest to the Moon, producing a bulge in the Earth's water surface. At the same time, centrifugal forces acting on the surface opposite to the Moon create a similar

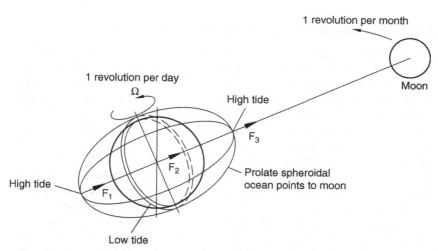

FIGURE 8.9   Schematic production of oceanic tidal bulges due to differential gravitational forces. $F_3 > F_2 > F_1$ (scales of deflection greatly exaggerated).

bulge of water. There are thus two bulges with two depressions between the bulges (Figure 8.9). These correspond to two high tides and two low tides, with a wavelength of half the circumference of the Earth. As the Earth makes one revolution in approximately 24 hours, the bulges tend to stay under the Moon as the Earth turns.

A point on Earth that is initially centered under a high tide passes through to a low tide, back to another high tide and another low tide and back to the original high tide position within one revolution of the Earth. However, while the Earth turns, the Moon is also orbiting the Earth and moves eastward along its orbit. After 24 hours, a point on Earth that was originally closest to the Moon corresponding to the bulge in the water surface is no longer directly under the Moon. The Earth has to turn an additional 12°, taking an additional 50 minutes to bring the starting point back in line with the Moon. This period defines the difference between a day and a lunar tidal day, with the lunar tidal day being defined as 24 hours and 50 minutes. This is why corresponding tides tend to arrive at any location about one hour later each day. The Sun also acts to produce a tidal wave, but because of the Sun's distant location the gravitational and centrifugal forces that cause tides are only 46% those of the Moon and have a correspondingly smaller though still significant influence.

Some atmospheric flows and phenomena occur on timescales that are short in comparison to the seasons. Other flow circulations persist as nearly permanent flow patterns. Meteorology is the study of weather and climate. The distinction between climate and weather is by means of scale. Weather refers to the state of the atmosphere at a local level and usually with a timescale ranging between minutes and months. Climate refers to the long-term behavior of the atmosphere in a specific region.

8.1.3 Structure and Composition of the Atmosphere

The atmosphere is the thin envelope of gases held to the Earth by gravitational attraction. The outer limit of the atmosphere is conventionally taken to have a radius of 1000 km from the Earth's surface. However, most atmospheric phenomena are governed by the flow and energy exchanges that occur in the region concentrated within 16 km of the Earth's surface at the equator and 8 km at the poles. This region contains the vast majority of atmospheric gases, due to the variation of density with height; approximately 99% of the mass of the atmosphere is contained within 40 km and 50% within 5.6 km of sea level. The layer of air and water that surrounds the air is relatively so thin that on a desk globe or printed book picture it would correspond to a coat of paint or a few pixels, respectively. All motions of air and water take place in this thin layer.

The vertical structure of the atmosphere and oceans is of significance in understanding and modeling air and water flows. The pressure of the atmosphere changes with height. In addition, the temperature also varies, but the energy exchanges involved result in a complex variation of temperature with

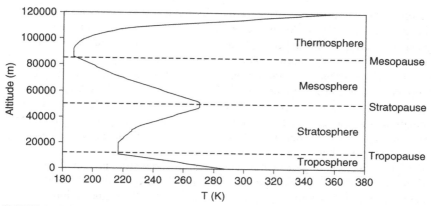

FIGURE 8.10 Vertical structure of the atmosphere (U.S. Standard Atmosphere, COESA, 1976).

height above the Earth's surface. The variation of temperature is used to classify the atmosphere into several distinct regions, including the troposphere, stratosphere, mesosphere, and thermosphere. These are illustrated in Figure 8.10.

The troposphere is characterized by a nearly linear decrease in temperature with height. The troposphere contains approximately 85% of the atmospheric mass, including the vast majority of the water vapor, dust, and pollutants. Typically the temperature in the troposphere decreases by 6.4 °C per 1000 m increase in altitude. The linear decrease of temperature with altitude is known as the lapse rate. The temperature variation is a result of the Earth's surface being heated by the incoming thermal radiation from the Sun, which in turn heats the air by conduction and convection. Pressure falls with height as the effect of gravity decreases. Wind speeds generally increase with height within the troposphere. Flow within this layer tends to be unstable. At the outer limit of the troposphere is a thin isothermal layer of atmospheric gas called the tropopause, and across this region the atmospheric temperature is actually constant despite the increase in height.

The stratosphere is characterized by a nonlinear increase in temperature caused by the concentration of ozone within this layer. The increase in temperature in the atmosphere is known as a temperature inversion. The ozone (O_3) absorbs incoming ultraviolet radiation and by heat exchange with the other surrounding gases causes the temperature rise. This region is characterized by low winds in the lower parts, but these rise with height away from the Earth's surface. Pressure continues to fall with height as the effect of gravity decreases and the air is dry with little moisture content. At the outer limit of the stratosphere is another thin isothermal layer of atmospheric gas called the stratopause where the atmospheric temperature is constant despite the increase in height.

The mesosphere is characterized by temperatures that fall rapidly with height as this region contains very little water vapor, cloud, dust, or ozone that act to absorb incoming radiation (principally short-wave radiation, e.g., X rays). The temperature within the mesosphere falls to about -90 °C and wind speeds can be high, up to 80 m/s. At the outer limit of the mesosphere is another thin isothermal layer of atmospheric gas called the mesopause where the atmospheric temperature is constant despite the increase in height.

The thermosphere, extending from about 90 km to 600 km above the Earth's surface is characterized by temperatures that increase rapidly with height as a result of an increasing proportion of atomic oxygen, which absorbs the incoming ultraviolet radiation from the Sun. Temperatures in this region are thought to rise to values of the order of 1750 °C. The portion of the Earth's atmosphere above 600 km is known as the exosphere.

Air is a mixture of gases whose composition varies from location to location in the atmosphere. Air can support a significant proportion of water vapor, before it becomes saturated. and also dust particles. The composition of standard dry air is listed in Table 8.2. The composition is site dependent and important factors include the surroundings, wind direction, time of day, and season. Nonstandard components may be present due to both natural processes and anthropogenic activities. Some nonstandard but common components include methane, nitrous oxide, sulfur dioxide, nitrous oxide, nitrogen dioxide, ammonia, and carbon monoxide. Although carbon dioxide is considered to be a standard component of air, its concentration will vary with location and season, and values for its concentration represent an approximation.

The relatively small thickness of the atmosphere allows some simplifications to be made in modeling. The decrease in gravity from the Earth's surface to an altitude of approximately 10 km is about 10^{-4} m/s^2, so gravity can be taken as constant for most atmospheric modeling applications. It is also sometimes possible to neglect the Earth's curvature and assume a planar disc geometry, as indicated in Section 8.1.1. This is comparable to a flat Earth rotating disc analogy.

Atmospheric water vapor is present in the atmosphere in variable amounts, a typical figure being 0.59%, but this figure varies significantly with location and local conditions. Water vapor comes primarily from evaporation from the oceans' surfaces. Water vapor is critically important for radiative heat transfer because it strongly absorbs in the infrared, for example, around $\lambda \approx 5$ μm to $\lambda \approx 8$ μm, the region of the spectrum in which Earth radiates energy to space. Water is of course also critical to life on Earth. Similarly CO_2 absorbs radiation at selective wavelengths such as $\lambda \approx 4$ μm and $\lambda \approx 14$ μm. CO_2 concentration in the atmosphere is controlled principally by photosynthesis, respiration, exchanges with the oceans, and anthropogenic activities. Whereas biosphere and ocean CO_2 emissions are globally of the order of 770 gigatonnes/year (Metz, 2001), anthropogenic fossil carbon dioxide emissions, including those from the production, distribution, and consumption of fossil fuels and as a by-product from cement production, based on 2000–2005 figures, are approximately 26.4 gigatonnes CO_2 per year (IPCC 4th Assessment Report, Climate Change 2007). Although H_2O, CO_2, and O_3 are present in only small proportions, they are very important as they are physically active in chemical and energy transfer

TABLE 8.2: Composition of dry air (Calvert, 1990)

Gas	Chemical symbol	Percentage of mixture by volume	Percentage by weight
Nitrogen	N_2	78.08	75.47
Oxygen	O_2	20.95	23.2
Argon	Ar	0.93	1.28
Carbon dioxide	CO_2	0.03	0.0590
Neon	Ne	1.8×10^{-3}	0.0012
Helium	He	5×10^{-4}	0.00007
Methane	CH_4	non-standard	non-standard
Krypton	Kr	1×10^{-4}	0.0003
Nitrous oxide	N_2O	non-standard	non-standard
Xenon	Xe	1×10^{-5}	0.00004
Hydrogen	H_2	$<1\times10^{-6}$	Negligible
		Total $\approx100\%$	

processes. Their small proportions also make them sensitive to major Earth events, such as volcanic action, solar activity, and anthropogenic activity.

For the first 50 km of the atmosphere, the mean free path of the molecules is so short and molecular collisions so frequent that the gas can be regarded as a continuum in local thermodynamic equilibrium. At altitudes above about 80 km, where the density is very low, the assumption of a continuum breaks down.

As noted earlier and indicated in Table 8.2, air is a mixture of gases, and for air below an altitude of about 50 km, the ideal gas law can be applied to its individual components. The concept of partial density is useful for mixtures of gases.

The partial density is the mass of the gas per unit volume if that gas alone occupied the volume. If ρ_v and ρ_d are the partial density of water vapor and dry air, respectively, then

$$p_v = \rho_v R_v T \tag{8.13}$$

$$p_d = \rho_d R_d T \tag{8.14}$$

Dalton's law of partial pressure states that the pressure, p, of a mixture of gases that do not interact together chemically is equal to the sum of the partial pressures. The sum of the partial pressures is given by adding together the pressures that each gas would exert if it alone occupied the volume of space concerned. The law is given by the expression:

$$p = \sum_{i=1}^{n} \rho_i R_i T \tag{8.15}$$

In the case of a mixture of dry air and water vapor,

$$p = \rho_d R_d T + \rho_v R_v T \tag{8.16}$$

As the amount of water vapor is comparatively small,

$$p \approx \rho_d R_d T \tag{8.17}$$

For a container partially filled with water with a free surface of air under equilibrium conditions, the rate of evaporation of water will equal the rate of condensation and the air is said to be saturated with water vapor. Provided there are enough condensation nuclei, there will be a mist. For these conditions the pressure is equal to the saturation vapor pressure, p_{sat} The saturation vapor pressure is a function of temperature only and increases rapidly with temperature, as illustrated in Figure 8.11 and can be approximated for typical atmospheric temperatures by Equation 8.18 (Marshall and Plumb, 2008).

$$p_{sat} = 611.2e^{-0.067T} \tag{8.18}$$

A more accurate determination of the saturation vapor pressure is given by the integrated form of the Clausius-Clapeyron relationship, Equation 8.19. This

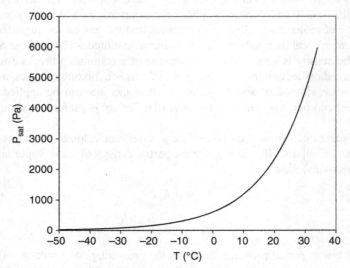

FIGURE 8.11 Variation of saturation vapor pressure with temperature.

equation enables the determination of vapor pressure of a liquid at any temperature if the heat of vaporization is known.

$$\ln\left(\frac{p_1}{p_2}\right) = \frac{\Delta H_{vapourization}}{\Re}\left(\frac{1}{T_1} - \frac{1}{T_2}\right) \tag{8.19}$$

where $\Delta H_{vaporization}$ is the enthalpy of vaporization, $\Re$ is the molar gas constant ($\Re = 8.314510$ J/mol K (Cohen and Taylor, 1987, 1999), and p_1 and p_2 are the pressures at two temperatures T_1 and T_2.

For $T = 288$ K, $p_{sat} = 16.7$ hPa $= 16.7$ mbar $= 1670$ Pa. The characteristic gas constant for water vapor is 461.39 J/kg K. From Equation 8.13, at $p_{sat} = 1670$ Pa and $T = 288$ K, $\rho_v = 0.0126$ kg/m^3. This is the maximum amount of water vapor per unit volume that can be held by the atmosphere at this temperature.

It is highly significant for the nature of the atmosphere that the saturation vapor pressure increases exponentially with temperature. As temperature in the lower regions of the atmosphere decreases with height, the moisture content also decays rapidly with height. As noted, the temperature of the atmosphere is about −50 °C at 10 km. At this temperature the saturation vapor pressure is almost zero, and the atmosphere cannot support any moisture. It is for this reason that most of the atmosphere's water is located in the lowest few kilometers. It should be noted also that the distribution of vapor around the Earth's atmosphere is not homogeneous with much more water vapor in the warm tropics than in cooler higher latitudes. As water vapor has significant absorption and emission characteristics in the infrared region the local vapor content of the atmosphere has an important impact on the local energy transfers in the atmosphere. Because it is warmer and the saturation vapor pressure is higher, air in the tropics tends to be much more moist than air over the poles and higher latitudes.

If a moist parcel of air moves vertically, it will tend to cool adiabatically. If the cooling is sufficient to saturate the parcel of air, some of the water vapor will condense on the condensation nuclei present to form a cloud. The corresponding release of latent heat will add buoyancy to the air parcel, thereby contributing to the convective fluid motion.

8.1.4 The Energy Cycle

The energy source for atmospheric flows is thermal radiation from the Sun. Thermal radiation is emitted by the Sun due to its temperature, and this emitted energy is intercepted by the Earth. The Earth also emits thermal radiation because of its temperature with an exchange between the Earth and space. In general, based on a historical basis of yearly averages, the Earth is neither warming nor cooling significantly. The Earth can be modeled as if it were in a state of equilibrium or energy balance, with the solar energy absorbed by the Earth equal to the energy released by the Earth by means of thermal

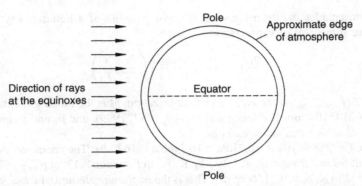

FIGURE 8.12 Schematic showing the intensity of solar thermal radiation at the equator and poles. (The vertical scale of the atmosphere is greatly exaggerated.)

radiation to space. It should be noted that this model ignores concerns regarding global warming owing to the accumulation of carbon dioxide in the atmosphere, ozone variations, and atmospheric water content. These may be influencing global temperatures by approximately 0.10 to 0.16 °C per decade, based on the IPCC (2007) figures for 1956–2005, although the figures in the various models and hypotheses vary. The assumption of energy equilibrium for the Earth is, however, acceptable as an approximation for the purpose of an Earth energy cycle model, given the scale of the energy transfers involved.

It should be noted, however, that the thermal radiation from the Sun is not uniform across the Earth's surface. This is illustrated for the equinoxes, March 21 and September 23, in Figure 8.12. Here it can be seen that the region of the Earth's surface around the equator is heated by the incoming solar radiation, while that around the poles receives very little incident radiation. The Earth also emits thermal radiation due to its temperature, and it radiates more energy from the poles than it receives as a result of solar radiation. The Earth is heated by solar thermal radiation at the equator and cooled by emission of thermal radiation to space at the poles. This temperature difference causes a flow heat from the equator to the poles and energy is transported by winds and ocean currents.

The energy exchanges between Earth and space are complex. Approximately 30% of the thermal radiation incident on the Earth is reflected into space by the outer atmosphere. Of the remaining 70% or so, a third is transformed into heat in the atmosphere and two-thirds on the Earth's surface. Most of the energy absorbed by the Earth is consumed in the evaporation of water and is stored in the atmosphere as latent heat, while the rest is transferred directly from the Earth to the atmosphere by conduction and convection. The latent heats of evaporation and solidification of water are highly significant as they provide a major source of energy storage and energy transportation. Temperature differences in the atmosphere result in the transformation of thermal energy into kinetic energy, which can be observed in winds and ocean currents. These

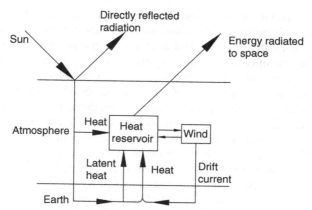

FIGURE 8.13   Energy balance between the Sun, Earth, and space.

fluid flow motions dissipate, and the frictional heat produced is returned to the atmosphere, which releases thermal radiation to space in the form of infrared radiation, thereby maintaining the energy balance. This model of energy exchanges is illustrated in Figure 8.13.

The rate at which the Sun emits energy is approximately $Q = 3.87 \times 10^{26}$ W. The flux of solar radiation at the Earth is called the solar constant, S_o, and depends on the distance between the Earth and the Sun. The solar constant is given by the inverse square law.

$$S_o = \frac{Q}{4\pi r^2} \tag{8.20}$$

Taking an average approximate value for the distance between the Earth and the Sun of $r = 150 \times 10^9$ m, the solar constant is approximately

$$S_o = 1367 \ W/m^2 \tag{8.21}$$

All matter will emit electromagnetic radiation due to its temperature. Heat transfer due to radiation between a small object and its surroundings can be modeled by

$$q = \varepsilon\sigma(T_s^4 - T_\infty^4) \tag{8.22}$$

where

q = heat flux (W/m^2)
ε = total surface emissivity
σ = the Stefan-Boltzmann constant (5.67051×10^{-8} Wm^{-2}K^{-4}; Cohen and Taylor, 1999)
T_s = temperature of the surface (K)
T_∞ = temperature of surroundings (K)

The term *thermal radiation* describes the energy that is emitted in the form of electromagnetic radiation at the surface of a body that has been thermally excited. Thermal radiation is emitted in all directions, and if it strikes another body, part may be reflected, part may be absorbed, and part may be transmitted through it. Absorbed radiation will appear as heat within the body. As it is an electromagnetic wave, thermal radiation can be transferred between two bodies, without the need for a medium of transport between them, as is the case with the Sun and the Earth. For example, if two bodies are at different temperatures and are separated by a vacuum between them, then heat will still be transferred from the hotter object to the colder by means of thermal radiation. Even in the case of thermal equilibrium, the temperature of both bodies being the same, an energy exchange will occur even though the net exchange will be zero.

The thermal radiation emitted by a surface is not equally distributed over all wavelengths. Similarly, the radiation incident, reflected or absorbed by a surface, may also be wavelength dependent. The wavelength dependency of any radiative quantity or surface property is referred to as spectral dependency. The adjective "monochromatic" is used to refer to a radiative quantity at a single wavelength.

The term *emissive power* is used for the thermal radiation, leaving a surface per unit surface area of the surface. This is qualified depending on whether it is summed over all wavelengths or whether it is emitted at a particular wavelength. Total hemispherical emissive power or just total emissive power, normally denoted by the symbol E, defines the emitted energy summed over all directions and all wavelengths. The total emissive power is dependent on the temperature of the emitting body, the type of material, and the nature of the surface features such as its roughness. The spectral emissive power, E_λ, defines the emissive power emitted at a particular wavelength.

When dealing with thermal radiation from and to real surfaces, it is useful to use the concept of an ideal surface as a comparison baseline. This ideal surface is given the name blackbody. A blackbody surface has the following properties:

- A blackbody absorbs all incident radiation regardless of wavelength and direction.
- For a given wavelength and temperature, no surface can emit more energy than a blackbody.
- Radiation emitted by a blackbody is independent of direction.

The spectral emissive power for a blackbody can be determined using Planck's law, Equation 8.23 (Planck, 1959).

$$E_{\lambda,b} = \frac{C_1}{\lambda^5 [\exp{(C_2/\lambda T)} - 1]} \, (\text{W/m}^3) \tag{8.23}$$

where
$E_{\lambda,b}$ = spectral emissive power for a blackbody (W/m^3)

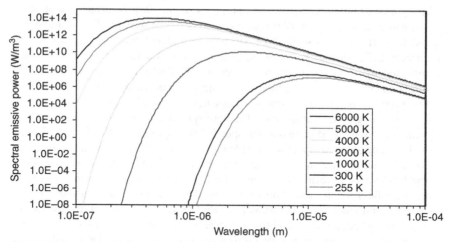

FIGURE 8.14   Energy emitted at different wavelengths for a blackbody at 6000 K, representative of the Sun's surface temperature and one at 255 K representative of the emission temperature of the Earth.

C_1 = the first radiation constant = $3.7417749 \times 10^{-16}$ W·m^2 (Cohen and Taylor, 1999)

λ = wavelength (m)

C_2 = the second radiation constant = 0.01438769 m·K (Cohen and Taylor, 1999)

T = absolute temperature (K)

This equation is plotted for a range of wavelengths and inculding those temperatures, representative of the Sun's surface temperature and the emission temperature of the Earth in Figure 8.14. This figure illustrates how the energy radiated for a given temperature is a function of the wavelength at which the energy is radiated. In the figure the wavelength has been plotted in μm and emissive power calculated in W/m^2/μm as wavelengths are commonly referred to using the unit μm. The first and second radiation constants in Equation 8.23 become $C_1 = 3.7417749 \times 10^8$ W·μm^4/m^2 and $C_2 = 14387.69$ μm·K accordingly.

For the case of the Sun, with a surface temperature of approximately 6000 K, 95% of the emitted energy is for wavelengths between 0.25 μm and 2.5 μm. Taking the Earth as a simple disc with radius 6.37×10^6 m, the solar power incident on the Earth is given by

$$Power = \pi a^2 S_o = 1.74 \times 10^{17} \text{ W} \tag{8.24}$$

In general, thermal radiation can be reflected, transmitted, or absorbed by a material. The spectral irradiation balance for the proportions of radiation reflected, transmitted, or absorbed by a material at a specific wavelength is given by

$$G_\lambda = G_{\lambda,ref} + G_{\lambda,abs} + G_{\lambda,tr} \tag{8.25}$$

In the case of the incident radiation of the Earth, the Earth's surface can be taken as opaque, and the quantity transmitted can be taken as zero. Thus some of the radiation given by Equation 8.24 will be reflected and only a proportion will be absorbed. The surface property that determines the fraction of incident radiation reflected by a surface is called reflectivity. Reflectivity depends on both the direction of the incident radiation and the direction of the reflected radiation; this property results in several different definitions of reflectivity in common use. An integrated average over the hemisphere associated with the reflected radiation can be used known as the spectral hemispherical reflectivity, if dealing with radiation at a specific wavelength or total hemispherical reflectivity, representing an integrated average over all wavelengths. Furthermore surfaces can be modeled as diffuse or specular. Diffuse reflection is said to have occurred if, regardless of the direction of the incident radiation, the intensity of the reflected radiation is independent of the reflection angle. If the reflection angle of the radiation is equal to the incident angle of the radiation, it is defined as specular radiation. In reality, no material is perfectly specular or diffuse, but specular surfaces can be approximated by mirrorlike or polished surfaces and diffuse surfaces by rough surfaces. In the case of geophysical applications, a water surface, if still, can take on the characteristics of a specular surface, but if the surface is roughened by waves and wind shear it takes on the characteristics of a diffuse surface. In geophysical applications the term *albedo* is used, rather than reflectivity, to define the proportion of radiation from the Sun diffusely reflected by a surface. For snow, ice, deserts, and clouds the albedo or reflectivity is high, 0.4 to 0.95, 0.2 to 0.7, 0.35 to 0.45, and 0.3 to 0.7, respectively, and for water bodies it is generally low, below about 0.08 (Marshall and Plumb, 2008)). Clouds, because of their high reflectivity, approximately double the reflectivity or albedo of the Earth from about 0.15 to 0.3. Clouds therefore have a very significant impact on the global energy balance.

For an opaque medium, the sum of absorptivity and reflectivity, Equation 8.26, is unity,

$$\alpha_r + \rho_r = 1 \tag{8.26}$$

For current planetary conditions, the albedo can be taken as approximately $\alpha_r = 0.3$. The solar radiation absorbed by the Earth is given by

$$Q = (1-\alpha_r)\pi a^2 S_o = 1.22 \times 10^{17} \text{ W} \tag{8.27}$$

For equilibrium conditions, the energy absorbed by the Earth must equal the energy radiated to space.

Approximating the spinning Earth as a body of uniform temperature, an estimate of the emission temperature of the Earth can be made using Stefan-Boltzmann's law.

$$q = \sigma(T_e^4 - 0) \tag{8.28}$$

$$Q = qA = \sigma 4\pi a^2 T_e^4 = (1-\alpha_r)\pi a^2 S_o = 1.22 \times 10^{17} \text{ W} \qquad (8.29)$$

$$T_e = \left[\frac{S_o(1-\alpha_r)}{4\sigma}\right]^{0.25} = 255 \text{ K} \qquad (8.30)$$

The emission temperature represents the temperature one would infer by representing the Earth as a radiative blackbody at a uniform temperature. The emission temperature calculated, 255 K, is about 33 K below the global average surface temperature of 288 K. The principal reasons for the low emission temperature are that radiation is absorbed within the atmosphere mainly by water vapour and because convective air currents carry heat both vertically and horizontally in the atmosphere.

Because the atmosphere comprises a variety of gases, some of which are opaque to solar radiation at certain wavelengths, it is important to consider the spectral characteristics. The atmosphere is transparent in the visible spectrum (0.38 μm to 0.75 μm), corresponding to a peak in the solar spectrum (4.83 μm). The atmosphere is almost totally opaque in the ultraviolet spectrum (400 nm to 10 nm); that is, these wavelengths of radiation do not pass through the atmosphere. The atmosphere has highly variable opacity across the infrared spectrum. At some wavelengths it is almost transparent, whereas at others it is nearly opaque. The dominant gaseous constituents in the atmosphere, nitrogen and oxygen, are transparent for the majority of the spectrum. The absorption of terrestrial radiation is dominated by the triatomic molecules. In the ultraviolet spectrum O_3 is dominant, and in the infrared, H_2O, CO_2 along with various others are dominant in controlling the absorption of radiation. Triatomic molecules are significant because they have rotational and vibrational modes that can readily be excited by infrared radiation. As the triatomic molecules have a significant effect on the transmission of radiation in the atmosphere and as they are present in relatively small quantities, the atmosphere's radiative balance is sensitive to changes in their proportions.

As described, radiation emanates from the Earth to space principally from the upper troposphere, rather than the Earth's surface. Much of the heat radiation from the Earth's surface is absorbed within the atmosphere. The Earth's surface is therefore heated by both direct solar radiation and radiation from the atmosphere. Equilibrium conditions are established in the atmosphere by a combination of radiative heat transfer and convective heat transfer processes, as illustrated in Figure 8.15. Convection can readily occur under these conditions because the ground surface is warm relative to the upper atmosphere and the associated unstable conditions allow for the formation of recirculating flows.

Convection relates to the mode of heat transfer that occurs in a fluid owing to the motion of the fluid itself. This motion can be induced by external means such as a pump or a moving surface, in which case it is referred to as forced convection. Alternatively, the motion can be induced by temperature differences

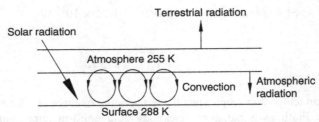

FIGURE 8.15   Equilibrium temperature established by radiative and convective heat transfer.

that give rise to density variations and instabilities in the flow with resultant bulk fluid motion; this mode is known as free or natural convection.

8.2. ATMOSPHERIC CIRCULATION

The equatorial surfaces of the Earth are nearly perpendicular to an Earth–Sun line. Each square meter at the equator intercepts the solar radiation available at the Earth's orbital distance from the Sun. This is approximately 1.367 kW/m². Away from the equator with increasing latitude, surfaces are progressively inclined, and each square meter of surface intercepts only a proportion of the available energy (proportional to $\cos\phi$). At the poles the intercepted energy approaches zero. The lower levels of energy intercepted at the poles result in lower temperatures here. The difference in temperature between the poles and the equator causes net energy flows away from the equator toward the poles. Simplistically, this temperature difference could produce a convective cell, with heat rising radially at the equator, moving away from the equator toward the poles, losing heat by radiation to space, and then sinking as it cools at the poles with the air moving back toward the equator at a lower radius, as illustrated in Figure 8.16.

The motion in Figure 8.16, however, ignores three highly significant factors present in governing atmospheric motion:

- flow instability
- mass conservation
- Coriolis phenomena

The distance between the equator to either pole is of the order of 10,000 km, which in comparison to the height of the troposphere (between 7 km and 11 km), is of the order of 1000 times larger. It is therefore highly unlikely that an upper heated layer of air could move from the equator to the poles without interacting in some way with the adjacent cooler layer moving toward the equator at a lower radius. If conservation of matter in a spherical control volume is considered, then for the simplistic convective cell of Figure 8.16 to function, air from the equators would need to be confined to an ever smaller volume as it nears the poles with either vastly increased speeds or increases in density. The tangential velocity at the equator is 464 m/s, while at the poles it is zero. As air

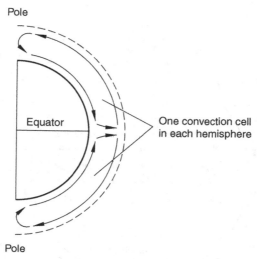

FIGURE 8.16 Simplistic atmospheric convective cells. (The vertical scale of the atmosphere is greatly exaggerated.)

moves from north or south from the equator, it moves to regions where the Earth's surface has smaller and smaller tangential velocities and Coriolis effects need to be accounted for accordingly.

Convection in long aspect ratio systems as illustrated in Figure 6.18 tends to result in a series of convective cells. The more complex flow pattern is a result of flow instability. Prediction of convective cells and their interaction is complex. In the Earth's atmosphere this is even more so, and instead observation has largely been responsible for developments in our understanding. These observations have revealed the existence of three principal circulations, as illustrated in Figure 8.17: Hadley, Ferrel, and polar cells.

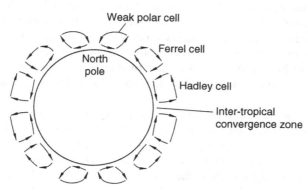

FIGURE 8.17 Meridional atmospheric flow circulations. (The vertical scale of the atmosphere is greatly exaggerated.)

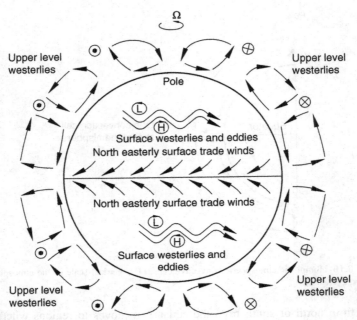

FIGURE 8.18 Bulk latitudinal flows.

Strong meridional (north–south) circulations called Hadley cells exist between the thermal equator (located near the geographic equator) and the latitudes of about 30° north and south of the equator. Hadley cells are nearly closed cells with little exchange of air with neighboring cells. Much weaker counter-rotating cells, called Ferrel cells, exist between latitudes of 30° and 60°. The flow in this band of latitude is far less stable. Much of the flow of energy toward the poles is actually a result of flow eddies of varying size and direction. The arctic and Antarctic cells form polar fronts extending from the poles to latitudes of about 60°, although the extent varies widely in winter. The views shown in Figure 8.17 represent solely a cross section of the bulk meridional flows in the troposphere. In addition, the atmosphere is subject to Coriolis forces and instabilities, governed by the normal rules of conservation, momentum, and energy, which are responsible for bulk latitudinal flows as illustrated in Figure 8.18.

The movement of air in the atmosphere may typically be vertical, rising or falling, or horizontal. Flows in the atmosphere tend to be stratified because of density and humidity differences. This suppresses the magnitude of vertical velocities, and flows tend to be mainly horizontal. Horizontal movements of atmospheric air on the Earth's surface are of course commonly known as wind. The convention for winds, as developed by early mariners, is to name a wind by the direction from where it comes. A westerly wind, for example, blows from

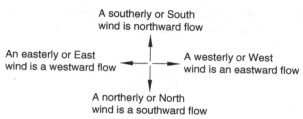

FIGURE 8.19   Nomenclature for describing the direction of wind flow.

the west toward the east. In oceanography, however, the convention is to name a flow of water by the direction to which it is going. An eastward flow of water, for example, flows to the east. Southerly, westerly, northerly and easterly winds are illustrated schematically in Figure 8.19.

A model for the flow dynamics in the atmosphere can be developed by consideration of the various phenomena involved. At the equator, air in contact with the Earth's surface will, due to the no-slip condition, acquire the Earth's tangential velocity. In addition, the air acquires thermal energy as a result of both an increase in temperature and latent heat of evaporation. As a result of its high thermal energy, it will begin to rise, and as it does so moisture in the air will condense, releasing its latent heat to the air, enabling the air to reach altitudes of about 12 km, and producing towering cumulonimbus clouds, frequent afternoon thunderstorms, and low pressure characteristic of the equatorial climate. This upper atmospheric air flows away from the equator, losing heat by radiation to space, to latitudes of about 30° north and 30° south. It then sinks back to the Earth's surface and flows back toward the equator. There is little exchange of air between the northern and southern hemispheres. As the upper layer of air flows away from the equator, it retains the equatorial tangential velocity. As a result, it is moving eastward faster than the local surface, forming an upper atmospheric westerly. As air sinks to the surface it is then subjected to increased shear stresses at the surface and slowed to approximately the same tangential velocity of the Earth at that latitude. As the air moves back toward the equator, it is moving toward regions of the Earth's surface that are moving faster. The air motion is east to west, relative to the Earth, and the winds formed are called easterlies. These easterlies are the tropical trade winds. The convergence of the trade winds from both the northern and southern hemispheres in the equatorial region forms the inter-tropical convergence zone (ITCZ). The trade winds pick up moisture and increased energy as a result of latent heat, as they cross the warm tropical oceans. The convergence of the trade winds at the equator completes the cycle, which then repeats with the energetic warm air rising again.

In meteorology it is customary to use pressure rather than height as the primary vertical coordinate because it is easier to measure pressure than height and because pressure is indicative of the overlying mass of fluid. If p is the vertical coordinate, then height becomes the dependent variable, $z(p)$, and represents the height of a pressure surface. From the hydrostatic balance and ideal gas equations,

$$\frac{\partial z}{\partial p} = -\frac{RT}{gp} \tag{8.31}$$

Integration of Equation 8.31 gives the geopotential height, $z(p)$, of a pressure surface, assuming a constant value for g, as

$$z(p) = \frac{R}{g} \int_{p}^{p_s} T \frac{dp}{p} \tag{8.32}$$

where p_s is the pressure of the pressure surface.

As noted previously, strong westerly winds blow in the upper troposphere and lower stratosphere of the midlatitudes, with the intensity of the wind increasing vertically. These winds result from the thermal gradient between the equator and poles and the rotation of the Earth. The term *thermal wind* is used to describe a wind that flows parallel to the temperature gradient in the troposphere, and a relationship can be developed defining why horizontal gradients of temperature must be accompanied by vertical gradients of flow. This relationship is known as the thermal wind equation.

From the hydrostatic balance and ideal gas equations,

$$\frac{\partial z}{\partial p} = -\frac{RT}{pg} \tag{8.33}$$

Differentiating with respect to y gives the temperature gradient $\partial T/\partial y$,

$$\frac{\partial^2 z}{\partial p \partial y} = -\frac{R}{pg} \frac{\partial T}{\partial y} \tag{8.34}$$

The velocity component u can be defined in terms of the horizontal gradient of geopotential height, Equation 8.32, by rearranging the geostrophic relationship in the form

$$\frac{\partial z}{\partial y} = -\frac{fu}{g} \tag{8.35}$$

Differentiating with respect to pressure gives

$$\frac{\partial^2 z}{\partial y \partial p} = -\frac{f}{g} \frac{\partial u}{\partial p} \tag{8.36}$$

From Equations 8.34 and 8.36,

$$\frac{\partial u}{\partial p} = \frac{R}{pf} \frac{\partial T}{\partial y} \tag{8.37}$$

Similarly

$$\frac{\partial v}{\partial p} = -\frac{R}{pf}\frac{\partial T}{\partial x} \tag{8.38}$$

The corresponding equations using height are

$$\frac{\partial u}{\partial z} = -\frac{g}{fT}\frac{\partial T}{\partial y} \tag{8.39}$$

and

$$\frac{\partial v}{\partial z} = \frac{g}{fT}\frac{\partial T}{\partial x} \tag{8.40}$$

The Hadley cell model provides a reasonable approximation for the flow observed in tropical regions, and the thermal wind associated with the equator to pole temperature gradient provides an approximate indication of the westerly winds. The combination of these models does not predict sufficient heat transfer toward the poles in comparison to that observed. Observations of our weather show that the midlatitudes are full of large-scale eddies, which arise from baroclinic instabilities; these generate meridional motion and significantly affect the meridional transport of heat. Relatively simple experiments can be set up to illustrated baroclinic instability using a rotating turntable and tanks (see Hide, 1958, 1964, 1966, 1967; Tucker, 1993; Marshall and Plumb, 2008). Figures 8.20 and 8.21 show the results from an experiment set up by Marshall

FIGURE 8.20   Characteristic baroclinic instabilities associated with thermal gradients and rotation for a rotating tank (Marshall and Plumb, 2008).

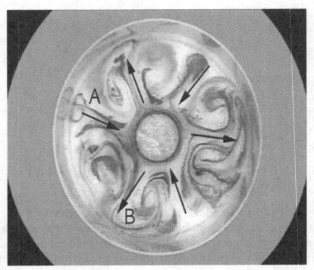

FIGURE 8.21 Characteristic baroclinic instabilities associated with thermal gradients and rotation for a rotating tank, top view (Marshall and Plumb, 2008).

and Plumb. Here a 700-mm-diameter tank has been filled with water to a depth of about 150 mm. A radial thermal gradient has been produced by placing a 150-mm ice bucket at the axis. The tank has then been rotated at 10 rpm to produce the characteristic baroclinic instabilities associated with a thermal gradient and rotation. The flow structure can be revealed by introducing ink traces.

The eddying motion illustrated in Figures 8.21 and 8.22 is characteristic of the large-scale eddying motion found in the midlatitudes. Vapor energy

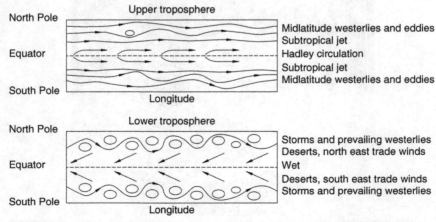

FIGURE 8.22 Characteristic flow in the lower and upper troposphere (after Marshall and Plumb, 2008).

exchange in the atmosphere can lead to localized regions of high and low pressure known as anticyclones and cyclones, respectively. These disrupt the simplistic flow model indicated in Figures 8.17 and 8.18, leading to the much more complex characteristic flow patterns with which we are familiar (see Figures 1.10 and 1.11).

The flow in the lower and upper troposphere of the atmosphere can be crudely modeled by a combination of Hadley cells, thermal wind, and baroclinic instabilities. leading to a meridional transfer of energy as illustrated in Figure 8.22.

8.2.1 Rossby Waves and Jet Streams

A belt of strong westerly winds are found in the upper troposphere. These follow a meandering path to form a complete loop around the Earth (see Figure 8.23). Evidence of strong winds in the upper troposphere first arose when World War 1 Zeppelins were blown off course and also when several interwar balloons observed to be traveling at speeds in excess of 200 km/hour. In World War II, aircraft flights above 8 km found the eastward flights much faster and their return flights much slower than expected, while north to south flights exhibited a tendency to be blown off course. The velocity of these winds is not uniform, and narrow bands of extremely fast moving air known as jet streams are found. The air flow in a jet stream can exceed speeds of 230 km/hour. The loop of winds illustrated in Figure 8.24 is known as a Rossby wave or Rossby waves. The number of waves forming the loop varies seasonally: There are usually four to six waves in summer and three in winter.

There are several recognizable jet streams including the polar front, the subtropical, the stratospheric night and equatorial easterly. The *polar front jet stream*, which exists in both hemispheres between latitudes of 40° and 60°,

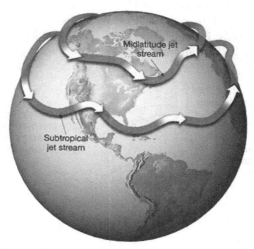

FIGURE 8.23   Jet Streams. Image courtesy of NASA.

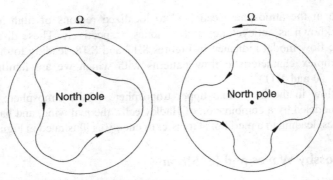

FIGURE 8.24 Rossby waves: (a) Winter, (b) summer.

forms the division between the Ferrel and Polar cells. Polar front jet streams vary in extent, location, and intensity and are mainly responsible for giving fine or wet weather on the Earth's surface. For the case of the northern hemisphere, if the jet stream moves south, it brings with it cold air, which descends in a clockwise direction to give dry, stable conditions associated with areas of high pressure (anticyclones). As a result of moving south, the jet stream will warm up, and this in turn provides the impetus for it to return northward. As it moves northward, it takes with it warm air, which rises in an anticlockwise direction to give the strong winds and heavy rainfall associated with areas of low pressure (depressions). The usual path of the polar front jet stream over Britain, for example is oblique, toward the northeast. This accounts for its frequent wet and windy weather. Occasionally, this path may be altered or blocked by a stationary anticyclone known as a blocking anticyclone (see Figure 8.25), producing extremes of climate such as the hot summers in Britain of 1976 and 2003, 2006 and the relatively cold January of 1987. The subtropical jet stream occurs in latitudes about 25° to 30° from the equator and forms the boundary between the Hadley and Ferrel cells. In comparison to the polar front jet streams, the winds in the subtropical jet stream are lower, and it meanders less but it follows

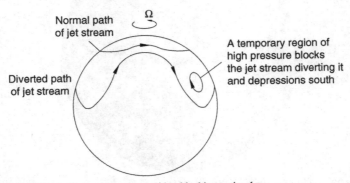

FIGURE 8.25 The polar front jet stream with a blocking anticyclone.

a similar west–east path. The easterly equatorial jet stream is seasonal, associated with the summer monsoon in the Indian subcontinent.

Use of the Rayleigh instability criteria shows that it is possible for an element of fluid to overshoot its undisturbed position, and unless it is heavily damped it will initiate an oscillatory wave motion called an inertial wave. Waves in fluid mechanics tend to be associated with the existence of a free surface (e.g., ship waves) or with the compressibility of the medium (e.g.. shock waves in supersonic flows). In general, waves occur in a physical system if the disturbed system is brought back into equilibrium by a restoring force. For example, for the case of a liquid with a free surface, surface waves are generated when gravity acts to bring fluid particles that are not in equilibrium back to their original bulk positions. Neighboring particles are affected by the resulting oscillation, and the disturbance spreads in a wave. In a rotating system, the Coriolis force functions as gravity does for surface waves, providing the restoring impetus.

Inertial waves can be produced in the laboratory using the apparatus illustrated in Figure 8.26. In this experiment, demonstrated by Oser (1958), water is contained inside a cylindrical vessel that can be rotated at a constant angular velocity. A small disc, suspended centrally in the vessel, is oscillated vertically at a frequency ω. Visualization of the flow structure can be achieved by means of aluminium flakes or discs suspended in the water with a light or laser sheet.

The vertical oscillations of the centrally mounted disc produce radial disturbances as the water flows around the edge of the disc. If the water was not rotating, inertial waves would not occur, but because the vessel and the fluid are rotating, then as identified by the Rayleigh instability criterion, inertial waves occur, with parcels of fluid overshooting their undisturbed positions and following an oscillatory motion. The waves move radially and axially, producing conical shear surfaces. A disc frequency of $\omega = 0$ produces a Taylor-Proudman column. As the disc frequency rises to $\omega = 2\Omega$, the vertex angles of the cones

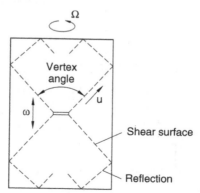

FIGURE 8.26 Inertial waves produced in a fluid contained in a rotating cylindrical vessel with an oscillating disc.

rise to $180°$. In this application, the local temporal acceleration is balanced by the Coriolis acceleration.

Geostrophic flow is only possible in spaces or containers in which a column of fluid, parallel to the axis of rotation, can maintain a constant height. While geostrophic flows dominate large-scale flow motions in the atmosphere and oceans, in some cases the geostrophic mode is not available owing to the lack of constant height, closed contours. A demonstration of this can be made following Pedlosky and Greenspan (1967), using a circular cylindrical vessel whose base is tilted relative to the horizontal at an angle of α. The cylinder and enclosed liquid are initially rotated about the central axis of the cylinder at a constant angular velocity. The velocity is then impulsively changed. No constant height geostrophic flow is possible because of the container's base, and an infinite number of low-frequency oscillations are generated that result in the formation of the interior flow field illustrated in Figure 8.27. The flow structure is comparable to that of a Rossby wave. When both the cylinder and the liquid are rotating at the same angular velocity, vorticity is uniform throughout the liquid. After a change in the angular velocity of the cylinder, the liquid rotates relative to the tilted base. As a vertical vortex tube travels from the shallower end to the deeper end, it is stretched and intensifies; on completing its circuit back to the shallower end, it is shortened, returning to its original vorticity. This process generates a transverse vorticity gradient. The volume of fluid in the shallow end is less than that in the deep end. As the liquid rotates relative to the container, in order to satisfy conservation of mass there is a continuous perturbation of the

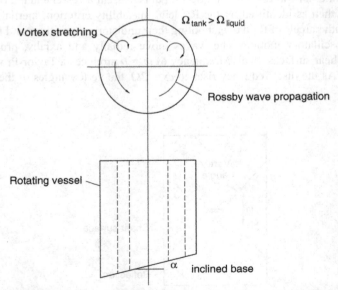

FIGURE 8.27 Inertial waves comparable to Rossby waves produced in a fluid contained in a rotating cylindrical vessel with an inclined base.

liquid velocity relative to the cylinder. These transverse perturbations relative to the transverse vorticity gradients generate the inertial waves illustrated.

Rossby (1939, 1940, 1945) developed a model for the movement of large-scale circulations in the middle troposphere based on the latitudinal variation of the vertical component of the Earth's vorticity (see also Dickenson, 1978). The derivation developed takes the form of consideration of a linear perturbation. Small-amplitude disturbances about a state of uniform latitudinal, west–east motion, are considered and analyzed. The atmosphere is assumed to be represented by a stably stratified fluid. The Coriolis parameter is considered to be constant except where differentiated with respect to latitude, when its derivative represents the gradient of the Earth's vorticity.

The vertical component of vorticity at ϕ is

$$\omega_z = f = 2\Omega\sin\phi \tag{8.41}$$

The vertical component of total vorticity is $f + \omega_z$

For an arbitrary displacement of a parcel of air with no externally applied torque, the sum of $f + \omega_z$ must remain constant. That is,

$$\frac{d(f + \omega_z)}{dt} = 0 \tag{8.42}$$

Rearranging and on substitution,

$$\frac{d\omega_z}{dt} = -\frac{df}{dt} = -\frac{dy}{dt}\frac{\partial f}{\partial y} = -u_y\frac{\partial f}{\partial y} = -\beta_w u_y \tag{8.43}$$

v = northward velocity, y = northward direction

$\beta_\omega = \partial f/\partial y$ is the rate of change of the Earth's atmospheric vorticity relative to latitude and is given by

$$\beta_\omega = \frac{2\Omega\cos\varphi}{a_{Earth}} \tag{8.44}$$

The Earth's westerly winds result from the equator–pole thermal gradient modified by the Coriolis force. At latitudes between about 30° and 60° north and south, the surface trade winds blow from the west to the east, albeit with temporary dislocations, and the patterns shift by approximately 10° according to the season.

The flow velocity components can be represented by

$$u = U_o + u' \tag{8.45}$$

and

$$v = v' \tag{8.46}$$

where:
u = velocity to the east (m/s)
v = velocity to the north (m/s)
U_o = trade wind (m/s)
u' and v' are perturbations (m/s)

$$\omega = \frac{\partial v}{\partial x} - \frac{\partial u}{\partial y} = \frac{\partial v'}{\partial x} - \frac{\partial u'}{\partial y} \tag{8.47}$$

where ω is the relative vorticity.

$$\frac{\partial \omega}{\partial t} + U_o \frac{\partial \omega}{\partial x} + u' \frac{\partial \omega}{\partial x} + v' \frac{\partial \omega}{\partial y} = -\beta_\omega v' \tag{8.48}$$

Neglecting products of small quantities,

$$\frac{\partial \omega}{\partial t} + U_o \frac{\partial \omega}{\partial x} = -\beta_\omega v' \tag{8.49}$$

If a wave exists, it can be represented as a function of $(x \pm c_{wave}t)$. Representing vorticity as $\omega(x \pm c_{wave}t)$ where c_{wave} is the eastward velocity of the wave,

$$\frac{\partial \omega}{\partial t} = -c_{wave} \frac{\partial \omega}{\partial x} \tag{8.50}$$

From Equations 8.49 and 8.50,

$$(U_o - c_{wave}) \frac{\partial \omega}{\partial x} = -\beta_\omega v' \tag{8.51}$$

Assuming the perturbation is only a small function of x, then from Equation 8.47

$$\omega = \frac{\partial v'}{\partial x} \tag{8.52}$$

Substitution gives

$$v' = \sin \frac{2\pi}{\lambda} (x - c_{wave}t) \tag{8.53}$$

where λ is the wavelength.

$$(U_o - c_{wave}) \frac{4\pi^2}{\lambda^2} \sin \frac{2\pi}{\lambda} (x - c_{wave}t) = \beta_\omega \sin \frac{2\pi}{\lambda} (x - c_{wave}t) \tag{8.54}$$

This can be stated in the form

$$c_{wave} = U_o - \frac{\beta_\omega \lambda^2}{4\pi^2} \tag{8.55}$$

where c_{wave} is the wave speed, U_o is the easterly wind, λ is the wavelength, and β_ω is given by Equation 8.44.

The wave speed is zero when

$$U = \frac{\beta_\omega \lambda^2}{4\pi^2} \qquad (8.56)$$

$$\lambda_{stationary} = 2\pi \sqrt{\frac{U_o}{\beta_\omega}} \qquad (8.57)$$

Alternatively, Equation 8.55 can be stated as

$$c_{wave} = U_o \left(1 - \frac{\lambda^2}{\lambda_{stationary}^2} \right) \qquad (8.58)$$

When $\lambda = \lambda_{stationary}$, the wave moves westward at a velocity U_o and is stationary over the Earth. If $\lambda > \lambda_{stationary}$, c_{wave} is negative and the wave has a net westward motion relative to the Earth, giving a classic Rossby wave with a very large wavelength traveling up-current, that is, westward.

A similar analysis can be applied to a rotating cylinder with a tilted base.

8.3. PRESSURE SYSTEMS AND FRONTS

In regions where one air mass is dominant, there is little seasonal variation in the weather, for example, as in the tropics and the poles. For regions such as Ireland and Britain where air masses constantly interchange, there is much greater seasonal and daily variation in weather. It is useful to be able to characterize air masses by means of pressure systems and fronts, and these are introduced in Sections 8.3.1 and 8.3.2. An anticyclone is a large pressure system and is considered in Section 8.3.3. A number of synoptic charts and satellite images are presented in Section 8.3.4 illustrating some typical characteristics of flow in the atmosphere.

8.3.1 Pressure Systems

A wind will flow in the atmosphere if there is a difference in pressure between two regions. The difference in pressure may be as a result of differences in temperature, gravity, vorticity, and Coriolis effects. An increase in temperature, for example, can cause a region of air to expand and become less dense and rise, generating an area of low pressure below. Conversely, a reduction in temperature produces an area of high pressure.

When both the Rossby and the Ekman numbers are small, this indicates that the viscous forces are negligible; this condition is referred to as the geostrophic approximation (see Section 2.5 and Equation 2.163). The Earth's rotation causes the Coriolis force to dominate over both inertial and viscous forces away from the Earth's surface, and a calculation of the Rossby and Ekman numbers gives from Equation 8.3, $Ro \approx 0.02$, and from Equation 2.161,

$$Ek = \frac{\nu}{\Omega L^2} = \frac{1.465 \times 10^{-5}}{7.27 \times 10^{-5} \times \left(6371 \times 10^3\right)^2} = 4.96 \times 10^{-15} \qquad (8.59)$$

Both the Rossby and Ekman numbers are small, relative to unity, and the geostrophic approximation, Equation 2.163, can be assumed to be valid. Substituting for the Coriolis terms in Equation 2.163 using the Coriolis parameter gives

$$f z \times w = -\frac{1}{\rho}\nabla p \qquad (8.60)$$

In component form using a Cartesian coordinate system, the eastward and northward velocity components are given, respectively, by Equations 8.61 and 8.62.

$$u = \frac{1}{fp}\frac{\partial p}{\partial x} \qquad (8.61)$$

$$v = \frac{1}{fp}\frac{\partial p}{\partial y} \qquad (8.62)$$

In a geostrophic flow, the pressure gradient is perpendicular to the flow direction. From knowledge of the pressure distribution, lines of constant pressure known as isobars can be constructed, and the isobars can be taken to be the lines along which the wind is blowing. This is illustrated in Figure 8.28. The existence of geostrophic winds was originally recognized in 1832 by Buys Ballot whose law states that if you stand in the northern hemisphere with your back to the wind, low pressure is always on your left and high pressure on your right. The approximation is suitable for winds in the upper troposphere where the effects of friction from the Earth's surface are small. This generalization

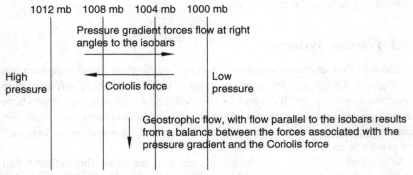

FIGURE 8.28 Geostrophic flow; the forces due to the pressure gradient and Coriolis force are balanced, and flow is parallel to the isobars.

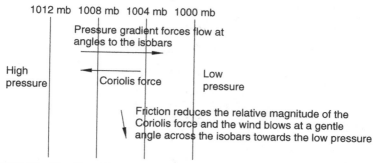

FIGURE 8.29 The effect of increased friction on geostrophic flow; the forces due to the pressure gradient and Coriolis force are no longer balanced, and flow blows at a gentle angle across the isobars toward the low pressure.

needs to be modified in the case of atmospheric flows where other factors such as the Coriolis force are acting. Friction between the Earth's surface and the flow reduces the effect of the Coriolis force. As the relative importance of the pressure gradient increases, the wind tends to blow across isobars toward the low pressure as indicated in Figure 8.29. The imbalance between the forces due to the pressure gradient and the Coriolis force increases near to the ground as friction increases and the effect of the Coriolis force diminishes. The effects of friction are smaller over water as the surface is smoother than that of land.

In geostrophic flow the vertical component, w, is zero. From Equations 8.61 and 8.62,

$$\frac{\partial u}{\partial x} + \frac{\partial v}{\partial y} = 0 \qquad (8.63)$$

Equation 8.63 implies that geostrophic flow is nondivergent. Comparison with the three-dimensional form of the steady-state continuity equation implies that $\partial w/\partial z = 0$. If $w = 0$ on, for example, a flat surface boundary then $w = 0$ everywhere, and the flow as defined by geostrophic motion will be horizontal everywhere.

The hecto Pascal is the official unit of atmospheric pressure: 1 hPa = 100 Pa. The hecto Pascal is more commonly referred to as a millibar, however. 1 mbar = 100 Pa = 1 hPa. Differences in atmospheric pressure are commonly represented on a map by lines of constant pressure called isobars. The usual practice is to present the pressure in millibars (mb).

$$1000 \text{ mb} = 1 \text{ bar} = 100000 \text{ Pa} \qquad (8.64)$$

$$1 \text{ mb} = 100 \text{ Pa} \qquad (8.65)$$

Isobars are typically drawn at 4-mb intervals. In general, the closer the isobars are on a map, the greater the pressure gradient and the stronger the wind. There

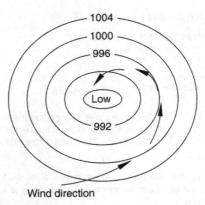

FIGURE 8.30 A low-pressure system.

are two basic pressure systems: a low pressure system and a high pressure system (see Figures 8.30 and 8.31). In a low-pressure system, winds tend to blow toward the center, with increasing strength as they near the center, but at an angle to the isobars and in an anticlockwise direction due to the effects of Coriolis forces and friction (in the case of the northern hemisphere). The winds in a low-pressure system tend to be strong because of the steep pressure gradient as indicated by the close proximity of the isobars. In a high-pressure system, the pressure gradient is less severe and winds tend to be more gentle, with descending air blown outward in a clockwise direction.

Equations 8.61 and 8.62 express pressure gradients at constant height. In order to apply the geostrophic equations to atmospheric observations, it is convenient to express them in terms of height gradients on a pressure surface. Figure 8.32 shows a surface of constant height, z_0, and another at constant pressure, p_0. The two surfaces intersect at point A, where the pressure will be p_0 and the height $z_A = z_0$. At constant height, the pressure gradient in the x direction is

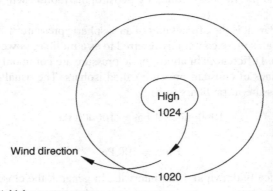

FIGURE 8.31 A high-pressure system.

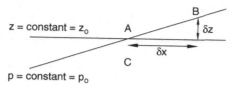

FIGURE 8.32 Constant height and constant pressure surfaces.

$$\left(\frac{\partial p}{\partial x}\right)_{z=constant} = \frac{p_C - p_o}{\delta x} \tag{8.66}$$

The gradient of height along the constant pressure surface is given by

$$\left(\frac{\partial z}{\partial x}\right)_{p=constant} = \frac{z_B - z_o}{\delta x} \tag{8.67}$$

As $z_A = z_0$ and $p_B = p_0$ in conjunction with the hydrostatic equation gives

$$\frac{p_C - p_o}{z_B - z_o} = \frac{p_C - p_o}{z_B - z_C} = -\frac{\partial p}{\partial z} = \rho g \tag{8.68}$$

Hence, from Equations 8.66 and 8.67,

$$\left(\frac{\partial p}{\partial x}\right)_{z=constant} = \rho g \left(\frac{\partial z}{\partial x}\right)_{p=constant} \tag{8.69}$$

$$\left(\frac{\partial p}{\partial y}\right)_{z=constant} = \rho g \left(\frac{\partial z}{\partial y}\right)_{p=constant} \tag{8.70}$$

The geostrophic flow equation using pressure coordinates becomes

$$\boldsymbol{u} = \frac{g}{f} \boldsymbol{z_p} \times \nabla_p z \tag{8.71}$$

where z_p is the upward unit vector in pressure coordinates and ∇_p is the gradient operator in pressure coordinates.

In component form using a Cartesian pressure coordinate system, the eastward and northward velocity components are given, respectively, by Equations 8.72 and 8.73.

$$u = -\frac{g}{f}\frac{\partial z}{\partial y} \tag{8.72}$$

$$v = \frac{g}{f}\frac{\partial z}{\partial x} \tag{8.73}$$

Density does not appear explicitly in Equations 8.72 and 8.73, and it is therefore possible to ignore its variation in evaluating a flow field from observations.

Example

If the 500-mbar pressure surface slopes down by a height of 800 m over a distance of 5000 km, at a latitude of 45°, calculate the wind speed.

Solution

From Equation 8.72,

$$u = \left| \frac{g}{f} \frac{\partial z}{\partial y} \right| = \frac{9.81}{1.03 \times 10^{-4}} \times \frac{800}{5000 \times 10^{3}} \approx 15 \; m/s$$

8.3.2 Fronts

If air remains stationary over an area for several days as it comes into thermal equilibrium it will tend to take on the temperature and humidity properties of that area. Stationary air in the Earth's atmosphere is mainly found in the high-pressure belts of the subtropics, such as the Azores and the Sahara, and the high latitudes, such as Siberia and northern Canada. The areas in which homogenous air masses develop in this context are called *source regions*. Source regions are usually classified according to the latitude in which they originate, which determines their temperature, and the nature of the surface over which they develop, which determines their moisture content.

In the case of the British Isles, the principal source regions that affect the climate are Arctic, Polar, and Tropical. These are given the designations Am, Pc, Pm, Tm, and Tc depending on whether the source region has developed over land or sea, with the postscripts m and c representing maritime and continental respectively.

When an air mass moves away from its source region, its temperature, humidity, and stability will alter according to the surface over which it passes. In the case of tropical air moving northward, it will cool down and become more stable because its density increases relative to the density of the air above it. Polar air moving southward warms and becomes increasingly less stable. When two air masses meet in an atmospheric flow, they do not readily mix due to differences in temperature and density. The region over which air masses meet is called a front. The frontal theory was originally developed by Vilhelm Bjerknes and his colleagues at the Bergen School in Norway around 1920 who recognized that air masses did not readily mix and called the boundary between two air masses a front, following an analogy with the division between opposing armies in World War I. Although the model did not originally include all the

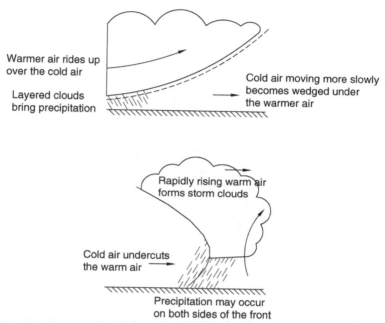

FIGURE 8.33   Schematic showing the formation of warm and cold fronts.

physical phenomena involved, it is nevertheless very useful for model atmospheric motions and forms the basis for meteorology.

A front can form, for example, if two bodies of air, but at different temperatures, are moving in the same direction. A front is formed as warm air rides up over colder air ahead of it, or if cold air undercuts and lifts the body of warm air. The term *warm front* is used to describe a region of advancing warm air that is forced to override cold air. A cold front describes a region of advancing cold air that is forced to undercut a body of warm air. In both cases the rising air cools, usually producing heaped clouds and precipitation. Fronts can be several hundred kilometers wide. (The formation of warm and cold fronts is illustrated in Figure 8.33.) Cold fronts move faster than warm fronts. They can generate an unstable state of the atmosphere when humid cold air is pushed upward. This instability causes violent storms in cold fronts that can produce thunderstorms, hail, and tornadoes. Such storms can occur in intense lows, which are called low-pressure storms.

Cold and warm fronts tend to move at different speeds, with warm air overtaking and rising above cold air or cold air undercutting and lifting warm air. A wave in the front can develop in these situations, with a wedge of warm air intruding into cooler air as it rises; a roughly circular area of low pressure forms at the apex of the wedge, and the system divides into separate warm and cold fronts forming a cyclone, low, or depression.

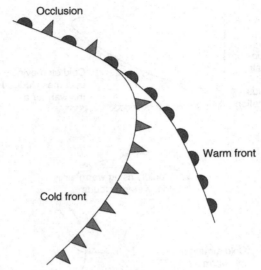

FIGURE 8.34 Front symbols.

Warm fronts are designated on weather maps by half circles. Cold fronts are shown by triangles. Usually, red semicircles are used for warm fronts and blue triangles for cold fronts. If a cold front overtakes a warm front in a depression, lifting the warm air away from the ground the front is called an occlusion or an occluded front. Alternating semicircles and triangles are used to indicate an occlusion. These symbols are illustrated in Figure 8.34.

The most notable type of front is the polar front, which is formed when moist tropical air (Tm) meets colder relatively drier polar air (Pm). A depression is an area of low pressure, and it is at a polar front that depressions form. Polar fronts form most readily over the oceans in midlatitudes. They then tend to move eastward, bringing rain and clouds to the western margins of continents as a result of the prevailing easterly flow of air.

The formation of a typical depression is illustrated using these symbols in Figures 8.35 and 8.36. In Figure 8.35 the embryo depression begins as a small wave on the front with, say, warm moist tropical air meeting colder drier polar air. It should be noted that the boundary between the air masses is across a region rather than at a sharp linear division. The convergence of the two air masses results in the warmer, less dense air rising in a spiral motion. The upward air movement causes an area of lower pressure at the Earth's surface.

In a mature depression, pressure continues to fall as more air in the warm sector is forced to rise. As pressure falls and the pressure gradient rises, radially inward winds increase in strength. Because of the Coriolis force, these anticlockwise winds come from the southwest (see Figure 8.30). As the relatively warm air of the warm sector continues to rise along the warm front, it eventually

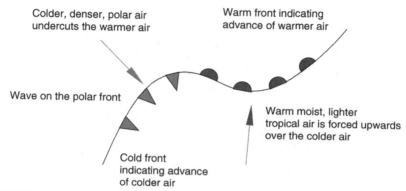

Colder, denser, polar air
undercuts the warmer air

Warm front indicating
advance of warmer air

Wave on the polar front

Warm moist, lighter
tropical air is forced upwards
over the colder air

Cold front
indicating advance
of colder air

FIGURE 8.35 Embryo depression.

cools to the dew point, and some of its vapor will condense, releasing large quantities of latent heat, and clouds develop. Continued uplift and cooling of the air mass will result in precipitation as the clouds become thicker and lower. A depression will begin to decay (Figure 8.37), when the cold front catches up with the warm front to form an occlusion or occluded front. By this stage, the warm tropical air supplying the uplift will have been exhausted, leaving no warm sector at ground level. As the uplift is reduced, so too do the amount of condensation and hence precipitation, resulting in maybe only one rainfall. Cloud cover will decrease, pressure will rise, and wind speeds lessen as colder air replaces the uplifted air and infills the depression.

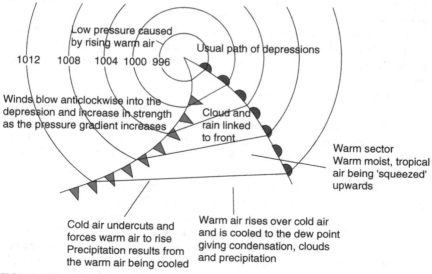

Low pressure caused
by rising warm air

Usual path of depressions

1012   1008   1004 1000 996

Winds blow anticlockwise into the
depression and increase in strength
as the pressure gradient increases

Cloud and
rain linked
to front

Warm sector
Warm moist, tropical
air being 'squeezed'
upwards

Cold air undercuts and
forces warm air to rise
Precipitation results from
the warm air being cooled

Warm air rises over cold air
and is cooled to the dew point
giving condensation, clouds
and precipitation

FIGURE 8.36 Mature depression.

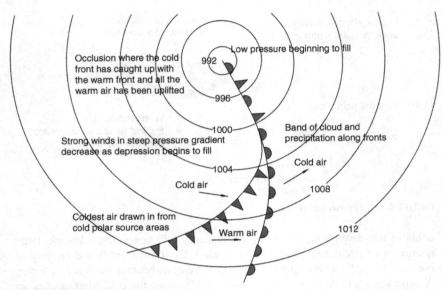

FIGURE 8.37 Decaying depression.

8.3.3 Anticyclones

An anticyclone is a large mass of subsiding air from the upper atmosphere which produces an area of high pressure on the Earth's surface. As the source of the air is from the upper atmosphere, which is cooler and hence has a lower saturation vapour pressure, the quantity of water vapour associated with it is limited. On descent the air warms at the dry adiabatic lapse rate and dry conditions result. Pressure gradients associated with anticyclones are gentle, producing low winds or calms. The wind blows gently radially outward and tends to rotate clockwise in the northern hemisphere. Anticyclones can be 3000 km in diameter and tend to be much larger than depressions. Once established, anticyclones can give several days, or under extreme conditions weeks, of settled weather.

The characteristic weather associated with an anticyclone varies with the season. In the British summer, for example, due to the relative absence of cloud, there is intense insolation that produces hot sunny days (up to 30 °C in southern England) and an absence of rain. Rapid heat loss overnight due to thermal radiation to space can lead to temperature inversions and the formation of morning dew and mist, although these conditions rapidly clear with the renewed insolation. Coastal areas may experience advection fogs as well as land and sea breezes, while highlands experience mountain and valley winds. If North Africa is the source of the air, then heat-wave conditions can result. Often after several days of increasing thermals resulting from an anticyclone, there is an increased risk of thunderstorms. In contrast, in the British winter, although sinking air will again give cloudless skies, there is relatively little insolation during the day due to the low angle of the Sun and resulting low temperatures. At night the already low temperatures combined with additional loss of thermal radiation results in even lower temperatures and the development of fog and frost. In the subsequent morning, the fog and frost may take some hours to disperse in the weak sunshine. Polar continental air with its source in Central Asia and a slow movement over the cold European land mass will be cold and stable until it reaches the North Sea. Here its lower layers will pick up heat and moisture, which can then cause heavy snowfalls on the east coasts of England and Scotland. A so-called blocking anticyclone can form when cells of high pressure detach themselves from the major high-pressure areas of the subtropics or poles. Once established, they can last several days and block eastward-moving depressions to create anomalous conditions such as extreme temperatures, rainfall, and sunshine. Examples include the British summer of 1995 and winter of 1987.

8.3.4 Weather Maps

A weather map or synoptic chart shows the weather for a particular area at a particular time. Weather maps are produced by the collection and analysis of data from weather stations. This data is plotted using internationally accepted symbols, a selection of which is presented in Table 8.3. Weather maps serve a number of purposes. A daily weather map as seen on television, phone display, or website gives a simple, clear impression of the weather. A synoptic chart shows selected meteorological characteristics for specific weather stations. An example of the use of the symbols is given in Figure 8.38. Detailed weather maps are also generated, giving details of the amounts and type of cloud, at various altitudes, dew point temperatures, and trends of pressure change.

The task in weather forecasting is to use the available data to estimate the speed and direction of the movement of the various principal air masses and any associated fronts and to predict the type of weather these will bring with them to a particular location. Considerable use is also made of satellite images and radar images. Figure 8.38 shows the synoptic chart for the British Isles and surrounding region, and the corresponding satellite images produced are shown in Figures 8.39 and 8.40. As of March 25, 2008, the British Isles lay between areas of low pressure centered over Scandinavia and high pressure across the Azores. This put the UK into a cold northwesterly flow of wind. A series of weather fronts can be seen to the west of Ireland. These would subsequently move slowly eastward, spreading less cold but also cloudier weather across the country. The corresponding synoptic chart and satellite images for the subsequent day are shown in Figures 8.41, 8.42, and 8.43, respectively.

The images shown in Figures 8.39 and 8.42 come from a satellite that remains above a fixed point on the Earth, known as a geostationary satellite. The visible images taken record visible light from the Sun reflected back to the satellite by cloud tops, land, and sea surfaces. They are equivalent to a black and white photograph from space. These visible images are able to show low cloud

TABLE 8.3 Pressure chart symbols. MetOffice

SYMBOL	NAME	DESCRIPTION
	Cold front	The leading edge of an advancing colder air mass. Its passage is usually marked by cloud and precipitation, followed by a drop in temperature and/or humidity.
	Warm front	The leading edge of an advancing warmer air mass, the passage of which commonly brings cloud and precipitation followed by increasing temperature and/or humidity.
	Occluded front (or 'occlusion')	Occlusions form when the cold front of a depression catches up with the warm front, lifting the warm air between the fronts into a narrow wedge above the surface. Occluded fronts bring cloud and precipitation.
	Developing cold/warm front (frontogenesis)	Represents a front that is forming due to increase in temperature gradient at the surface.
	Weakening cold/warm front (frontolysis)	Represents a front that is losing its identity, usually due to rising pressure. Cloud and precipitation becomes increasingly fragmented.
	Upper cold/warm front	Upper fronts represent the boundaries between air masses at levels above the surface. For instance, the passage of an upper warm front may bring warmer air at an altitude of 3000 m, without bringing a change of air mass at the surface.
	Quasi-stationary front	A stationary or slow-moving boundary between two air masses. Cloud and precipitation are usually associated.
—— 996 ——	Isobars	Contours of equal mean sea-level pressure (MSLP), measured in hectopascals (hPa). MSLP maxima (anticyclones) and minima (depressions) are marked by the letters H (High) and L (Low) on weather charts.

Thickness lines	Pressure decreases with altitude, and thickness measures the difference in height between two standard pressure levels in the atmosphere. It is proportional to the mean temperature of this layer of air and is therefore a useful way of describing the temperature of an air mass. Weather charts commonly show contour lines of 1000 hPa to 500 hPa thickness, which represent the depth (in decameters, where 1 dam = 10 m) of the layer between the 1000 hPa and 500 hPa pressure levels.
Trough	An elongated area of relatively low surface pressure. The troughs marked on weather charts may also represent an area of low thickness (thickness trough), or a perturbation in the upper troposphere (upper trough). All are associated with increasing cloud and risk of precipitation.
Convergence line	A slow-moving trough, which is parallel to the isobars and tends to be persistent over many hours or days.

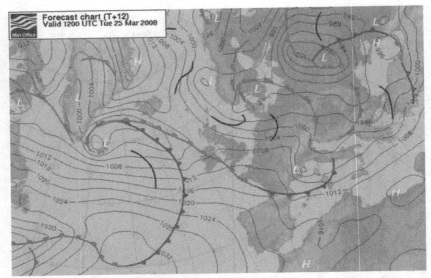

Forecast chart (T+12)
Valid 1200 UTC Tue 25 Mar 2008

FIGURE 8.38   Synoptic chart for Europe for March 25, 2008. Courtesy of the Met Office.

EVEH31 MSG 0.8 micron Visible Image 25 Mar 2008 1100 UTC

FIGURE 8.39   Visible satellite image for March 25, 2008. Courtesy of the Met Office.

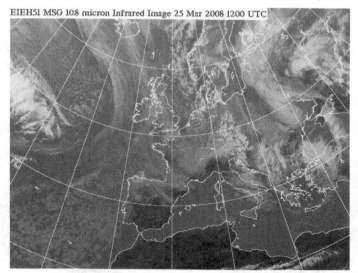

EIEH51 MSG 10.8 micron Infrared Image 25 Mar 2008 1200 UTC

FIGURE 8.40   Infrared satellite image for March 25, 2008. Courtesy of the Met Office.

better than infrared images, as low cloud is more reflective than the underlying
land or sea surface. The infrared satellite images presented in Figures 8.40 and
8.43 show the infrared radiation emitted directly by cloud tops, land, and sea
surfaces. The warmer the object, the more intensely it emits infrared radiation.
The intensity of infrared radiation has been converted to a gray scale in these

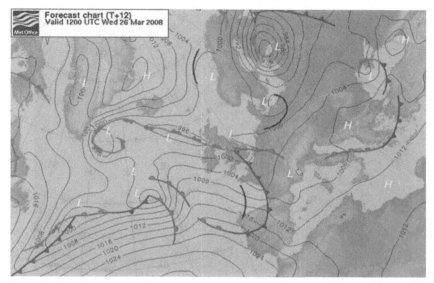

Forecast chart (T+12)
Valid 1200 UTC Wed 26 Mar 2008
Met Office

FIGURE 8.41   Synoptic chart for Europe for March 26, 2008. Courtesy of the Met Office.

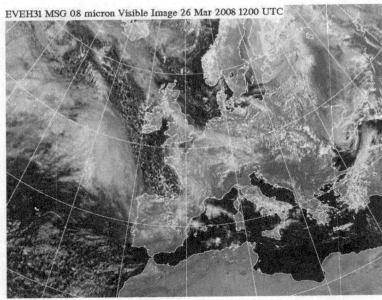

EVEH31 MSG 0.8 micron Visible Image 26 Mar 2008 1200 UTC

FIGURE 8.42   Visible satellite image for March 26, 2008. Courtesy of the Met Office.

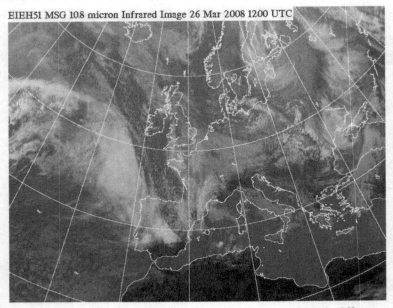

EIEH51 MSG 10.8 micron Infrared Image 26 Mar 2008 1200 UTC

FIGURE 8.43   Infrared satellite image for March 26, 2008. Courtesy of the Met Office.

images, with cooler temperatures indicated by lighter tones and higher temperatures indicated by darker tones. The light areas of cloud show where the cloud tops are cooler and therefore help identify the location of features such as weather front and shower clouds.

8.4.   INTENSE ATMOSPHERIC VORTICES

A variety of intense atmospheric vortices occur, including whirlwinds, dust devils, tornadoes, and water spouts, large-scale vortex storms, and hurricanes, also known as cyclones and typhoons. All of these rotate relative to the Earth, usually with their axis nominally vertical. Whirlwinds, dust devils, tornadoes, large-scale vortex storms, and hurricanes differ in their size, duration, how they form, how they are maintained, and how they dissipate and end.

8.4.1 Thunderstorms

A thunderstorm can form as an isolated convection cell. Alternatively, a series of storms can form into a cluster of storms that evolve into a line or lines of storms known as squall lines. The energy associated with a thunderstorm is derived from condensation of moisture in rising, unstable moist air currents. Strong vertical currents in the convecting central region of the storm extend to the stratosphere. The rising air generates a low-pressure system, drawing air in from all directions affecting the region up to hundreds of kilometers away. The final flow structure resembles an inviscid vortex with a sink flow. The classic and largest thunderstorm examples produce an upper anvil cloud. The converging air tends to produce cyclonic rotation with tangential velocities approximating an inviscid vortex. The vertical winds produce toroidal vortices that are concentrated at the forward-moving edge of the anvil cloud. The horizontal toroidal vortices sometimes interact with the major vertical vortex of the storm to produce what is known as a rotor cloud or squall cloud.

8.4.2 Tornadoes and Waterspouts

A tornado is a columnar vortex, formed in the atmosphere with a diameter of up to 500 m and a height of about 3 km. The highest wind velocities near the Earth's surface occur in tornadoes and can be 200 or 300 m/s. The duration of a tornado tends to be of the order of minutes but can be hours such as one on March 8, 1925 in the United States that lasted 3 hours. Tornadoes occur worldwide, and names for them also include *trombe* in France, *Trombe* or *Windhose* in Germany, and *Smerch* in Russia.

The vorticity required for tornadoes to form is normally derived from cyclones. Vorticity in the mesorange and synoptic range is mostly cyclonic. As a result, tornadoes tend to rotate counterclockwise, although anticyclonic tornadoes also occur with clockwise rotation. In the United States, tornadoes tend to form in cold fronts and squall lines near the jet stream, particularly in spring and in afternoon hours. Tornadoes are also observed in hurricanes, usually north of the center. Tornadoes have also been observed to form during the eruption of a volcano (Thorarinsson and Vonnegut, 1964). In a squall line, winds originating from opposite directions converge on each other and are twisted by Coriolis effects. With the vertical upward motion, a circulation at an altitude of about 5 km is generated, which tends to be about 5 to 15 km across, called a tornado cyclone. Unstable cold air in the parent cloud sinks downward and concentrates to form the tornado vortex. The formation of a tornado is illustrated by means of the following model.

1. Air strongly heated from below rushes upward to form a convection cell in which air moves vertically upward at up to 50 m/s inside a storm cloud. The winds high up in the storm cloud start the air rotating at high speed.
2. More air flows inward, and the rotation extends downward, narrowing as it gets close to ground level. The rotational speed of the air increases as it narrows to conserve angular momentum. Low pressure in the vortex causes ware vapor, if present, to condense, making the vortex funnel visible.
3. Friction between the flow and the surrounding atmosphere and the Earth's surface gradually dissipates the energy of the tornado, and as the supply of moist rising air dwindles, the forcing convective currents in the cloud weaken and the tornado disperses.

Tornadoes are characterized by a funnel-shaped cloud that broadens with height and blends into the parent cloud. The funnel tends to be about 800 to 1500 m high. Both inside and outside of the funnel, a secondary flow is superposed on the horizontal rotation, and the flow pattern generated is in the form of a spiral. Near the ground is an Ekman layer where the flow is radially inward toward the

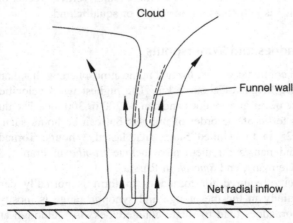

FIGURE 8.44 Schematic for the meridional flow in a tornado.

TABLE 8.4: Classification of tornadoes according to the Fujita scale

F Scale Number	Wind speed (mph)	Characteristic damage
F0	40-72	Light damage
F1	73-112	Moderate damage
F2	113-157	Considerable damage
F3	158-206	Severe damage
F4	207-260	Devastating damage
F5	261-318	Incredible damage

axis. The meridional flow structure is illustrated in Figure 8.44. Velocities in the boundary layer at ground level have been estimated to be up to approximately 150 m/s and above the boundary layer, at a height of about 40 m above ground level, up to approximately 200 to 300 m/s. Vertical velocities of up to 70 m/s can occur. Tornadoes can be classified using the Fujita scale with a number between F0 and F5 as listed in Table 8.4 (Fujita,1971). A tornado with a tangential velocity of 300 m/s is sometimes popularly known in the United States as a supertornado. The tangential velocity profile corresponds

FIGURE 8.45 Waterspout. Image courtesy of the National Oceanic and Atmospheric Administration.

FIGURE 8.46 The family of four waterspouts in this picture developed out of a rapidly growing thunderstorm. Lake Huron, September 9, 1999. Image courtesy of NOAA, Glover.

approximately to that of a Rankine vortex (see Section 3.2.3), and the flow regime is highly turbulent.

If an intense atmospheric vortex occurs over water, a waterspout can be formed, as illustrated in Figures 8.45 and 8.46. In essence, a waterspout is the visible columnar vortex formed if a tornado passes over water. However, waterspouts can also be formed by weaker whirlwinds. Water spouts are particularly common around the Florida Keys but can occur anywhere over water.

8.4.3 Whirlwinds and Dust Devils

Small whirlwinds consist of an intense, typically vertical, rotating column of air. They are often formed downstream of a building, hill, cliff, or other obstruction. The forward kinetic energy of the sluggish flow in the boundary layer over the surface of the obstruction is not sufficient to overcome the adverse pressure gradient causing flow reversal and separation. The free stream will separate from the surface, and vorticity in the boundary layer will form a vortex in the wake downstream of the body. If there is enough momentum in the flow, then at appropriate Reynolds numbers the vortices will be shed from the region. If the vortex core has one end on a solid surface, such as the ground, then a radially inward secondary flow will form along the surface. The radially inward flow supplies the low-pressure core, providing a vertical component of motion. Whirlwinds of varying intensity will tend to form near every surface in a wind, but they are only evident in the presence of leaves or dust. Helmholtz's theorem stating that a vortex cannot end in a fluid is true for an inviscid, constant density fluid. For a viscous fluid an open-ended vortex can be initiated, but it cannot persist unless it is energized by some external source of energy or momentum. The energy for a whirlwind comes from wind, and the whirlwind will tend to dissipate in a few seconds due to viscous dissipation of energy at the ground, the introduction of low vorticity air into the core, and diffusion and dissipation with the air above. The direction of rotation for a whirlwind depends on the side of the obstacle off which it was formed. Whirlwinds can rotate clockwise or anticlockwise; they are too small and occur too quickly for Coriolis effects to be significant.

Dust devils are atmospheric vortices with a rotating core between 1 m and 10 m in diameter with a column between 50 m and several hundred meters tall. They can persist for several minutes to more than an hour and become visually apparent as a relatively gentle or meandering column of dust or sand moving slowly across a desert or plane. A dust devil results from a concentration of vorticity from wind shear between wind and the Earth's surface, or between opposing winds. If the vortex formed is initially horizontal, it will try to assume a more stable orientation by attaching itself to the ground. Favorable winds and thermal convection can intensify the vortex by longitudinal stretching. In a similar fashion to whirlwinds, dust devils are small enough for Coriolis effects not to be significant enough for them necessarily to be cyclonic. However, cyclonic motion provides a source of angular momentum for the larger and more persistent dust devils.

8.4.4 Hurricanes and Typhoons

Hurricanes and typhoons are intense atmospheric vortices. Both are cyclones, with the term *hurricane* used for tropical cyclone in the Atlantic and a *typhoon* describing a tropical cyclone in the Pacific. They are characterized by very high winds and torrential rainfall (of the order of hundreds of millimeters in a day). Hurricanes and typhoons form over warm tropical seas, between latitudes of 10° and 20°, when the water temperature is above 26 °C. Once formed, they follow a path away from the equator, typically westward but following an erratic path. They tend to grow in intensity while they remain over warm water; they swing poleward on reaching land where their energy is rapidly dissipated but not without usually causing substantial destruction.

Hurricanes tend to be approximately several hundred kilometers to 1000 km in diameter and can persist for up to several weeks, although the typical life span is 7 to 14 days, with surface winds up to 80 m/s (180 mph). Only when the storm has matured does the characteristic central eye develop. Mature hurricanes are illustrated in Figures 8.47 and 8.48. The cross-sectional structure of a typical tropical cyclone is illustrated in Figure 8.49. Concentrated around the eye of the hurricane is a ring of thunderstorms known as the eyewall. A series of rainbands surround the eye in concentric circles of increasing radius. Within the

FIGURE 8.47   Hurricane Katrina, August 28, 2005. Image courtesy of GOES Project, NASA-GSFC.

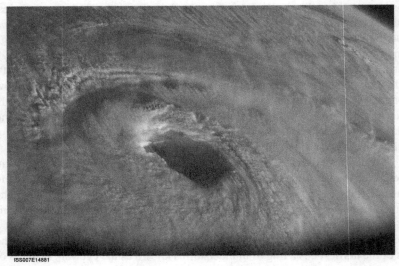

FIGURE 8.48   Hurricane Isabel, September 15, 2003, Image courtesy of NASA.

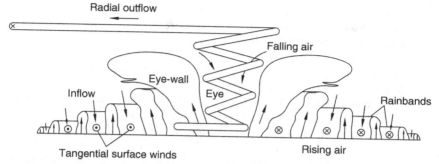

FIGURE 8.49   Cross section of a northern hemisphere hurricane.

eyewall and also in the rainbands, warm moist air rises. In the eye and around the rainbands, air from higher in the atmosphere descends toward the surface. The rising moist air will cool, and water vapor will condense to form precipitation. The descending air in the eye warms and dries, producing a calm cloud-free area in the eye. The low pressure at and near the ocean surface in the heart of the hurricane draws in air from the surrounding regions. The spiralling inward winds accelerate as they approach the eye, transferring more heat and moisture from the ocean surface. The stronger the convection cells in the thunderstorms become, the more precipitation they will produce. As the rate of precipitation increases, so does the amount of heat released to the surrounding atmosphere, providing additional energy for fueling the storm.

Hurricanes are a direct result of Coriolis phenomena. Air converging on a thermally induced sink concentrates angular momentum. Consider a convective cell at a latitude of 30° north. Air from the south approaches the cell with larger tangential velocity than that of the Earth at a latitude of 30° north. As the air approaches the cell, it moves eastward relative to the Earth. Air approaching from the north has a smaller tangential velocity and moves westward. Because of the Coriolis effect, westerlies tend to move to the south and easterlies move to the north. The net effect is a cyclonic vortex with a sink. The formation of a tropical cyclone tends to follow a similar sequence of events (Vanyo, 2001).

1. There is a region of thermally unstable air over the central region or eastern region of an ocean between a latitude of about 10° and 20°.
2. The water temperature is equal to, or above, 26.5 °C.
3. The lower and middle troposphere is warm and moist and near to saturation. Such conditions tend to occur in late summer when the thermal equator is displaced from the equator.
4. Convective cells start. Heavy precipitation throughout the troposphere, with the associated release of latent energy, leads to growth of thunderstorms.
5. Radial inflow of air across a regions acts to coalesce a group of thunderstorms in one mesoscale giant thunderstorm. The radially

inflowing air is warmed by the oceans, acquiring latent heat by evaporation. The high winds and waves lead to an increase in the evaporation rate from the ocean.

6. The normal vorticity ($f = 2\Omega sin\phi$) at the latitude concerned is concentrated by the converging air. The rate of rotation accelerates, and this can be further augmented by wind shear.

7. After about a week of forming, the tangential velocities near the middle of the cyclone reach hurricane strength (115 km/hour)

8. Until the eye of the hurricane forms, the central region tends toward rigid body rotation. The central angular velocity eventually becomes so large that the radial pressure gradient is not sufficient to maintain the centripetal acceleration. The rotating inner wall enlarges its diameter until equilibrium is reached between the pressure differential and the centripetal acceleration and the eye of the hurricane is formed.

9. In a well-defined eye, the low-pressure core will extend to the stratosphere and will draw dry nonrotating stratospheric air downward. As the air descends, it will be heated adiabatically by compression, and its relative humidity will decrease. An eye can be up to 10 to 100 km in diameter, and the air within it can be dry, warm, and almost motionless. The air in the eye slowly diffuses outward into the wall of the eye.

10. The maximum tangential velocities occur just outside of the eye. Rainfall can reach levels of the order of 500 mm per day. As condensation provides a mechanism for the release of latent energy of evaporation from the ocean surface, this energy also maintains upward convection and therefore radial inflow, which in turn maintains the high tangential velocities by concentrating the vorticity through Coriolis effects.

A free vortex and central sink provide an approximate model for a thunderstorm and tropical cyclone. There are, however, significant differences. In the free vortex, the tangential velocity varies inversely with radius, while in a hurricane or thunderstorm, outside the central core the tangential velocity varies with the −0.5 or −0.7 power of radius ($u_\phi \propto (1/r)^{0.5}$ to $u_\phi \propto (1/r)^{0.7}$). Furthermore, the model has a sink at the axis, while in reality the flow is extracted across a large central area in the vortex, and in addition the flow is three-dimensional and time dependent.

A crude estimate can be made for the energy associated with a hurricane based on the rotational energy of the swirling air. A hurricane can be modeled as a rotating disc of diameter 700 km, compatible with Hurricane Katrina, of height 10,000 m. The rotational speed of the mass of rotating air can be estimated from the tangential speed at the outer radius. Taking an extreme value of 80 m/s at a radius of 350 km, the angular velocity can be estimated as 2.29 rad/s. The moment of inertia of a rotating disc is given by

$$I = \frac{1}{2}ma^2 \qquad (8.74)$$

where m is the mass of the disc and a is the outer radius. The mass of air in a hurricane can be estimated from an approximate value for the density and the volume of the rotating air, given in the case of a disc by $\pi a^2 L$, where L is the height of the body of rotating fluid. Here the density will be taken as 1.2 kg/m^3.

The rotational energy of a rotating rigid object is given by

$$E = \frac{1}{2} I \Omega^2 \tag{8.75}$$

Substituting for the mass and moment of inertia, the rotational energy of the hurricane is given by

$$E = \frac{\pi}{4} a^4 L \rho \Omega^2 \tag{8.76}$$

With $a = 350 \times 10^3$ m, $L = 10000$ m, $\Omega = 2.29 \times 10^{-4}$ rad/s, the rotational energy is

$$E = \frac{\pi}{4} \left(350 \times 10^3\right)^4 \times 10^4 \times 1.2 \times \left(2.29 \times 10^{-4}\right)^2 = 7.39 \times 10^{18} \text{ J} \tag{8.77}$$

This is a very high value of energy, and to put it in context, it can be compared with that of one of the most destructive manmade items, an atomic bomb. The explosive "yield" of the U235 "Little Boy" bomb dropped on Hiroshima in World War II has been estimated to be equivalent to approximately 15 kilotons of TNT. This is equivalent to $E = 15 \times 10^3 \times 4.185 \times 10^9 \approx 6 \times 10^{13}$ J of energy. The energy associated with a hurricane, modeled as outlined above, is thus $7.39 \times 10^{18}/6 \times 10^{13} \approx 120,000$ times more than that of an atomic bomb.

8.5. OCEANIC CIRCULATION

The oceans can be represented as a stratified fluid on the rotating Earth's surface, and the principal oceanic circulations result from wind shear and buoyancy-driven processes. Although there are similarities between oceanic and atmospheric circulations and the atmosphere and oceans form a coupled system with one influencing the other, there are important differences. These differences arise from the differences in fluid properties and also because the oceans are laterally bounded, except in the southern ocean. Indeed, the heat capacity of the oceans is so large relative to the atmosphere that relatively small variations in oceanic flow are capable of causing major atmospheric variations. The momentum associated with oceanic flows is also much larger than that associated with atmospheric phenomena; therefore oceanic flows are even more persistent than atmospheric circulations.

Estimates of the total quantities of atmospheric gases and oceanic water are useful in comparing their relative heat capacities. The oceans are estimated to contain about 97% of the Earth's water. The average depth of the oceans is 3.73 km. The oceans account for approximately 70.8% of the Earth's surface. Taking

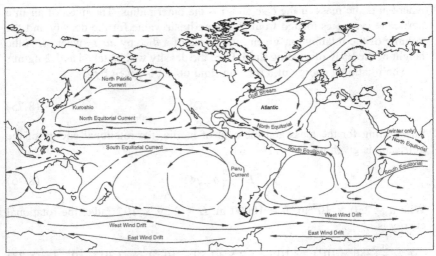

FIGURE 8.50 Seasonally averaged ocean currents. Image after the Office of Naval Research.

a typical density for seawater as 1024 kg/m³ gives an estimate for the total mass of sea water to be

$$m = \rho V = \rho \left[\frac{4\pi(r_o^3 - r_i^3) \times 0.708}{3} \right] = 1024 \left[\frac{4\pi(6367500^3 - 6363800^3) \times 0.708}{3} \right]$$
$$\approx 1.3 \times 10^{21} \text{ kg}$$

(8.78)

By comparison the total mass of air, taking an atmospheric pressure of 101325 Pa and the surface area of the Earth to be 5.1×10^{14} m², gives an estimate for the total mass to be

$$m = 5.1 \times 10^{14} \times 1.01325 \times 10^5 / 9.81 = 5.2 \times 10^{18} \text{ kg} \qquad (8.79)$$

The specific heat capacity of air is approximately 1005 J/kg K, while that of water is approximately 4000 J/kg K. The heat capacity of the oceans is therefore of the order of 1000 times ($m_{Water} \, C_{p,Water} / m_{air} \, C_{p,air}$) that of the atmosphere.

Whereas the general circulation of the atmosphere results from thermal convection, ocean circulations are produced by both drift currents, which are generated by the prevailing winds, and by convection flows generated as a result of temperature and density differences. Global circulations of oceanic flows are limited by the physical constraints of the ocean basins formed by the continents. Some of the principal currents and circulations resulting from these are illustrated in Figure 8.50.

Drift currents are oceanic flows generated by viscous friction action between the water surface and the prevailing wind. The principal oceanic drift currents are illustrated in Figure 8.50. Many of these have been named, the major ones being the following:

- The West Wind Drift
- The East Wind Drift
- The North Equatorial Current
- The South Equatorial Current
- The Peru Current
- The Kuroshio Current
- The Gulf Stream

The shearing effect of the trade winds and westerlies generate the North Atlantic circulation. The north equatorial current, which is driven by the trade winds, forms the southern border of the circulation. It bifurcates before the West Indies and flows partly into the Gulf of Mexico and partly along the Bahamas. The two parts rejoin off the coast of Florida to form the Gulf Stream, which flows off the coasts of Georgia and the Carolinas with a flow velocity of up to 3 m/s at the surface and a width of approximately 100 km. The Gulf Stream leaves the coast of North America at Cape Hatteras, flowing toward Europe where it divides ,with part flowing past the western coasts of the British Isles and the coast of Norway and part flowing south past the coast of Portugal and North Africa, where it then starts a new cycle with the flow being driven by the trade winds. The Kurushio current off the coast of Japan forms the western boundary for the North Pacific circulation. At Cape Nojima, south of Tokyo, the Kurushio leaves the Asiatic continent and flows toward North America. In a manner similar to Europe being warmed by the Gulf Stream, so too is Alaska warmed by the Kurushio.

These currents flow in large rotating loops that are called gyres. In the northern hemisphere the gyres tend to spin in a clockwise direction, and in the southern hemisphere they tend to spin in a counterclockwise direction as result of the Earth's rotation and the Coriolis force, which, for the anticlockwise rotation of the Earth causes a flow to tend to turn to the right. As it is not just the Coriolis force that controls flow, but also wind shear and buoyancy effects as well as conservation of mass, it is possible for a gyre to rotate in the opposite direction to the general rule, and indeed some of the smaller gyres do.

The ocean currents play a significant role in bringing relatively warm water and thus heat from low latitudes to higher latitudes. The model of oceanic flow presented so far is of circulations of flow. In addition to wind shear effects, density plays a significant role. In the Atlantic, for example, warm, high-salinity water flows northward in the Gulf Stream along the east coast of North America. Some of this water continues northeastward in the North Atlantic current toward Iceland and Norway. Off the coast of Greenland, a proportion of the surface water cools, becoming denser, and sinks. A further portion of surface water continues into the Arctic Ocean before also cooling, becoming denser, and sinking. Together these sinking plumes off Greenland and in the Arctic form deep water that plays an important role in global oceanic circulation. An indication of a two-layer model for oceanic circulation is illustrated in Figure 8.51.

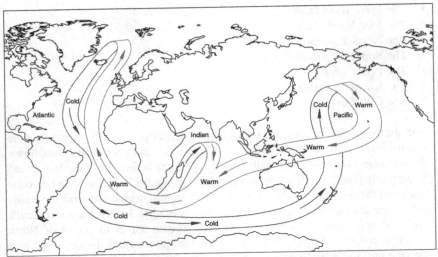

FIGURE 8.51 Two-layer oceanic circulation model. Image after NASA (2004).

Oceanic flows are driven mostly by viscous drag from the prevailing winds and deflected by Coriolis forces and ocean boundaries. Localized flow phenomena also occur, which tend to have shorter scales and time constants.

If the Earth did not rotate, wind shear on a water surface would push the surface layer of liquid in the same direction as the wind. By viscous shear, this layer would in turn drag the next layer below along and thereby set a current in motion with successive layers of fluid induced to move by the one above it—a bit like a pack of cards. The rotation of the Earth, however, significantly influences the flow induced in a body of water by wind shear. Ekman (1903) developed a model for how surface currents are developed by wind stress. He found that water near the ocean surface should flow at an angle of 45° to the right of the wind direction in the northern hemisphere and 45° to the left of the wind direction in the southern hemisphere and that the angle of net transport of water varies with depth from the surface. After a depth of about 100 m the induced flow is in the opposite direction to the wind.

Neglecting all terms except Coriolis terms and frictional effects and modeling wind shear using μ_e, an eddy viscosity, which when multiplied by the velocity gradient gives the wind shear stress τ_e in the direction of the flow, the equations of motion to be solved for a fluid bounded by a horizontal free surface are

$$\frac{\partial}{\partial z}\left(\mu_e \frac{\partial u}{\partial z}\right) + \rho f v = 0 \tag{8.80}$$

$$\frac{\partial}{\partial z}\left(\mu_e \frac{\partial v}{\partial z}\right) - \rho f u = 0 \tag{8.81}$$

The component of the Coriolis force in the z direction is balanced by a simple pressure gradient and does not contribute to the resulting flow dynamics.

If it is assumed that the eddy viscosity is independent of depth and that the ocean is not shallow, then the solution to Equations 8.80 and 8.81 is given by

$$u = \frac{\tau_e}{\mu k \sqrt{2}} e^{kz} \cos \left(kz - \frac{\pi}{4} \right) \tag{8.82}$$

$$v = \frac{\tau_e}{\mu k \sqrt{2}} e^{kz} \sin \left(kz - \frac{\pi}{4} \right) \tag{8.83}$$

where

$$k = \sqrt{\frac{f}{2\nu}} \tag{8.84}$$

and

$$\tau_e = \mu \frac{\partial u}{\partial z} \tag{8.85}$$

A steady flow near the surface with velocity components u and v is predicted by Equations 8.82 and 8.83, with a maximum magnitude of $\tau_e/(\mu K \sqrt{2})$. An outcome of these equations is that the velocity of the fluid at the free surface has a direction of 45° to the right of the applied stress. As the depth below the free surface increases, the direction of the flow velocity rotates until at a depth $z = \pi/k$, known as the penetration depth, the flow direction is actually opposite to that at the surface. This seemingly counterintuitive result is entirely due to the Coriolis

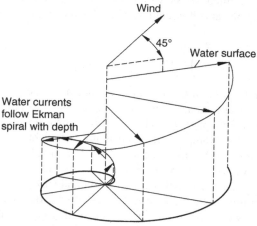

FIGURE 8.52   Ekman spirals.

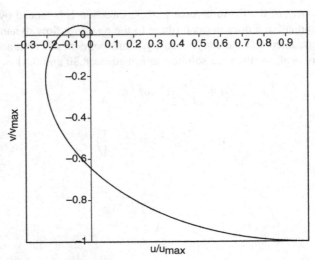

FIGURE 8.53 Ekman spiral.

force. The velocity of the flow reduces with depth below the surface, and at $z = \pi/k$ the magnitude is $e^{-\pi}$, which is about 4% of the surface value.

The variation of flow angle is illustrated in Figure 8.52, revealing the graceful nature of the velocity vector which turns with depth. The projection of the horizontal velocity components at different heights onto a horizontal surface is called an Ekman spiral, as illustrated in Figure 8.53. Integration of the flow from $z = 0$ to $z = -\infty$ provides a measure of the net water transport, and this is at an angle of approximately 90° to the flow direction.

From Equation 8.86 it can be seen that the net flux in the direction of the applied stress is zero. This is to be expected because any motion in the direction of the applied stress would give rise to a Coriolis force in an orthogonal direction, and this could not be balanced by any other external force.

$$\int_{-\infty}^{0} (u + iv)\,dz = -\frac{i\tau_e}{2\mu k^2} \tag{8.86}$$

Real ocean boundary conditions often do not match the idealized conditions of the Ekman wind shear model, and water movements induced by wind will often differ considerably from predictions based on an Ekman flow theoretical model. In shallow water, for example, the depth is not sufficient for the full Ekman spiral to develop, and the angle between the horizontal wind direction and surface-water movements can be as little as 15°. As the water depth increases, the angle between the horizontal wind direction and surface-water movements increases and approaches the theoretical value of 45°. Batchelor (1967) notes that the principal weakness in the application of this type of model to ocean currents lies in the assumption that the tangential shear stress across horizontal planes is due to viscosity. Instead, the notion of an effective viscosity that takes into account surface roughness is commonly employed, and values for the effective kinematic viscosity can range from about 1×10^{-6} m^2/s to 0.1 m^2/s.

8.6. CONCLUSIONS

Flow in the atmosphere can be characterized by bulk meridional circulations and latitudinal flow induced by thermal gradients and the effects of rotation. The bulk flows are significantly modified on a local scale by large-scale eddies that arise from baroclinic instabilities which generate meridional motion. Flow in the oceans can also be characterized by bulk circulations that result from wind shear and buoyancy driven processes.

The thin nature of the atmosphere and oceans relative to the Earth's radius allows many geophysical flows to be modeled using a two-dimensional Cartesian coordinate system, thereby neglecting the Earth's curvature, but accounting for the Coriolis force using the Coriolis parameter.

It is the combination of phase changes between ice, liquid, and vapor, interactions between phenomena with length scales from centimeters to thousands of kilometers, the transfer of radiation through the semitransparent medium of variable composition that forms the atmosphere, and the influence of rotation that makes understanding and modeling geophysical flows such a challenge. The observed flow phenomena often do not match our expectations and intuition, and it is through the insight revealed by detailed modeling that significant satisfaction in the study of geophysical flow can result.

REFERENCES

Batchelor, G.K. *An introduction to fluid dynamics.* Cambridge University Press, 1967.

Calvert, J.G. Glossary of atmospheric chemistry terms. *Pure and Applied Chemistry,* 62, No. 11, (1990), pp. 2167–2219.

COESA (U.S. Committee on Extension to the Standard Atmosphere). U.S. standard atmosphere. U.S. Government Printing Office, Washington DC, 1976.

Cohen, E.R., and Taylor, B.N. The fundamental physical constants. *Physics Today,* BG5-BG9, 1999.

Cohen, E.R., and Taylor, B.N. The 1986 adjustment of the fundamental physical constants. *Reviews of Modern Physics,* 59 (1987), pp. 1121–1148

Dickenson, R.E. Rossby waves—long period oscillations of oceans and atmospheres. *Annual Review of Fluid Mechanics,* 10 (1978), pp. 159–195

Ekman, V.W. On the influence of the Earth's rotation on ocean currents. *Arkiv för. Matematik, Astronomi och Fysik,* 2 (1905), pp. 1–52

Fujita, T.T. Proposed characterisation of tornadoes and hurricanes by area and intensity. Satellite and Mesometeorology Research Project Report 91, University of Chicago, 1971.

Greenspan, H.P. *The theory of rotating fluids.* Cambridge University Press, 1968.

Hide, R. An experimental study of thermal convection in a rotating liquid. *Phil. Trans. Royal Society* A., 250 (1958), pp. 441–478

Hide, R. The viscous boundary layer at the free surface of a rotating baroclinic fluid. *Tellus*, 16 (1964), pp. 523–529

Hide, R. Review article on the dynamics of rotating fluids and related topics in geophysical fluid dynamics. *Bulletin of the American Meteorological Society*, 47 (1966), pp. 873–885

Hide, R. Detached shear layers in a rotating fluid. *Journal of Fluid Mechanics*, 29 (1967), pp. 39–60

Intergovernmental Panel on Climate Change. IPCC Climate Change 2007. The Physical Science Basis, Technical Summary, 2007. www.ipcc.ch/ipccreports/ar4-wg1.htm

IPCC: Summary for Policymakers. In: *Climate Change 2007: The Physical Science Basis. Contribution of Working Group I to the Fourth Assessment Report of the Intergovernmental Panel on Climate Change*. Solomon, S., Qin, D., Manning, M., Chen, Z., Marquis, M., Averyt, K.B., Tignor, M., and Miller, H.L. (eds.). Cambridge University Press, Cambridge, United Kingdom and New York, NY, USA, 2007.

Marshall, J., and Plumb, R.A. *Atmosphere, ocean and climate dynamics*. Elsevier, 2008.

Metz, N. Contribution of Passenger Cars and Trucks to CO_2 , CH_4, N_2O, CFC and HFC Emissions. SAE Paper 2001-01-3758, 2001.

Oser, H. Experimetelle untersuchung über harmonische schwinungen in rotierenden flüssigkeiten. *ZAMM*, 38 (1958), p. 386.

Pedlosky, J., and Greenspan, H.P. A simple laboratory model for the oceanic circulation. *Journal of Fluid Mechanics*, 27 (1967), pp. 291–304

Planck, M. *The theory of heat radiation*. Dover, 1959.

Rossby, C.G., et al. Relation between variations in the intensity of the zonal circulation of the atmosphere and the displacements of the semi-permanent centres of action. *Journal of Marine Research*, 2 (1939), pp. 38–55

Rossby, C.G. Planetary flow patterns in the atmosphere. *Quarterly Journal of the Royal Meteorological Society*, 66 (1940), pp. 68–87

Rossby, C.G. On the propagation of frequencies and energy in certain types of oceanic and atmospheric waves. *Journal of Meteorology*, 2 (1945), pp. 187–204

Thorarinsson, S., and Vonnegut, B. Whirlwinds produced by the eruption of Surtsey volcano. *Bull. Am. Met. Soc.*, 45 (1864), p. 440.

Tucker, P.G. Numerical and experimental investigation of flow structure and heat transfer in a rotating cavity with an axial throughflow of cooling air. DPhil Thesis, University of Sussex, UK, 1993.

Vanyo, J.P. Rotating fluids in engineering and science. Dover Publications, 2001.

Properties of Air

The density in Table A1.1 has been determined from the ideal gas law

$$\rho = \frac{p}{(\Re/M)T} \tag{A1.1}$$

The molar gas constant has been taken as $\Re = 8.314510$ J/mol K (Cohen and Taylor, 1987, 1999).

The molecular mass of air has been taken as $M = 28.9644 \times 10^{-3}$ kg/mol (COESA, 1976).

Hence from Equation A1.1,

$$\rho = \frac{101325}{(8.31451/0.028966)T} \tag{A1.2}$$

The viscosity data has been sourced from Touloukian, Saxena, and Hestermans (1975) with an estimated uncertainty of 2%.

The ideal gas specific heat at constant pressure data, for $250\,K \leq T \leq 1000\,K$, has been obtained from empirical equations presented in Touloukian and Makita (1970) and is given here as Equations A1.3 and A1.4, assuming a conversion factor of 1 cal = 4.1868 J.

For 250 K to 600 K

$$c_p = 1023.2 - 0.176021T + 4.02405 \times 10^{-4}T^2 - 4.87272 \times 10^{-8}T^3 \tag{A1.3}$$

For temperatures between 600 K and 1500 K

$$c_p = 874.334 + 0.322814T - 3.58694 \times 10^{-5}T^2 - 1.99196 \times 10^{-8}T^3 \tag{A1.4}$$

The estimated uncertainty for c_p is 0.5% (Touloukian and Makita, 1970).

For air in the real gas state, Hilsenrath et al. (1955) give the following relationships, Equations A1.5 and A1.6, for the specific heat capacity at constant pressure.

For temperatures between 260 K and 610 K

$$c_p = 1045.36 - 0.316178T + 7.08381 \times 10^{-4}T^2 - 2.70521 \times 10^{-7}T^3 \tag{A1.5}$$

Rotating Flow, DOI: 10.1016/B978-0-12-382098-3.00009-3

TABLE A1.1 Properties of air at $p = 101325$ Pa, $\mathscr{R} = 8.314510$ J/mol K

T (K)	ρ (kg/m^3)	μ (Pa s)	ν (m^2/s)	c_p (J/kg K)	k (W/m K)	Pr
80	4.4122	5.520×10^{-6}	1.251×10^{-6}			
90	3.9219	6.350×10^{-6}	1.619×10^{-6}			
100	3.5298	7.060×10^{-6}	2.000×10^{-6}	1028.0	0.00922	0.787
150	2.3532	1.038×10^{-5}	4.411×10^{-6}	1011.0	0.01375	0.763
200	1.7649	1.336×10^{-5}	7.570×10^{-6}	1006.0	0.01810	0.743
210	1.6808	1.392×10^{-5}	8.282×10^{-6}			
220	1.6044	1.447×10^{-5}	9.019×10^{-6}			
230	1.5347	1.501×10^{-5}	9.781×10^{-6}			
240	1.4707	1.554×10^{-5}	1.057×10^{-5}			
250	1.4119	1.606×10^{-5}	1.137×10^{-5}	1003.6	0.02226	0.724
260	1.3576	1.657×10^{-5}	1.221×10^{-5}	1003.8	0.02305	0.722
270	1.3073	1.707×10^{-5}	1.306×10^{-5}	1004.1	0.02384	0.719
273	1.2930	1.722×10^{-5}	1.332×10^{-5}	1004.1	0.02407	0.718
280	1.2606	1.757×10^{-5}	1.394×10^{-5}	1004.4	0.02461	0.717
288	1.2256	1.795×10^{-5}	1.465×10^{-5}	1004.7	0.02523	0.715
290	1.2172	1.805×10^{-5}	1.483×10^{-5}	1004.8	0.02538	0.715
293	1.2047	1.819×10^{-5}	1.510×10^{-5}	1004.9	0.02561	0.714
300	1.1766	1.853×10^{-5}	1.575×10^{-5}	1005.3	0.02614	0.713
303	1.1649	1.867×10^{-5}	1.603×10^{-5}	1005.5	0.02636	0.712
310	1.1386	1.900×10^{-5}	1.669×10^{-5}	1005.9	0.02687	0.711
320	1.1030	1.946×10^{-5}	1.764×10^{-5}	1006.5	0.02759	0.710
330	1.0696	1.992×10^{-5}	1.862×10^{-5}	1007.2	0.02830	0.709
340	1.0382	2.037×10^{-5}	1.962×10^{-5}	1008.0	0.02900	0.708
350	1.0085	2.081×10^{-5}	2.063×10^{-5}	1008.8	0.02970	0.707
353	0.9999	2.094×10^{-5}	2.094×10^{-5}	1009.1	0.02991	0.707
360	0.9805	2.125×10^{-5}	2.167×10^{-5}	1009.7	0.03039	0.706
370	0.9540	2.168×10^{-5}	2.273×10^{-5}	1010.7	0.03107	0.705
373	0.9463	2.181×10^{-5}	2.305×10^{-5}	1011.0	0.03127	0.705
380	0.9289	2.211×10^{-5}	2.380×10^{-5}	1011.7	0.03173	0.705

Table A1.1 (*Continued*)

T (K)	ρ (kg/m^3)	μ (Pa s)	ν (m^2/s)	c_p (J/kg K)	k (W/m K)	Pr
390	0.9051	2.252×10^{-5}	2.488×10^{-5}	1012.9	0.03239	0.704
400	0.8824	2.294×10^{-5}	2.600×10^{-5}	1014.1	0.03305	0.704
410	0.8609	2.335×10^{-5}	2.712×10^{-5}	1015.3	0.03371	0.703
420	0.8404	2.375×10^{-5}	2.826×10^{-5}	1016.6	0.03437	0.703
430	0.8209	2.415×10^{-5}	2.942×10^{-5}	1018.0	0.03503	0.702
440	0.8022	2.454×10^{-5}	3.059×10^{-5}	1019.5	0.03568	0.701
450	0.7844	2.493×10^{-5}	3.178×10^{-5}	1021.0	0.03633	0.701
460	0.7673	2.532×10^{-5}	3.300×10^{-5}	1022.6	0.03697	0.700
470	0.7510	2.570×10^{-5}	3.422×10^{-5}	1024.3	0.03761	0.700
480	0.7354	2.607×10^{-5}	3.545×10^{-5}	1026.0	0.03825	0.699
490	0.7204	2.645×10^{-5}	3.672×10^{-5}	1027.8	0.03888	0.699
500	0.7060	2.682×10^{-5}	3.799×10^{-5}	1029.7	0.03951	0.699
510	0.6921	2.718×10^{-5}	3.927×10^{-5}	1031.6	0.0402	0.698
520	0.6788	2.754×10^{-5}	4.057×10^{-5}	1033.6	0.0408	0.698
530	0.6660	2.790×10^{-5}	4.189×10^{-5}	1035.7	0.0414	0.698
540	0.6537	2.825×10^{-5}	4.322×10^{-5}	1037.8	0.0420	0.698
550	0.6418	2.860×10^{-5}	4.456×10^{-5}	1040.0	0.0426	0.698
560	0.6303	2.895×10^{-5}	4.593×10^{-5}	1042.3	0.0432	0.698
570	0.6193	2.929×10^{-5}	4.730×10^{-5}	1044.6	0.0438	0.699
580	0.6086	2.969×10^{-5}	4.879×10^{-5}	1047.0	0.0444	0.700
590	0.5983	2.997×10^{-5}	5.009×10^{-5}	1049.4	0.0450	0.699
600	0.5883	3.030×10^{-5}	5.150×10^{-5}	1051.9	0.0456	0.699
610	0.5786	3.063×10^{-5}	5.293×10^{-5}	1053.4	0.0462	0.698
620	0.5693	3.096×10^{-5}	5.438×10^{-5}	1055.9	0.0468	0.699
630	0.5603	3.128×10^{-5}	5.583×10^{-5}	1058.5	0.0473	0.700
640	0.5515	3.161×10^{-5}	5.731×10^{-5}	1061.0	0.0479	0.700
650	0.5430	3.193×10^{-5}	5.880×10^{-5}	1063.5	0.0484	0.702
660	0.5348	3.224×10^{-5}	6.028×10^{-5}	1066.0	0.0490	0.701
670	0.5268	3.256×10^{-5}	6.180×10^{-5}	1068.5	0.0496	0.701

(*Continued*)

Table A1.1 (*Continued*)

T (K)	ρ (kg/m^3)	μ (Pa s)	ν (m^2/s)	c_p (J/kg K)	k (W/m K)	Pr
680	0.5191	3.287×10^{-5}	6.332×10^{-5}	1071.0	0.0501	0.703
690	0.5116	3.318×10^{-5}	6.486×10^{-5}	1073.5	0.0507	0.703
700	0.5043	3.349×10^{-5}	6.642×10^{-5}	1075.9	0.0513	0.702
710	0.4971	3.379×10^{-5}	6.797×10^{-5}	1078.3	0.0518	0.703
720	0.4902	3.409×10^{-5}	6.954×10^{-5}	1080.7	0.0524	0.703
730	0.4835	3.439×10^{-5}	7.112×10^{-5}	1083.1	0.0530	0.703
740	0.4770	3.469×10^{-5}	7.273×10^{-5}	1085.5	0.0535	0.704
750	0.4706	3.498×10^{-5}	7.433×10^{-5}	1087.9	0.0541	0.703
760	0.4644	3.528×10^{-5}	7.596×10^{-5}	1090.2	0.0546	0.704
770	0.4584	3.557×10^{-5}	7.759×10^{-5}	1092.5	0.0552	0.704
780	0.4525	3.586×10^{-5}	7.924×10^{-5}	1094.9	0.0558	0.704
790	0.4468	3.615×10^{-5}	8.091×10^{-5}	1097.1	0.0563	0.704
800	0.4412	3.643×10^{-5}	8.257×10^{-5}	1099.4	0.0569	0.704
810	0.4358	3.672×10^{-5}	8.426×10^{-5}	1101.7	0.0575	0.704
820	0.4305	3.699×10^{-5}	8.593×10^{-5}	1103.9	0.0580	0.704
830	0.4253	3.727×10^{-5}	8.764×10^{-5}	1106.2	0.0586	0.704
840	0.4202	3.755×10^{-5}	8.936×10^{-5}	1108.4	0.0592	0.703
850	0.4153	3.783×10^{-5}	9.110×10^{-5}	1110.6	0.0597	0.704
860	0.4104	3.810×10^{-5}	9.283×10^{-5}	1112.8	0.0603	0.703
870	0.4057	3.837×10^{-5}	9.457×10^{-5}	1114.9	0.0608	0.704
880	0.4011	3.864×10^{-5}	9.633×10^{-5}	1117.1	0.0614	0.703
890	0.3966	3.891×10^{-5}	9.811×10^{-5}	1119.2	0.0619	0.704
900	0.3922	3.918×10^{-5}	9.990×10^{-5}	1121.3	0.0625	0.703
910	0.3879	3.945×10^{-5}	1.017×10^{-4}	1123.4	0.0630	0.703
920	0.3837	3.971×10^{-5}	1.035×10^{-4}	1125.5	0.0635	0.704
930	0.3795	3.997×10^{-5}	1.053×10^{-4}	1127.5	0.0639	0.705
940	0.3755	4.023×10^{-5}	1.071×10^{-4}	1129.5	0.0644	0.706
950	0.3716	4.049×10^{-5}	1.090×10^{-4}	1131.6	0.0649	0.706
960	0.3677	4.075×10^{-5}	1.108×10^{-4}	1133.6	0.0654	0.706
970	0.3639	4.100×10^{-5}	1.127×10^{-4}	1135.5	0.0658	0.708

Table A1.1 (*Continued*)

T (K)	ρ (kg/m³)	μ (Pa s)	ν (m²/s)	c_p (J/kg K)	k (W/m K)	Pr
980	0.3602	4.126×10^{-5}	1.146×10^{-4}	1137.5	0.0663	0.708
990	0.3565	4.152×10^{-5}	1.165×10^{-4}	1139.4	0.0668	0.708
1000	0.3530	4.177×10^{-5}	1.183×10^{-4}	1141.4	0.0672	0.709
1100	0.3209	4.440×10^{-5}	1.384×10^{-4}	1160	0.0732	0.704
1200	0.2941	4.690×10^{-5}	1.594×10^{-4}	1177	0.0782	0.706
1300	0.2715	4.930×10^{-5}	1.816×10^{-4}	1195	0.0837	0.704
1400	0.2521	5.170×10^{-5}	2.051×10^{-4}	1212	0.0891	0.703
1500	0.2353	5.400×10^{-5}	2.295×10^{-4}	1230	0.0946	0.702
1600	0.2206	5.630×10^{-5}	2.552×10^{-4}	1248	0.100	0.703
1700	0.2076	5.850×10^{-5}	2.817×10^{-4}	1266	0.105	0.695
1800	0.1961	6.070×10^{-5}	3.095×10^{-4}	1286	0.111	0.692
1900	0.1858	6.290×10^{-5}	3.386×10^{-4}	1307	0.117	0.691
2000	0.1765	6.500×10^{-5}	3.683×10^{-4}	1331	0.124	0.685
2100	0.1681	6.720×10^{-5}	3.998×10^{-4}	1359	0.131	0.683
2200	0.1604	6.930×10^{-5}	4.319×10^{-4}	1392	0.139	0.694
2300	0.1535	7.140×10^{-5}	4.652×10^{-4}	1434	0.149	0.687
2400	0.1471	7.350×10^{-5}	4.998×10^{-4}	1487	0.161	0.679
2500	0.1412	7.570×10^{-5}	5.362×10^{-4}	1556	0.175	0.673

For temperatures between 610 K and 900 K

$$c_p = 1002.72 - 0.163092T + 5.69907 \times 10^{-4}T^2 - 2.68261 \times 10^{-7}T^3 \quad \text{(A1.6)}$$

The thermal conductivity values for 80 K to 1500 K are from Touloukian, Liley, and Saxena (1970).

Data for c_p and k for $T \geq 1100$ K are from Poferl, Svehla, and Lewandowski (1969).

Data for c_p and k for T=100 K to 200 K are from Hilsenrath et al. (1955).
The Prandtl number is defined by

$$Pr = \frac{\mu c_p}{k} \quad \text{(A1.7)}$$

Values for μ and k at 273 K, 288 K, 293 K, 303 K, 353 K, and 373 K have been determined by linear interpolation using the data from the nearest decade.

REFERENCES

COESA (U.S. Committee on Extension to the Standard Atmosphere). U.S. standard atmosphere. U.S. Government Printing Office, Washington DC, 1976.

Cohen, E.R., and Taylor, B.N. The 1986 adjustment of the fundamental physical constants. *Reviews of Modern Physics*, 59 (1987), pp. 1121–1148.

Cohen, E.R., and Taylor, B.N. The fundamental physical constants. *Physics Today*, BG5-BG9 (1999).

Hilsenrath, J., Beckett, C.W., Benedict, W.S., Fano, L., Hoge, H.J., Masi, J.F., Nuttall, R.L., Touloukian, Y.S., and Wooley, H.W. Tables of thermal properties of gases. *U.S. NBS Circular* 564 (1955).

Poferl, D.J., Svehla, R.A., and Lewandowski, K. Thermodynamic and transport properties of air and the combustion products of natural gas and of ASTM-A-1 fuel with air. *NASA Technical Note* D-5452 (1969).

Touloukian, Y.S., Liley, P.E., and Saxena, S.C. *Thermophysical properties of matter*. Volume 3: *Thermal conductivity, non-metallic liquids and gases*. p. 512. IFI/Plenum, 1970.

Touloukian, Y.S., and Makita, T. *Thermophysical properties of matter*. Volume 6: *Specific heat, non-metallic liquids and gases*, p. 293. IFI/Plenum, 1970.

Touloukian, Y.S., Saxena, S.C., and Hestermans, P. *Thermophysical properties of matter*. Volume 11: *Viscosity*, p. 611. IFI/Plenum, 1975.

The Vector Cross Product

The cross product, also called the vector product, is an operation on two vectors **A** and **B** denoted by **A** × **B**. In a three-dimensional Euclidean space, using the usual right-handed coordinate system, the cross product of two vectors produces a third vector that is perpendicular to the plane in which the first two lie. That is, for the cross of two vectors, **A** and **B**, we place **A** and **B** so that their tails are at a common point. Then, their cross product, **A** x **B**, gives a third vector, say **C**, whose tail is also at the same point as those of **A** and **B**. The vector **C** points in a direction perpendicular (or normal) to both **A** and **B**. The direction of **C** depends on the right-hand rule as illustrated in Figures B.1–B.3. The magnitude of the cross product is equal to the area of the parallelogram that the vectors span.

It should be noted that the order of vectors in and equation is important and

$$\mathbf{A} \times \mathbf{B} = -\mathbf{B} \times \mathbf{A} \tag{B.1}$$

In rotating systems, the direction of rotation is important. The direction of the angular velocity vector can be determined by use of the right-hand grip rule. Using this rule, the fingers are curled in the direction of rotation. The thumb then indicates the direction of the vector (see Figure B.2).

The momentum equation (see Chapter 2, Equation 2.88) can be stated in the form

$$\rho \left(\frac{D\mathbf{w}}{Dt} \right)_{rotating\ frame} = \rho \left[\frac{\partial \mathbf{w}}{\partial t} + (\mathbf{w} \cdot \nabla)\mathbf{w} + 2\mathbf{\Omega} \times \mathbf{w} + \mathbf{\Omega} \times (\mathbf{\Omega} \times \mathbf{r}) \right]$$
$$= -\nabla p + \mu \nabla^2 \mathbf{w} + \mathbf{F} \tag{B.2}$$

where $2\mathbf{\Omega} \times \mathbf{w}$ represent the Coriolis terms.

If, using the notation indicated in Figure B.1,

A becomes the angular velocity, **Ω**
B becomes the flow vector, **w**

Then **A** x **B** is the cross product and in this case gives the direction of the Coriolis terms, $2\mathbf{\Omega} \times \mathbf{w}$. This is illustrated in Figure B.3.

Rotating Flow, DOI: 10.1016/B978-0-12-382098-3.00010-X

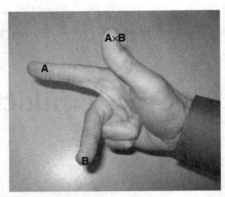

FIGURE B.1 Cross product for two vectors **A** and **B**

FIGURE B.2 Right hand grip rule

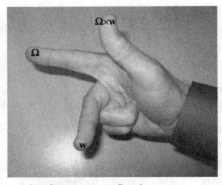

FIGURE B.3 The cross product for two vectors Ω and **w**

This approach can be used to determine the direction of the Coriolis force for a wind in the Earth's atmosphere. The Earth's direction of rotation is anticlockwise. The direction of the angular velocity vector can now be determined using the right-hand grip rule. This gives the angular velocity vector, Ω, pointing away from the north pole. Looking down on the Earth from outer space, the angular velocity vector, represented in the right-hand rule by the index finger, will therefore be pointing up from the north pole. For any wind direction, w, represented by the middle finger in the right-hand rule, in the northern hemisphere, the Coriolis force will be at right angles to the wind direction and to the right.

Flow in atmospheric pressure systems requires consideration of the pressure gradient and the effects of friction and is considered in Chapter 8, Section 8.3.1.

Glossary

Adiabatic A thermodynamic process in which no heat transfer across the boundary occurs.

Ageostrophic Ageostrophic flows are not dominated by the interaction of horizontal pressure gradients and Coriolis effects.

Albedo The proportion of radiation from the Sun diffusely reflected by a surface.

Annulus The space formed between two cylinders of different radii.

Anticyclone A large mass of subsiding air that produces an area of high pressure on the Earth's surface.

Anticyclonic For atmospheric flows, if the wind swirls clockwise in the northern hemisphere or anticlockwise in the southern hemisphere, then it is called anticyclonic flow.

Baroclinic A fluid where density depends on both pressure and temperature is called a baroclinic fluid.

Barotropic A fluid where density depends on pressure is called a barotropic fluid.

Batchelor flow A type of flow that occurs in an enclosed rotor-stator disc system, where fluid pumped radially outward in the rotor boundary layer moves axially across toward the stationary disc in a boundary layer on the cylindrical outer shroud. In between the rotor and stator disc boundary layers there is an inviscid core of fluid that rotates at some fraction of the rotor angular velocity.

Bernoulli's equation An equation that provides the relationship between static pressure, elevation, and velocity. Bernoulli's equation can be used to determine the exchange between static pressure and velocity in a flow.

Boundary layer The region of fluid flow adjoining a solid boundary in which viscous effects are important, characterized by large velocity gradients and shear stresses.

Boundary layer approximation If the effects of viscous or turbulent stresses are only important near a solid boundary, then the governing equations can be simplified to give a set of equations that can be readily solved to model, for example, the flow in the boundary layer of a rotating disc.

Boundary layer thickness The perpendicular distance from a solid surface to the outer edge of a boundary layer. Typically defined as the point at which the local velocity attains 0.99 times the "external" velocity.

Buoyancy wave Any wave where one portion of a fluid is buoyant or partially buoyant in another portion of the fluid or adjacent fluid.

Characteristic length A representative length of a physical system, such as the radius of a vortex.

Circular Couette flow Flow between two long concentric cylinders, with one or more of the cylinders rotating about their common axis.

Rotating Flow, DOI: 10.1016/B978-0-12-382098-3.00011-1

Circulation The line integral of the tangential velocity component about any closed curve fixed in the flow.

Climate The long-term behavior of the atmosphere in a specific region.

Cold front A region of advancing cold air that is forced to undercut a body of warm air.

Concentric annulus The space formed between two concentric cylinders of different radii.

Core This term is used in the context of a rotor-stator or rotor-rotor wheelspace to refer to a region of fluid where the flow can be assumed to be inviscid.

Coriolis force A force that acts on a body in motion in a rotating reference frame.

Coriolis parameter A factor used to account for the effects of the Coriolis force when modeling a flow using a two-dimensional Cartesian coordinate system.

Couette flow The flow of viscous fluid between two parallel plates or surfaces is commonly known as Couette flow (in honor of M. Couette who studied the flow of fluid between rotating cylinders [1890]).

Critical speed The angular velocity above which laminar flow breaks down and flow instabilities begin to grow, leading to the formation of secondary flow vortices.

Cyclonic If the wind swirls anticlockwise in the northern hemisphere or clockwise in the southern hemisphere, then it is called cyclonic flow.

Depression An area of low pressure.

Diabatic Involving transfer of heat.

Disc pumping The radial outflow resulting from a disc rotating in a fluid.

Drift current Motion of a body of water due to the viscous action of the prevailing wind on its surface.

Easterlies Wind, especially a prevailing wind, that blows from the east. The trade winds in the tropics and the prevailing winds in the polar regions are easterlies.

Efflux The flow emerging from a wheelspace.

Egress The flow emerging from a wheelspace.

Ekman layer Viscous region between a geostrophic zone and a solid boundary in a rotating fluid. The steady boundary layer associated with a rotating disc is commonly referred to as an Ekman (or Ekman-type) layer.

Ekman number Ratio of the effects of viscous force to Coriolis force.

Ekman spiral A polar diagram showing the variation of the velocity vector across the thickness of an Ekman layer.

Entrained flow The quantity of fluid supplied to meet the demands of conservation of mass.

Entrainment The supply of fluid to meet the demands of conservation of mass.

Eulerian In the Eulerian description of a fluid flow, the motion of fluid is described in terms of changes occurring at a fixed point (or observer) in space.

Forced vortex In a forced vortex all the fluid rotates with uniform angular velocity about an axis, and there is no relative movement between fluid elements. A forced vortex is also known as a flywheel vortex.

Form drag The force component exerted on a body by a moving fluid in the direction of the free stream remote from the body due to the effects of the static pressure variations around the body caused by flow separation.

Frame of reference The specification of a coordinate system that is fixed in space. Alternatively, for convenience, a moving frame of reference can be specified that translates or rotates.

Free disc A single disc rotating in an initially stationary fluid.

Free disc pumping The radial outflow resulting from a free disc rotating in a fluid.

Free stream The specification of conditions at a location outside or away from a boundary layer.

Free vortex A region of concentrated rotation where fluid elements moving in concentric circles do not rotate on their own axes and the tangential velocity is inversely proportional to the radius. A free vortex is also known as an irrotational flow vortex, a line vortex, or a potential vortex.

Frictional drag The force component exerted on a body by a moving fluid in the direction of the free stream remote from the body due to the action of the shear stress at the surface of the body.

Front The region where two atmospheric air masses meet.

Gap ratio The ratio of the axial gap between a coaxial rotor and a stator disc, and the outer radius.

Geostrophic flow A flow in which both the Rossby and the Ekman numbers are small indicating that the viscous forces are negligible. Geostrophic flows are dominated by the interaction of horizontal pressure gradients and Coriolis effects.

Grashof number This number provides an indication of the ratio of buoyancy to viscous forces acting on a fluid.

Gyre The approximately circular rotational current present in all oceans.

Helmholtz theorems A vortex tube cannot end in a fluid. It must extend to infinity, end at a solid boundary, or form a closed loop within the fluid. Vortices are assumed to remain attached to the same fluid elements as the fluid moves.

Homogeneous Having common physical properties throughout the medium.

Humidity The content of water vapor in a gas.

Inertial frame A frame of reference that is not accelerated so that Newton's first law is valid.

Ingestion Strictly, this means "taking in," a process that can lead to contamination of the flow in a cavity by an inflow of external fluid.

Ingress Inflow or, in the case of a rotor-stator wheelspace, ingestion of external fluid into the cavity.

Insolation The thermal radiation received from the Sun per unit surface area of the Earth.

Instability The tendency for a flow to undergo transition from one type of motion to another.

Integral equations The partial differential equations for continuity and momentum can be integrated across the boundary layer, giving the so-called integral equations.

Inviscid An inviscid flow solution is where the effects of viscosity have been ignored.

Irrotational A flow that has zero vorticity is called an irrotational flow.

Irrotational flow A flow for which no fluid element rotation occurs and the vorticity is zero everywhere

ITCZ The convergence of the trade winds from both the northern and southern hemispheres in the equatorial region forms the inter-tropical convergence zone (ITCZ).

Jet stream Fast flowing atmospheric currents found at an altitude of approximately 10 km.

Lagrangian In the Lagrangian description of fluid flow, attention is placed on the trajectory of a specific particle or body.

Laminar flow A flow where individual fluid elements follow paths that do not cross those of neighboring fluid elements.

Lapse rate The linear rate at which the temperature of air decreases with altitude.

Latitude An angular measurement in degrees from the equator, ranging from $0°$ at the equator to $90°$ north at the north pole and $90°$ south at the south pole.

Linear equations Also known as the Ekman layer equations. At high rotational speeds, where the Coriolis terms are much greater than the nonlinear inertial terms, the boundary layer equations can be simplified to give a set of linear equations called the Ekman layer equations.

Longitude Lines of longitude, also known as meridians, run from pole to pole. By international agreement, the meridian line through Greenwich, England, is given the value of $0°$ of longitude. This meridian is referred to as the Prime Meridian. Longitude values indicate the angular distance between the Prime Meridian and points east or west of it on the surface of the Earth.

Meridional This means along a meridian, that is, along a north–south line.

Mesopause A thin isothermal layer of atmospheric gas at the outer limit of the mesosphere where the atmospheric temperature is actually constant despite the increase in height.

Mesorange Atmospheric and oceanic motion with a lifetime or period of between 1 and 48 hours (or even longer in an ocean) with a wavelength or diameter of between 20 and 500 km.

Mesosphere A layer of the atmosphere characterized by temperatures that fall rapidly with height as this region contains very little water vapor, cloud, dust, or ozone that act to absorb incoming thermal radiation.

Meteorology The study of atmospheric phenomena.

Microrange Atmospheric and oceanic motion with a lifetime or period of about 1 hour with a wavelength or diameter of about 20 km.

Moment coefficient A coefficient derived from the definition of the moment produced by the drag on a rotating disc.

Newtonian fluid A fluid in which the shear stresses are proportional to the velocity gradient and obey Newton's law of fluid viscosity.

No-slip condition There is no relative velocity between a rigid surface and the fluid immediately next to it.

Oblate spheroid A surface obtained by rotating an ellipse about its minor axis.

Ocean gyre The approximately circular rotational current present in all oceans.

Prandtl number This number provides an indication of the relative thicknesses of the momentum and thermal boundary layers.

Pressure asymmetry A variation in the pressure with circumferential coordinate.

Pressure surface A constant pressure surface or pressure surface is a surface where the pressure is equal everywhere on that surface. It is also known as an isobaric surface.

Pumped flow The radial outflow resulting from a disc rotating in a fluid.

Quiescent Undisturbed.

Rayleigh-Bénard convection A flow structure normally associated with fluid contained between two stationary horizontal plates with the lower plate heated. Instability tends to occur for Rayleigh numbers above a critical value, and a buoyancy-induced flow structure is produced, comprising an array of contra-rotating vortices.

Reduced pressure The pressure that arises from the effect of motion of the fluid.

Relative humidity The amount of water vapor in a gas at a given temperature expressed as a percentage of the maximum amount of water vapor that the gas could hold at that temperature.

Rigid body rotation The rotation of a fluid element so that there is no relative motion between the fluid elements.

Rim seal A circumferential form or feature located near the outer radius of a rotor-stator disc wheelspace, on either or both of the stationary and rotating discs, used to limit the exchange of flow between the wheelspace and the external environment.

Rossby number Ratio of the effects of inertial force to Coriolis force.

Rotating cavity The space formed between two coaxially located co-rotating or contra-rotating discs. Also known as a rotor-rotor cavity.

Rotor-stator cavity The space formed between a rotating disc and a stationary disc. Also known as a wheelspace.

Siderial day The time required for one complete revolution of the Earth with reference to any star or to the vernal equinox at the meridian, equal to 23 hours, 56 minutes, 4.09 seconds in units of mean solar time.

Solar day The period between one noon and the next.

Source region An area defining the region from which a body of homogeneous fluid has originated. In the case of atmospheric flows, if air has remained static over a particular area, it will come into thermal equilibrium with that area.

Specific humidity The mass of water vapor in a gas normally expressed, in the case of air, in grams of water per kilogram of air.

Spheroid A surface obtained by rotating an ellipse about one of its principal axes. If the ellipse is rotated about its minor axis, the surface obtained is called an oblate spheroid.

Stewartson flow A type of flow in a rotor-stator wheelspace in which the tangential velocity in the rotor boundary layer reduces from the rotor speed to zero away from the boundary layer with no core rotation.

Stratopause A thin isothermal layer of atmospheric gas at the outer limit of the stratosphere where the atmospheric temperature is actually constant despite the increase in height.

Stratosphere A layer of atmospheric air characterized by a nonlinear increase in temperature caused by the concentration of ozone within this layer.

Streamline A streamline is defined as an imaginary line drawn within a fluid flow so that the tangent to the line at any point always lies in the direction of the velocity at that point.

Superposed flow Supply of fluid to a cavity additional to any entrained flow.

Surface force A short-range force that acts on a fluid element through physical contact between the fluid element and its surroundings.

Swirl Rotation of a region of the flow.

Swirl fraction Ratio of the local tangential velocity and the tangential speed of the disc evaluated at the outer radius of the disc.

Swirl ratio The ratio of the tangential velocity of the core flow to the tangential velocity of the disc.

Synoptic range Atmospheric and oceanic motion with a lifetime or period of about 48 hours (or even longer in an ocean), with a wavelength or diameter of about 500 km.

Taylor column When the Rossby and Ekman numbers approach zero, fluid motion is restricted to two-dimensional motion in planes perpendicular to the axis of rotation, and fluid cannot flow over a protuberance but must instead circumscribe it, producing a region of fluid rigidly attached to a protuberance known as a Taylor column.

Taylor number A dimensionless group of various parameters relevant to the flow in an annulus with rotation of either or both surfaces. The critical Taylor number indicates the onset of Taylor vortices in the flow.

Taylor vortices A series of co-rotating fluid vortices found to occur between concentric cylinders under certain conditions with relative rotation of the inner and outer cylinders.

Taylor-Proudman theorem This theorem states that for a system with solid boundaries perpendicular to the axis of rotation the flow is restricted to two-dimensional motion in planes perpendicular to the axis of rotation.

Temperature inversion The non-linear increase in temperature with height in the atmosphere.

Thermal equator A circumferential ring encircling the Earth defined by the locations having the highest mean annual temperature at each longitude.

Thermocline A region of rapid vertical variation in density.

Thermosphere A layer of the atmosphere characterized by temperatures that increase rapidly with height as a result of an increasing proportion of atomic oxygen.

Trade winds The prevailing winds in the tropics blowing from the high-pressure region in the horse latitudes (between 30 and 35° latitude north and south) toward the low-pressure area around the equator.

Tropic of Cancer One of the five major latitude circles. It is the northernmost latitude, currently 23° 26′ 22″ north, at which the Sun can appear directly overhead at noon.

Tropic of Capricorn One of the five major latitude circles. It is the southernmost latitude, currently 23° 26′ 22″ south, at which the Sun can appear directly overhead at noon.

Tropopause A thin isothermal layer of atmospheric gas at the outer limit of the troposphere where the atmospheric temperature is actually constant despite the increase in height.

Troposphere The layer of atmospheric air nearest to the Earth's surface characterized by a nearly linear decrease in temperature with height.

Turbulent flow A flow characterized by random fluctuations in the velocity components so that the instantaneous velocity varies with time at any given position.

Turbulent flow parameter A parameter based on the dimensionless mass flow rate and rotational Reynolds number that provides a useful means of characterizing flow regimes in enclosed rotating disc systems, indicating whether the flow is dominated by rotational flow effects or radial flow effects.

Typhoon A tropical cyclone forming in the Pacific.

Vortex A region of concentrated rotation in a flow; flow where the streamlines form concentric or elliptical circles.

Vorticity A measure of the local rate of rotation of a fluid particle or element; a vector quantity corresponding to the rotation of a fluid element as it moves in a flow field.

Warm front A region of advancing warm air that is forced to override cold air.

Weather The state of the atmosphere at a local level and usually with a timescale ranging between minutes and months.

Westerlies The prevailing winds in the midlatitudes between 30° and 60° north and south. A westerly wind blows from the west toward the east.

Wheelspace The region of space between two circular co-axial discs, one or more of which may be rotating.

Wind The horizontal movement of atmospheric air.

Windage The power required to overcome viscous drag due to rotation.

Zonal Along a latitude circle.

Index

Printed in the United States
By Bookmasters